BELOWGROUND RESPONSES TO RISING ATMOSPHERIC CO_2

Developments in Plant and Soil Sciences

VOLUME 60

The titles published in this series are listed at the end of this volume.

Belowground Responses to Rising Atmospheric CO_2: Implications for Plants, Soil Biota, and Ecosystem Processes

Proceedings of a workshop held at the University of Michigan Biological Station, Pellston, Michigan, USA, May 29–June 2, 1993

Edited by

P. S. CURTIS, E. G. O'NEILL, J. A. TEERI, D. R. ZAK and K. S. PREGITZER

Reprinted from *Plant and Soil*, Volume 165, No. 1 (1994)

Springer-Science+Business Media, B.V.

A C.I.P. Catalogue record for this book is available from
the Library of Congress.

ISBN 978-90-481-4415-0 ISBN 978-94-017-0851-7 (eBook)
DOI 10.1007/978-94-017-0851-7

Printed on acid-free papers

Contents

ECOSYSTEM PROCESSES

MODELLING

Preface

There is an increasing recognition that interdisciplinary studies are required to improve scientific understanding of the many complex environmental problems on earth. These studies will require an expansion in communication and collaboration among researchers in numerous, and often previously disconnected, disciplines. One example of this need is the research required to understand the potential responses of terrestrial ecosystems to rising levels of atmospheric CO_2. Many investigators now realize that the responses of soil biota and soil processes will be important in determining the nature of terrestrial ecosystem responses to elevated atmospheric CO_2. This realization has resulted in new collaborations and the development of multidisciplinary research teams with a greater range of expertise than has been the case in the past. This type of organizational framework is defining the shape of much of the current research on the potential ecological impacts of elevated atmospheric CO_2.

We convened a workshop with the goal of increasing the integration of aboveground and belowground perspectives in elevated CO_2 research. Scientists representing a broad range of disciplines met to discuss our current understanding of belowground responses to elevated atmospheric CO_2 and to consider the implications for plants, soil biota and ecosystem processes.

The workshop was held 29 May - 2 June, 1993, at the University of Michigan Biological Station at Douglas Lake, Michigan, USA. Participants presented results of original research which are reported in the contributions to this volume. We also divided into four working groups to consider the broader workshop objectives. The findings of the working groups are presented in the Executive Summary. We would like to thank the editors of *Plant and Soil* and Kluwer Academic Publishers for facilitating the timely appearance of this volume. We also are most grateful to the organizations listed herein that provided financial support for the workshop.

Peter S. Curtis
Elizabeth G. O'Neill
James A. Teeri

Workshop Organizing Committee

Plant and Soil **165**: 1–6, 1994.

Belowground responses to rising atmospheric CO_2: Implications for plants, soil biota and ecosystem processes

Executive summary

Peter S. Curtis[1], Elizabeth G. O'Neill[2], James A. Teeri[3], Donald R. Zak[4] and Kurt S. Pregitzer[5]
[1]*Department of Plant Biology, The Ohio State University, Columbus, OH 43210, USA,* [2]*Environmental Sciences Division, Oak Ridge National Laboratory, P.O. Box 2008, Oak Ridge, TN 37831-6034, USA,* [3]*Biological Station, University of Michigan, Pellston, MI 49769, USA,* [4]*School of Natural Resources and Environment, University of Michigan, Ann Arbor, MI 48109, USA and* [5]*School of Forestry, Michigan Technological University, Houghton, MI 49931, USA*

Introduction

As atmospheric CO_2 concentrations increase there almost certainly will be alterations in the flux of C from plants to the soil. It is likely that such alterations will be accompanied by changes in the allocation of C to organic structures and to soil processes. These changes have the potential for further altering the structure and function of terrestrial ecosystems. While there has been increasing recognition of the importance of soil-mediated responses to global climate change, the nature and magnitude of these responses are not well understood.

In an effort to expand our understanding of belowground responses to rising atmospheric CO_2, a workshop was organized that resulted in the contributions that are contained in this volume. The primary objectives of the workshop were to:

1) Assess the present level of understanding of belowground responses to elevated atmospheric CO_2 and identify key areas where fundamental knowledge was lacking;
2) Evaluate existing and proposed methods for investigating belowground responses to rising CO_2; and
3) Identify high priority research topics aimed at improving our understanding of belowground responses to elevated CO_2 and the consequences for whole ecosystems.

Recognizing the wide range of processes and biota spanned by considering "belowground responses", these objectives were addressed separately by four working groups (Appendix I), covering: 1) Root biology, 2) Soil biota, 3) Ecosystem processes, and 4) Modeling. Working groups were asked to consider belowground systems from a hierarchical perspective (sensu O'Neill et al., 1986); i.e., to determine how belowground processes and organisms might constrain ecosystem response to CO_2 enrichment. This, of course, would require that belowground effects be considered in the context of negative and positive feedbacks, and the significance of these feedbacks or constraints to terrestrial ecosystems. Presented here is a summary of key areas of uncertainty and an evaluation of research methodologies from each working group, followed by a statement of research priorities across groups.

Root biology

The extent to which the belowground represents a significant constraint to ecosystem response to CO_2 enrichment ultimately depends upon the allocation of C to coarse or fine roots, i.e., to storage vs. resource capture functions. Allocation to coarse roots should have little short-term effect on belowground function, and would tend to buffer responses of ecosystems to rising CO_2. On the other hand, allocation to fine roots could have an immediate and significant effect on belowground structure and function and extend to the ecosystem level through changes in soil organic matter (SOM), effects on microbial communities and biomineralization, and eventual effects on plant community composition.

Significant changes in C fluxes into SOM could occur with increases in atmospheric CO_2 concentrations. Questions of concern include coarse and fine root production, fine root longevity and tissue chemistry, the magnitude and role of exudation, and the contribution of mycorrhizae to these processes.

Increased C allocation to fine roots may increase the total viable rhizosphere and bulk soil microbial biomass, or induce shifts in microbial and mycorrhizal community structure (see below). Such shifts have the potential to alter patterns of biomineralization and soil formation, change nutrient pool sizes, and affect the formation of SOM. We have very little information on root-rhizosphere interactions at elevated CO_2.

Finally, differential responses among plant species in the dynamics of C allocation belowground with rising CO_2 may lead to changes in plant community composition and hence to fundamental changes in ecosystem-level processes. Indeed, changes in species composition due to CO_2 effects on plant-plant interactions should have a greater impact on processes such as SOM formation and nutrient cycling than will CO_2-induced changes in plant tissue chemistry. Traditionally, emphasis has been placed on root physiological responses when assessing the impact of plant-plant interactions on community composition. The degree to which root morphology determines competitive outcomes among plants is equally important and not well understood. The architecture of root systems, vertical and horizontal distribution of roots, and the plasticity of these features may constrain the potential for adjustment and the ability of plants to more efficiently exploit soil resources.

Methodologically, improved quantification of fine root turnover, the source of much of the C flowing to soil microbes, is essential. Wider application of non-destructive, repetitive video sampling using minirhizotron and related technologies was recommended. Measurement of fine root dynamics in a multi-species setting remains problematic. Molecular methods and new radio- and/or stable-isotope tracer methods must be developed to better quantify the nature, amounts, and ultimate fate of compounds exuded by roots under field conditions. We are also lacking reliable methods for quantifying mycorrhizae and measuring their function in soil.

Soil biota

Ecosystem processes that are mediated by soil biota (e.g., biomineralization) are constrained by energy (C) input to the soil. This input is likely to increase as CO_2 concentrations in the atmosphere rise, either through increased root production or exudation, or through increased aboveground production and litterfall. The magnitude of changes in quality or quantity of C input with rising CO_2 will determine whether soil biota in turn represent a significant constraint to ecosystem functioning in a changing climate.

We do not know if present constraints exerted by soil biota will increase in importance as atmospheric CO_2 rises. Soil organisms vary in their sensitivity to changes in C input and in their impact on ecosystem processes. We need to identify which biotic groups regulate these processes, assess the importance of species diversity and functional redundancy in regulation, and determine the scales over which soil processes operate.

Soil biota may be affected by rising CO_2 at the organism, population or community levels. Although all soil biota may ultimately feel the impact of changing C inputs to soil, several organisms were identified as critical to understanding soil system responses. Mycorrhizae are of key interest due to their importance in linking roots to soil communities, their role in establishing and maintaining soil structure, and their dependence on plant-derived C. Future experiments should avoid studies without normal levels of mycorrhizal infection. Responses of organisms involved in nutrient transformations also deserve high priority. Finally, plant pathogens merit attention because of their potential to alter ecosystem structure and impact human society. For soil biota in general we may see differential interspecific or even intraspecific responses to increased C input and consequent changes in belowground community structure and process rates. If environmental pressures force ecosystem migration, we need to know whether migration rates of soil organisms and communities will prove to be a critical constraint to plant movement.

Detecting responses of soil communities and assessing the impact of those responses on the ecosystem are particularly difficult because the stressor (elevated CO_2) is several levels removed from the community of interest. Thus, the most important conceptual problems with regard to soil biota relate to scale and heterogeneity. Soil processes operate on a multitude of temporal and spatial scales; we must identify the scales over which each soil process is important. There is a primary need to determine if, and to what degree, changes in C input will influence the soil biota in both the short term (transient) and the longer term (years to decades). We must also learn to deal with the almost insurmountable degree of heterogeneity that exists in most soil systems. The latter will require continued development of statistical techniques designed to characterize and understand heterogeneity, rather than eliminate it.

Methodology may prove to be a significant obstacle to our ability to measure responses of soil communities to CO_2 enrichment. Historically, emphasis has been placed either on counting numbers of organisms or on measuring process rates (e.g., nitrogen mineralization, soil respiration). In order to understand the role of soil biota in ecosystem response to CO_2 enrichment, it will be necessary to link these two approaches. For example, the basic assumption of potential soil biota response to elevated CO_2 depends upon whether or not C input from roots significantly limits soil population size or activity and whether population size (of microbes, of nematodes, of soil arthropods, etc.) or activity determines process rates. Yet, aside from recognizing the existence of the "rhizosphere effect", we cannot agree on the magnitude of root C inputs nor have we determined whether this input significantly stimulates soil processes or influences soil community composition.

Ecosystem processes

The ecosystem processes working group identified four specific gaps, in order of priority, relative to our understanding of ecosystem responses to CO_2 enrichment. First, there is a primary need to understand the amount and fate of C fluxes belowground, including soil organic matter dynamics, soil biota biomass and turnover, and soil C efflux. Soil communities regulate C flux and partitioning in the soil; at present we have a limited understanding of the nature of this regulation and how C entering the soil is partitioned into biomass, respiration, and long-lived C pools. We need to know which different ecosystem types will change their source or sink potential for atmospheric C. What, for example, can we say about belowground responses in C-rich vs. C-poor systems, in mesic vs. arid systems, in managed vs. unmanaged systems, or in wetlands, where anaerobic conditions may predominate?

Second, issues of scaling, both spatial and temporal, must be resolved. This is especially critical in assessing belowground responses because of the wide range of scales over which belowground processes operate. Third, more information is needed on the potential feedback mechanisms that could mediate whole-ecosystem responses to environmental change.

Finally, we recognize that population- and community-level phenomena exert a strong influence on ecosystem structure and function. Competitive relationships, colonization/recruitment, disturbance response and recovery, and plant-animal interactions are all potentially sensitive to CO_2-induced change. We know little about how these ecological processes operate in belowground systems or how they influence relationships at higher scales.

Addressing these needs will require changes in experimental approach and design, rather than new techniques. We recommend that, 1) belowground questions be given high priority and provisions for addressing these questions be included at the initial design stage of any ecosystem-level project, 2) whole-ecosystem elevated CO_2 experiments be established (with an appropriate representation of biome diversity), 3) site selection criteria for ecosystem-level studies be determined, and 4) experiments at the ecosystem level continue for time periods appropriate to the goals of belowground research. For many sites, a minimum of ten years exposure may be needed to allow ecosystem structure and nutrient cycling processes to become established. For most ecosystems, short-term experiments measure acclimation rather than equilibrium responses.

Modeling

Models are useful tools for organizing our thinking and for identifying critical gaps in our understanding of processes. In exploring CO_2 effects on belowground processes, we must mix experiments and models, with timely exchanges of information between the two approaches. At present, there is a need for increased feedback between ecosystem-level CO_2 studies and ecosystem modeling efforts. For modelers, additional empirical work would be of great benefit in the areas of control over belowground C allocation, root phenology, root inputs to soil C, and soil organic matter turnover.

There are currently two principal modeling schemes that have been used to allocate photosynthate belowground. One method is to assume that root growth is proportional to aboveground growth (allometric), and the other is to assume that allocation takes place so as to balance the plant's C and N requirements (balanced). Neither of these schemes is mechanistic and both have serious limitations in modeling plant response to elevated CO_2. Models using allometric allocation would require an arbitrary change to the proportionality constant under elevated CO_2 to account for changes in root:shoot ratio. However, plants may undergo changes in this ratio during ontogeny, and the values of the proportionality constant would have to be determined experimentally for the principal plant species in each ecosystem. Models using balanced allocation mimic some responses of herbaceous plants and young seedlings to elevated CO_2 reasonably well, but this may be limited to periods of exponential growth and they do not account for disparate root:shoot ratios of different growth forms. For model development to progress, it is imperative to develop a more general understanding of the nature of allocation and the way it is influenced by elevated CO_2.

In addition to the fundamental issue of potential changes in amount of C translocated belowground with rising CO_2, changes in the timing of root growth and C use are important to ecosystem functioning and are inadequately incorporated into current ecosystem models. The timing of root size and activity in the dominant species of, for example, a northern hardwood forest, may affect ecosystem hydrology, nutrient uptake and storage, C storage and fluxes, and soil faunal activity. In spite of its significance to ecosystem processes, little is known of root phenology in many important ecosystems, particularly compared to aboveground phenology.

Most models with a soil C component must use incomplete information on the amount of root inputs, which include fine root turnover, mortality of coarse roots, and root exudates. Given that fine root turnover, as estimated by some methods, is as large or larger than aboveground inputs to SOM, this is a critical area in modeling ecosystem C and nutrient cycling, and estimating potential C sequestration in soil with rising CO_2. The dynamics of root decomposition are almost unknown. Empirically, attention must be paid not only to bulk kinetics during decomposition, but also to changes in root tissue chemistry and associated nutrient dynamics.

The dynamics of bulk SOM are captured quite well by current models under a wide range of climate conditions, although it is uncertain how well they will perform under novel conditions such as elevated atmospheric CO_2. This uncertainty stems from the largely correlative nature of our understanding of the interactions between physical, chemical, and biological factors and the stability of SOM. Because of the complexity of these interactions, a complete mechanistic description will probably not be useful to modelers, however. Rather, these processes must be captured in some larger concept, such as litter quality, that could be tested for its generality.

Research recommendations

There was unanimous agreement that measurement of C flux dynamics was the single most important research priority with regard to belowground responses to rising CO_2, beginning with a reasonable estimate of C input through increased root turnover and exudation. In order of priority, the following research goals were recommended:

1) Quantify C input to the soil (exudation, fine root production/turnover, litter) and partitioning of C among different soil pools with CO_2 enrichment.
2) Determine effects on interactions (especially plant-pathogen, plant-symbiont), potential for alterations in soil biota community structure, changes in diversity of belowground communities, soil community migration/dispersal, and most importantly, feedback to higher organizational levels.
3) Measure rates of microbially-mediated nutrient transformations, and assess potential for long-term changes in nutrient availability.
4) Investigate interactions of increased CO_2 concentrations with other environmental variables, such as soil nutrient status, changes in amount or temporal patterns of water availability, and potential increases in annual or seasonal temperatures.

Finally, several issues were considered to be high priority due to their importance to all of the above research questions. The related questions of organism activity vs. number and turnover vs. standing crop must be resolved, as they have very different implications for ecosystem response to elevated CO_2. Scaling and heterogeneity must be dealt with – determination of appropriate scales of measurement and identification of scale-dependent processes must precede any serious effort to answer the questions raised above.

Acknowledgements

Funding for this workshop was provided in part by the U. S. Environmental Protection Agency, the Electric Power Research Institute and the University of Michigan Biological Station. The contribution of the Environmental Sciences Division of Oak Ridge National Laboratory* is gratefully acknowledged. The organizers would also like to thank Beth Wolfe, Karen

Popham, Lisa Readmond, Diana Randlett, Tammy Long and April Jones for administrative and logistic support before and during the workshop. ESD Publication No. 4322.
*Managed by Martin Marietta Energy Systems, Inc., under contract DE-AC05-84OR21400 with the U.S. Department of Energy.

References

O'Neill R V, DeAngelis D L, Waide J B and Allen T F H 1986 A Hierarchical Concept of Ecosystems. Monographs in Population Biology 23. Princeton University Press, Princeton, New Jersey. 254 p.

Appendix I

Working groups

Root biology - Chair: Hugo H. Rogers, USDA-ARS, Auburn, AL, USA

- David Atkinson, University of Aberdeen, Aberdeen, Scotland, UK
- Glen Berntson, Harvard University, Cambridge, MA, USA
- Peter S. Curtis, The Ohio State University, Columbus, OH, USA
- David M. Eissenstat, University of Florida, Lake Alfred, FL, USA
- H. William Hunt, Colorado State University, Ft. Collins, CO, USA
- Susan Kalisz, Michigan State University, Hickory Corners, MI, USA
- Jonathan Lynch, Penn State University, University Park, PA, USA
- Kurt Pregitzer, Michigan State University, East Lansing, MI, USA
- David Ringelberg, Center for Environmental Biotechnology, Knoxville, TN, USA
- Johan Swinnen, Institute for Soil Fertility Research, Wageningen, The Netherlands
- Rett Weber, Old Dominion University, Norfolk, VA, USA
- David C. White, Center for Environmental Biotechnology, Knoxville, TN, USA

Soil biota - Chair: Carole C. Klopatek, USFS, Arizona State University, Tempe, AZ, USA

- Michael F. Allen, San Diego State University, San Diego, CA, USA
- Robert Fogel, University of Michigan, Ann Arbor, MI, USA
- James D. Lewis, Duke University Durham, NC, USA
- John Lussenhop, The University of Illinois, Chicago, IL, USA
- Jack Morgan, USDA-AIRS, Fort Collins, CO, USA
- Eldor A. Paul, Michigan State University, East Lansing, MI, USA
- Brett Runion, Auburn University, Auburn, AL, USA
- Paul T. Rygiewicz, USEPA, Corvallis, OR, USA
- Steven R. Shafer, USDA-NCSU, Raleigh, NC, USA
- Chris Vogel, The Ohio State University, Columbus, OH, USA
- Julie Whitbeck, Stanford University, Stanford, CA, USA
- Donald R. Zak, University of Michigan, Ann Arbor, MI, USA

Ecosystem processes - Chair: Clenton Owensby, Kansas State University, Manhattan, KS, USA

- Jay Arnone, University of Basil, Basil, Switzerland
- Frank P. Day, Old Dominion University, Norfolk, VA, USA
- Bert Drake, Smithsonian Environmental Research Center, Edgewater, MD, USA
- Bruce Hungate, University of California, Berkeley, CA, USA
- Dale W. Johnson, Desert Research Institute, Reno, NV, USA
- Mark Kubiske, Michigan Technological University, Houghton, MI, USA
- Bruce D. McVeety, Batelle-Pacifc Northwest Laboratories/USEPA, Corvallis, OR, USA
- Richard J. Norby, Oak Ridge National Laboratory, Oak Ridge, TN, USA
- Elizabeth G. O'Neill, Oak Ridge National Laboratory, Oak Ridge, TN, USA

Steven A. Prior, USDA-ARS, Auburn, AL, USA

G. Phillip Robertson, Michigan State University, East Lansing, MI, USA

Modeling - Chair: Stan D. Wullschleger, Oak Ridge National Laboratory, Oak Ridge, TN, USA

Jeffrey S. Amthor, Lawrence Livermore National Laboratory, Livermore, CA, USA

Mark Johnson, Oregon State University, Corvallis, OR, USA

Paul Kemp, Duke University, Durham, NC, USA

Steven W. Leavitt, University of Arizona, Tucson, AZ, USA

W. M. Post, Oak Ridge National Laboratory, Oak Ridge, TN, USA

David Tingey, USEPA, Corvallis, OR, USA

David Vieglas, NASA, Kennedy Space Center, FL, USA

Participants-at-Large

James A. Teeri, University of Michigan, Ann Arbor, MI, USA

Roger A. Dahlman, U.S. DOE, Washington, DC, USA

Diana Randlett, University of Michigan, Ann Arbor, MI, USA

Tammy Long, University of Michigan, Ann Arbor, MI, USA

ROOT BIOLOGY

Plant and Soil **165**: 9–20, 1994.

Issues and perspectives for investigating root responses to elevated atmospheric carbon dioxide

Richard J. Norby
Environmental Sciences Division, Oak Ridge National Laboratory Building 1506, Oak Ridge, TN 37831, USA

Key words: carbon storage, CO_2 enrichment, fine roots, global change, root-to-shoot ratio, root growth

Abstract

A thorough assessment of how plants and ecosystems will respond to increasing concentrations of atmospheric CO_2 requires that the responses of root systems and associated belowground processes be understood. Static measures of root-to-shoot ratio have not been satisfactory for describing the integrated responses of plants to CO_2-enriched atmospheres, but research with a process orientation has suggested that elevated CO_2 can stimulate root growth or root activity and provide a positive feedback on plant growth. There are, however, critical questions concerning the relevance of root data from short-term studies with potted plants when scaling to questions about plants in the field. Data on root responses to CO_2 enrichment in the field are fragmentary, but they allow us to more clearly define research questions for further investigation. Three perspectives for analyzing the significance of root responses as a component of the overall response of the terrestrial biosphere to increasing atmospheric CO_2 are suggested: (1) roots as a platform for nutrient acquisition and a mediator of whole-plant response to CO_2; (2) carbon storage in roots as a component of whole-plant carbon storage; and (3) effects of CO_2 enrichment on root turnover and the implications for carbon storage as soil organic matter. The relative importance of these different perspectives will vary depending on the ecosystem of interest and the larger-scale issues being considered.

Roots and global change issues

Analyses of plant responses to elevated concentrations of atmospheric CO_2 have focused largely on processes occurring above the ground. When questions are framed about plants in a CO_2-enriched atmosphere, root systems, being "out of sight", have often been "out of mind" as well. This is not surprising, since aboveground plant structures are generally of greater economic interest and are certainly easier to study than roots and belowground processes. Photosynthesis is particularly well studied, as it is the process by which the primary interaction between a green plant and atmospheric CO_2 occurs. But despite great advances in our understanding of photosynthetic responses to elevated CO_2 at levels ranging from molecular genetics to biochemistry to environmental physiology and micrometeorology, it is becoming increasingly clear that an understanding of photosynthesis alone is not sufficient to answer the important questions about terrestrial responses to a changing atmosphere. A whole-plant perspective is needed to understand the critical feedbacks and adjustments that occur within a plant and between plant and soil. Whether the questions concern carbon sequestration in vegetation and ecosystems or process level responses to a complex suite of environmental variables, root systems must be a significant component of the analysis.

The importance of understanding root responses has recently been highlighted by the challenge of the "missing carbon" within the global carbon cycle. Analyses of the annual exchanges of carbon between the atmosphere, terrestrial vegetation, and oceans (Post et al., 1990; Siegenthaler and Sarmiento, 1993; Tans et al., 1990) have failed to account for all of the carbon injected into the atmosphere each year by fossil fuel combustion. The physicist Freeman Dyson made the connection between the "missing carbon" and roots in a lecture at Oxford University in 1990 (Dyson, 1992):

"It is known from growth-chamber experiments that a common response of plants to an enrichment of carbon dioxide in the air is an increased root-to-shoot ratio ... Whether it occurs, or how much it occurs, in the world outside, we do not know. But it is at least plausible that the 15 percent increase of atmospheric carbon dioxide due to human activities has produced a worldwide shift in root-to-shoot ratios and a consequent growth of the carbon reservoir in soil. This, rather than accelerated growth of trees in undestroyed forests, may be the process that accounts for the bulk of the missing carbon".

There are many studies that have investigated at least some aspects of root responses to elevated CO_2. Recent reviews (Stulen and den Hertog, 1993; Rogers et al., 1994) have comprehensively compiled the data on those responses, and new field-oriented research projects are underway that will substantially augment that data base. It is the aim of this paper to review the responses of roots to elevated CO_2 while considering whether the data can be used to address the important questions concerning the terrestrial biosphere and atmospheric CO_2. Those questions, with an increased focus on biotic and abiotic interactions and feedbacks within ecosystems (Mooney, 1991), should direct the development of a new conceptual framework for studying root responses to elevated CO_2.

Responses of small plants

Root-to-shoot ratio

Most of the data on root responses to elevated CO_2 describe root mass of agronomic plants and, to a lesser extent, small tree seedlings and other perennials that had been grown in pots for weeks or months in ambient or elevated CO_2. The general conclusion that can be drawn from such studies is that root mass increases in elevated CO_2, as does shoot mass. This statement, however, does not specify whether there is a direct, specific effect of CO_2 on roots or rather a coordinated, whole-plant acceleration of ontogeny. The question then becomes whether root mass increases more or less than shoot growth. That is, does CO_2 enrichment affect carbon allocation patterns such that root growth specifically is altered. This question is most often answered by the simple calculation of the ratio of root dry mass to shoot (or stem) dry mass (root-to-shoot ratio) when the plants are harvested.

The data on effects of CO_2 concentration on root-to-shoot ratio are confusing and ambiguous, as evidenced by the wide range of generalizations present in review articles. Oechel and Strain (1985) state that "studies of root growth show that with increasing atmospheric CO_2 most plants allocate proportionately more of the extra carbon belowground, causing an increase in root-to-shoot ratios ..." Bazzaz (1990) also accepted that there generally is an increase in allocation to roots, but he suggested that the response was most apparent when nutrients and water are limiting. Eamus and Jarvis (1989) expressed the role of other resources more strongly: "... when nutrients have been supplied in adequate amounts sufficiently often, an increase in CO_2 concentration resulted in a decrease or in no change in the root:shoot ratio." From a horticultural point of view, Idso et al. (1988) concluded that for plants whose primary yield component is produced above-ground, "... it appears unlikely that atmospheric CO_2 enrichment has any significant effect on root:shoot ratios".

Experimental data can be presented to support any of these differing generalizations. The compilation of all of the published data, however, suggests that on average, root-to-shoot ratio changes little in elevated CO_2. Among crop plants the range of response in root-to-shoot ratio with CO_2 doubling was an 8.5% decrease to a 6.4% increase, except in sweet potato (34.9% increase) (Cure, 1985). Wullschleger et al. (1994) compiled 398 observations of 73 tree species in elevated CO_2 and used this data set to examine trends in root-to-shoot ratio. From a subset of 224 observations, the log-transformed mean response in root-to-shoot ratio was only a 6% increase for plants grown in elevated CO_2 (S. D. Wullschleger, personal communication; Norby et al., 1994). Although the generalizations presented above might suggest that the availability of other resources could alter the response of root-to-shoot ratio to CO_2, the mean response in this data base was independent of soil water or nutrient status (S. D. Wullschleger, personal communication). Regardless of the average response of root-to-shoot ratio, large shifts in the ratio have been the dominant response to CO_2 enrichment in some experiments (e.g., Norby et al., 1986).

This analysis of root-to-shoot ratio of plants in elevated CO_2 does not lead to a simple or satisfactory conclusion. The statement that is best supported probably is that of Stulen and den Hertog (1993), as echoed by Rogers et al. (1994): "The assumption that a larger proportion of the extra dry matter produced under CO_2 enrichment is allocated to roots needs critical reconsideration." Stulen and den Hertog (1993) list several reasons why determination of root-to-shoot ratio is subject to experimental error, including problems in defining the root-shoot boundary, quantitative recovery of roots (especially fine roots) and accounting for root decomposition, control of the nutrient and water regime in a pot, ontogenetic effects, and accumulation of non-structural carbohydrates. In addition to evaluation of these experimental problems, part of any reconsideration of root-to-shoot ratio should include thought about why the parameter is of interest. Root-to-shoot ratio may well be a useful index of carbon partitioning within the framework of a specific experiment. But in better defining the research questions pertaining to global change issues, we may find that root-to-shoot ratio of small, potted plants provides little useful information for addressing the larger-scale issues concerning the integrated responses of plants and ecosystems to elevated CO_2.

A process orientation

In studies of annual plants or short-term studies of seedlings of woody perennials, our interest in the responses of roots to CO_2 enrichment derives primarily from their role in water and nutrient uptake. (An exception are those crop plants for which the roots have economic value.) Root systems of CO_2-enriched plants can provide critical feedbacks to whole-plant response by altering the capacity of the plant to acquire sufficient water and nutrients from the environment. For example, our research program at the Oak Ridge National Laboratory on responses of trees to elevated CO_2 centered around a hypothesis that elevated CO_2 would stimulate belowground processes, thereby circumventing nutrient limitations to growth (Norby et al., 1986, 1994). Static measures of root-to-shoot mass ratio do not help us understand the processes of resource acquisition by roots. A more useful framework considers CO_2 effects on root growth and soil exploration, as well as root activity and associated microbial activity (Zak et al., 1993). The concept is that nutrient availability could increase in elevated CO_2 either because of increased size of the root system or increased activity per unit root, although as will be discussed later, there are different implications for these two alternatives as the scale of interest increases. The relevant research topics include root length and branching, periodicity in root growth, distribution between different sizes or type of roots, nutrient uptake by roots, respiration, exudation, and symbiotic activity. This framework represents a process orientation that should prove more useful in understanding the role of roots in plant response to CO_2 enrichment.

Root growth

Experiments primarily with agronomic plants have shown that elevated CO_2 can cause significant changes in plant root morphology that can alter the capacity of a root system to explore the soil (Rogers et al., 1992). Several studies with woody plants have divided root systems by size class to determine the separate effects of CO_2 on fine roots and larger, coarse roots. Fine roots were the most responsive plant component to CO_2 enrichment in *Quercus alba* (Norby et al., 1986) and *Castanea sativa* seedlings (El Kohen et al., 1992), but in *Liriodendron tulipifera*, fine roots increased in mass with CO_2 enrichment to the same extent as coarse roots (Norby and O'Neill, 1991). In a model wet-tropical assemblage, the effect of CO_2 enrichment on root mass decreased with increasing root diameter. Mass of roots less than 1 mm diameter was 63% greater in elevated CO_2, roots from 1 to 3 mm diameter were enhanced by 37%, and larger roots were enhanced by 16% (Körner and Arnone, 1992). Separation of root responses by size class may be particularly important in tree species. The metabolically-active fine roots are of paramount importance in nutrient uptake and microbial interactions. The coarse, woody roots may represent most of the dry mass of the root system and are important in carbon and nutrient storage, but they are less important metabolically. In many experiments with herbaceous plants or small tree seedlings, the plants may be too small or the experiment of insufficient duration for any significant differentiation in root size to occur.

Fine root production is of particular interest to considerations of physiological adjustments to increased CO_2 concentrations because of the presumption that plants with more fine roots can exploit a larger volume of soil, perhaps increasing the supply of water and nutrients to the plant. While the direct effects of elevated CO_2 on stomatal aperture (Eamus and Jarvis, 1989) may reduce whole-plant water use, water shortages will continue to be an important limitation to plant growth in many habitats. The balance between fine roots and leaf area may be particularly interesting. In *Liriodendron*, the ratio of fine root mass to leaf area increased with increasing CO_2, and this was interpreted as a compensatory response that could increase whole-plant water-use efficiency (Norby and O'Neill, 1991).

Increased fine root mass, however, does not necessarily imply a larger volume of soil occupied or exploited by the root system. A more direct assessment is provided by measures of root length and branching patterns. Del Castillo et al. (1989) reported that

CO_2 increased the number of soybean roots but not their elongation rate, from which they inferred that the soil volume explored by the root system would not increase, but a given volume of soil would be explored more thoroughly. Berntson and Woodward (1992) reached a different conclusion by measuring in situ the root system architecture of *Senecio vulgaris* (Fig. 1). Elevated CO_2 resulted in more highly branched, longer root systems that foraged through larger volumes of soil. Furthermore, the reductions in branching and root length in ambient CO_2 resulting from low water availability were compensated for by CO_2 enrichment. While there were few aboveground measurements from which to assess whether changes in root system structure were a specific response to CO_2 concentration or an indirect response to altered ontogeny, the authors concluded that under elevated CO_2 the intensity of foraging root systems might be unchanged while the extent of foraging is increased.

Root activity

An increased supply of photosynthate to roots could stimulate various metabolic or microbial activities in the roots or associated rhizosphere. The first question to address, then, is whether there is increased supply of photosynthate to roots of plants grown in elevated CO_2. Norby et al. (1987) used $^{14}CO_2$ to determine C allocation in *Pinus echinata* seedlings growing in ambient or elevated CO_2. Proportionately more ^{14}C was allocated to fine roots in the CO_2-enriched seedlings. Lekkerkerk et al. (1990), however, reported no differences in ^{14}C distribution to roots in wheat plants. Elevated CO_2 increased carbohydrate concentrations in *Pinus taeda* seedlings by 68% and compensated for the decline in carbohydrates caused by water stress (Tschaplinski et al., 1993). The concentration of non-structural carbohydrates (mainly starch) in roots of cotton plants was two- to four-fold higher in elevated CO_2 (Wong, 1990). While these increased carbohydrate concentrations are not necessarily indicative of increased photosynthate allocation to roots, they may be important as carbon reserves (Chomba et al., 1993) or in the response of roots to water stress (Tschaplinski et al., 1993).

If there is excess carbohydrate allocated to roots, the expected result could include higher rates of respiration or increased exudation of soluble organic compounds. Whipps (1985), however, reported that the percentage of ^{14}C translocated to maize roots that was subsequently released through respiration or exudation was not significantly affected by the CO_2 concentration in which the plants were grown. In the *Pinus echinata* study (Norby et al., 1987) exudation was marginally higher in elevated CO_2 and root respiration was lower. The apparent effect on exudation was attributed to a greater fine root mass rather than increased exudation per unit root. Lekkerkerk et al. (1990) concluded that as a percentage of the ^{14}C translocated to roots, relatively less was retained by the roots in elevated CO_2. The remainder was respired by roots or by the soil microbial community, presumably after exudation or sloughing of ^{14}C material. Körner and Arnone (1992) assumed that fine root turnover or exudation was higher in elevated CO_2 in their model tropical systems because soil xylanase and protease activities were higher. The possible effects of increased carbon supply to roots on microbial biomass or activity, as well as on mycorrhizal and nitrogen fixing relationships, are discussed by Norby et al. (1994) and O'Neill (1994).

Ion uptake across root membranes is an energy-demanding process that might respond to an increased supply of carbohydrate to root systems, but there have been few investigations on responses of ion uptake processes to elevated CO_2. Luxmoore et al. (1986) used a growth analysis approach to show that in *Pinus virginiana* seedlings specific nutrient uptake (i.e., nutrient uptake per unit root and time) was not altered by CO_2 enrichment. There have also been detailed mechanistic studies of ion uptake in response to soil CO_2 partial pressure (Thibaud et al., 1990).

Relevance of short-term experiments

Evidence from short-term experiments with small plants supports the suggestion that elevated CO_2 concentrations can in some cases stimulate root growth or root activity, providing a positive feedback on plant growth by improving resource availability. For example, in *Quercus alba*, the increase in fine root mass and associated mycorrhizal and rhizosphere activity was thought to be the mechanism by which P availability in the soil increased, thereby avoiding a P deficiency in the CO_2-enriched seedlings (Norby et al., 1986). Similar increases in N availability have not been observed, except in the case of N_2-fixing systems (Norby, 1987; O'Neill, 1994). Increased root and associated belowground activity in the CO_2-enriched tropical systems of Körner and Arnone (1992) apparently increased nutrient availability, but this resulted in nutrient loss from the system rather than in increased uptake by the plants. Stimulation of root growth in

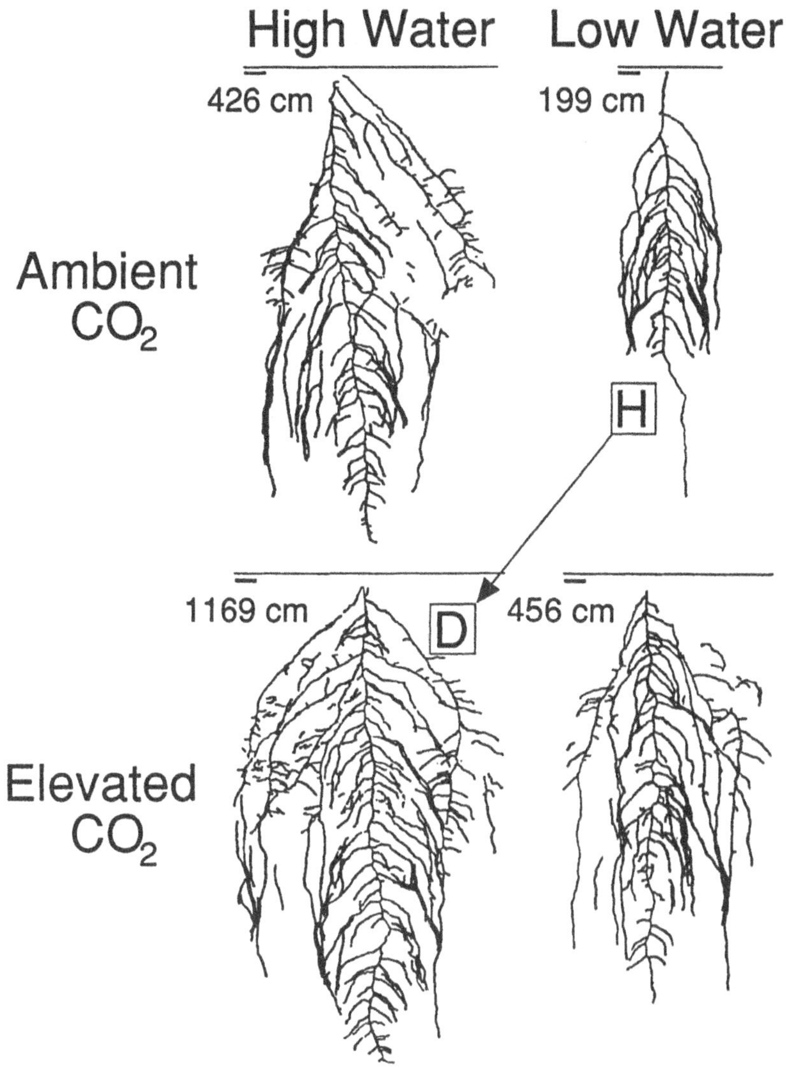

Fig. 1. Representative tracings of root systems of *Senecio vulgaris* plants grown in ambient of elevated CO_2 and high or low water availability (Berntson and Woodward, 1992). The long horizontal bar above each root is the maximum horizontal spread of the root system, with the shorter bar representing 1 cm. The number below the bars gives the total root length. CO_2 enrichment resulted in a more dichotomously branched (D) and longer root system that foraged through larger volumes of soil. Low water availability reduced branching and root length, resulting in a herringbone pattern (H), but this was compensated for by CO_2 enrichment. Figure courtesy of Glenn Berntson, Harvard University.

CO_2-enriched *Liriodendron tulipifera* seedlings, particularly the ratio of fine roots to leaf area, was associated with higher whole-plant water-use efficiency, which occurred in the absence of consistent effects of CO_2 on leaf-level stomatal conductance and transpiration efficiency (Norby and O'Neill, 1991).

Can the data from short-term experiments and our interpretations of those data be applied to plants growing in the field? In most cases the primary rationale for conducting the experiments was the hope that they would provide insight — if not direct predictions — about the responses of plants growing in the field in the CO_2-enriched atmosphere of the future. (The exceptions are considerations of responses to CO_2 in greenhouses for commercial production; Porter and Grodzinski, 1985). Hence, it is particularly important to consider whether the typical design of the short-term experiment produces relevant and appropriate data for larger-scale perspectives. The first consideration is whether the use of an artificial rooting medium contained in a pot creates an artifact large enough to prevent extrapolation to the field. Small pots can alter root growth and morphology, and CO_2 effects on root deployment can interact with pot characteristics (Berntson et al., 1993). Photosynthesis and other aboveground responses to elevated CO_2 are altered when root growth is restricted by small pots (Thomas and Strain, 1991). The effect of root restriction on growth and reproductive yield may be reversed by the addition of supplemental nutrients (McConnaughay et al., 1993), but nutrient additions to an inert rooting medium can create serious artifacts in plant nutrition and growth if growth rate and nutrient flux are mismatched (Brown and Higginbotham, 1986). Sophisticated approaches can maintain steady-state plant nutrition by matching nutrient flux density to relative growth rate (Pettersson and McDonald, 1992), but in the field steady-state nutrition is achieved by root growth (and concomitantly exploited soil volume and nutrient flux density) increasing in relation to shoot growth (Ingestad, 1982). Substituting nutrient additions for increased soil exploration may allow for a detailed examination of particular plant metabolic processes in response to CO_2 enrichment, but the responses of root growth and whole-plant carbon allocation cannot then be applied directly to a field situation in which root-soil interactions are important.

Some artifacts can be avoided by using unamended native soil as the rooting medium (e.g., Norby et al., 1986; Norby and O'Neill, 1989, 1991; El Kohen et al., 1992), although soil heterogeneity, microbial processes, and aeration are still disrupted, and maintenance of appropriate water supply is difficult. While the responses to CO_2 in unamended soil may be confounded with nutrient interactions, this is not necessarily a problem when the research questions are specifically focused toward responses in nutrient-limited habitats. Regardless, no pot can contain a tree for several growing seasons without affecting root growth when the roots of the field-grown trees extend for a meter or more in every direction. The results of CO_2 enrichment experiments with potted plants can suggest useful hypotheses about root responses to be tested in a more natural setting, but the responses should be used cautiously if the purpose is to predict the responses in the field.

Another problem is the difficulty in interpreting static measures of root mass when root growth is episodic or when significant root death and sloughing occur. While harvests of plant shoots can be scheduled in relation to observable phenological events (e.g., completion of leaf expansion), temporal patterns of root growth remain invisible without specialized equipment. The relative effects of CO_2 enrichment on root *versus* shoot growth are likely to be different just prior to a flush of root growth compared to after the completion of the growth flush. The difference between *Cedrus atlantic* and *Pinus nigra* in their chronological patterns of root growth in relation to shoot growth was an important determinant of the overall response to CO_2 (Kaushal et al., 1989). Poorter et al. (1988) also emphasized the importance of ontogenetic shifts in relative root growth. Root-to-shoot ratio of *Plantago major* was at first higher and later lower in elevated CO_2 compared to the ratio in plants in ambient CO_2, with the shift attributed to the increase in self-shading of leaves with increasing plant size. With this sort of variation, extrapolations from static measures of relative root mass to the field situation have little meaning.

A related complication in using short-term studies to predict responses of plants over several years is the role of roots in storage and remobilization of carbohydrates and nutrients. These issues usually are not relevant and therefore not considered in most short-term experiments. The possibility of remobilization of stored reserves complicated the analysis of CO_2 effects in *Quercus alba* seedlings because of the species' cyclic, episodic growth pattern (Norby and O'Neill, 1989). Wong (1990) argued that differential accumulation of non-structural carbohydrates in different CO_2 concentrations obscures the effect of CO_2 on growth and could lead to misinterpretations, particularly of the root-to-shoot ratio. Sasek and Strain (1988) suggested that their 60-day experiment with *Pueraria lobata* (kudzu) underestimated the effect CO_2 enrichment might have had on root-to-shoot ratio over a full growing season, since photosynthates are allocated to root growth and storage in the latter part of the growing season. Increased storage in overwintering roots could be expected to support increased shoot growth in the next growing season, a response to CO_2 not readily observable in short-term experiments with potted plants. The most difficult issues to address in relating the short-term pot experiment to the field situation are those requiring scaling across levels of ecological complexity. Patterns of resource acquisition within a competitive stand may differ from that under non-competitive conditions (Bazzaz and McConnaughay, 1992), but little is known about how CO_2 affects root growth in the presence of interspecific competition for soil resources. An increase in fine root proliferation that was observed in a pot experiment, for example, may not occur in the field if the rooting space is already occupied by another root system. Hence, pot experiments may be most useful for identifying the responses of small-scale processes to CO_2 enrichment rather than for describing root growth itself.

In extrapolating process data from a pot to a complex ecosystem, it is important to consider whether the response was a result of a specific stimulation of a process, such as increased exudation per unit root, or a result instead of increased root growth providing a larger platform for the process to occur. Although an increase in total exudation may be important within the simple ecosystem defined by the pot, if the response occurred only because there were more roots, then we have learned nothing new that would alter our predictions of plant response in the field. That is, if exudation scales linearly with root growth (as apparently was the case in the studies cited above), only the effect of CO_2 on root growth need be known to evaluate the potential consequences of increased exudation. Furthermore, if root density reaches a maximum value in a developing ecosystem (similar to the maximum leaf area after canopy closure), then any effects of CO_2 on exudation, nutrient uptake, mycorrhizal density, symbiotic N_2 fixation, or other such process would not continue to be affected by CO_2 unless the rates per unit root were stimulated.

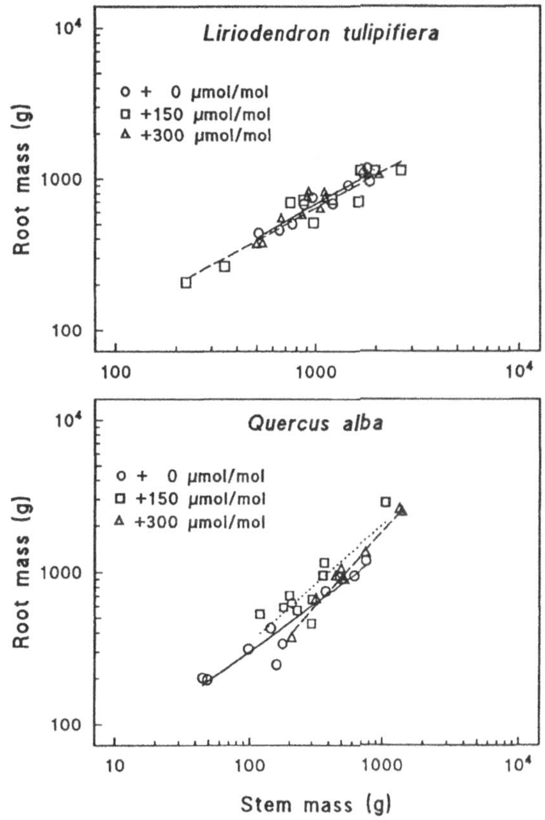

Fig. 2. Allometric analysis of root and stem mass of *Liriodendron tulipifera* and *Quercus alba* saplings grown for 3–4 growing seasons in open-top field chambers at different atmospheric CO_2 concentrations (Norby et al., 1992, 1993). In *Q. alba* the slopes of the log-log regression increased with increasing CO_2 concentration ($p < 0.09$).

Root responses in the field

Given the difficulty in applying the information about root responses from experiments with small, potted plants to answer larger-scale questions about plants and ecosystems in a CO_2-enriched atmosphere, it is especially important that root responses are measured in field experiments. Unfortunately, the number of experiments on elevated CO_2 conducted in the field is small, and none have focused specifically on root responses. The problems with studying roots in the field are well known: excavation is tedious, labor-intensive, and prone to error; non-destructive observations are difficult, so static measures cannot easily be related to root phenology; most approaches to measuring root processes such as respiration or exudation are likely to alter the process itself. Nevertheless, there exist enough data on root responses to CO_2 in the field to allow us to explore the important concepts and better define hypotheses and future research priorities.

The experiments that will be drawn upon in this discussion are the free-air CO_2 enrichment (FACE) of cotton in Mississippi and Arizona, in which Rogers et al. (1992) measured many root parameters and associated soil microbial responses; open-top chamber exposure of *Liriodendron tulipifera* (yellow-poplar) and *Quercus alba* (white oak) saplings for 3 to 4 years (Norby et al., 1992, 1993), in which root mass, fine root density, mycorrhization, and CO_2 efflux were measured; long-term exposure of *Citrus aurantium* (sour orange) trees under horticultural conditions (Idso and Kimball, 1991, 1992), in which fine root density was measured; and one-year exposure of *Populus grandidentata* (bigtooth aspen) (Zak et al., 1993), in which total root mass and associated soil responses were measured. New experiments with *Pinus taeda*, *Pinus ponderosa*, *Pseudotsuga menziesii*, and temperate grasslands have an increased focus on belowground responses, and should be providing valuable data in the future.

Root mass

Carbon stored in large, woody, perennial roots is a component of ecosystem carbon storage, just as is the carbon stored in woody stems, yet it is largely invisible and therefore largely ignored in analyses of global carbon budgets. To assess the potential of trees in a CO_2-enriched atmosphere to store additional carbon based only on their stem growth responses is implicitly to assume that roots and stems respond similarly. To address this problem, we excavated the tap roots of *Liriodendron* and *Quercus* saplings from the soil within the open-top chambers in which the trees were grown under different atmospheric CO_2 concentrations (Norby et al., 1992, 1993). The *Liriodendron* tap roots extended approximately 0.6 m deep, and the *Quercus* roots were as deep as 1.2 m. Lateral roots, which extended 1–2 m from the trunk, could not be excavated completely, so their mass was estimated from their diameter at the point of attachment to the tap root, using a regression analysis developed from a subset of lateral root systems that were completely excavated. *Liriodendron* root mass was 10% greater in trees grown in ambient + 150 μmol/mol CO_2, and 15% greater in ambient + 300 μmol/mol, compared to trees grown in ambient air. These differences were similar to the differences in whole-tree mass and were not close to statistical significance. In contrast, *Quercus* root mass was 77% greater in +150 and 136% greater in +300 compared to ambient-grown trees, significantly different at $p < 0.05$. Root-to-shoot ratio in these data sets is not particularly meaningful because the ratio decreased with increasing plant mass, thus confounding any specific effects of CO_2 on root growth with ontogenetic shifts coinciding with CO_2 effects on whole-plant growth. The allometric approach, in which root mass of individual plants is plotted against stem mass on log scales (Fig. 2), suggests that there were no effects of CO_2 on relative root growth in *Liriodendron*, but root growth was enhanced relative to stem growth in *Quercus*. The slopes of the regressions for *Quercus* are 0.64 in ambient CO_2, 0.76 in ambient + 150 μmol/mol, and 0.97 in ambient + 300 μmol/mol ($p < 0.09$). Specific effects of CO_2 enrichment on *Quercus* root mass are particularly important with regard to carbon storage because root mass was more than twice that of stems, whereas in *Liriodendron*, root mass was about 60% of stem mass.

Carbon storage in roots is probably not an important issue in the analysis of responses of agronomic systems to CO_2 enrichment (the root-to-shoot mass ratio of cotton is less than one-tenth that of *Quercus alba* saplings), but changes in rooting could influence other aspects of crop performance. In the FACE study of cotton in Arizona (but not as clearly in the similar Mississippi experiment), root length and mass density were greater in elevated CO_2, especially in the upper profile (Rogers et al., 1992). Taproot diameter, mass, and volume (but not length) increased, as well as the number, length, and mass of lateral roots. The responses were considered to be potentially important for seedling establishment, especially under stress conditions, but perhaps less significant under conditions of unlimited water and nutrient supply.

Fine roots

The importance of considering fine roots separately was emphasized in several previously discussed seedling experiments. In the field experiment with *Quercus* and *Liriodendron* saplings, fine root density of both species increased with increasing CO_2, and in *Liriodendron*, the increase greatly exceeded any effects of CO_2 on other plant components (Norby et al., 1992). However, the percentage of total root mass accounted for by fine roots was less than half the comparable percentage in seedlings (cf. Norby and O'Neill, 1991), and the increase in fine root mass had a minimal impact on whole-tree mass. Because of ontogenetic shifts in plant morphology, specific effects of elevated CO_2 on fine root mass will become a relatively less important component of whole-plant carbon storage as trees grow, and large effects of CO_2 on total root mass observed in seedlings may not translate into an equivalent effect in larger trees if the CO_2 effect is due primarily to increases in the fine root component. This analysis emphasizes the importance of separating functionally distinct components of root systems to increase the value of the data at different scales.

Regardless of the limited implications of fine root mass to carbon storage in the plant, fine roots are undoubtedly important physiologically for nutrient and water uptake, as a platform for interactions with soil microbes, and for carbon cycling in ecosystems. In addition to the *Liriodendron* and *Quercus* study, other field studies have shown fine roots to increase in elevated CO_2. In the cotton study, fine root density increased in elevated CO_2 at most depths to 90 cm, but was increased more significantly in the upper 45 cm of soil. The effect of CO_2 also tended to be larger as the distance from the row center increased, suggesting faster or more prolific spread of roots in elevated

CO_2 (Rogers et al., 1994). In *Citrus* trees fine root mass to a depth of 1.2 m increased 180% in elevated CO_2, identical to the increase in calculated aboveground mass (Idso and Kimball, 1991). Subsequent extractions of soil cores to 0.4 m led to an estimate of a 130% increase in fine root mass in the upper profile (Idso and Kimball, 1992). The relationship of fine root mass to trunk cross-sectional area through time was the same in ambient and elevated CO_2. Together these results suggest that CO_2 enrichment had no specific effect on fine root production in *Citrus* trees. Buried root ingrowth bags were used to measure root production in a tallgrass prairie ecosystem (Owensby et al., 1993b) and a brackish marsh ecosystem (Curtis et al., 1990). Root production in the tallgrass system was 130% greater in elevated CO_2 chambers in a wet year and 17% greater in a dry year, compared to that in ambient CO_2 chambers. Root production in the marsh community dominated by the C_3 sedge *Scirpus olneyi* was increased 83% in elevated CO_2.

None of these studies have been able to measure root mortality and decomposition in order to estimate total fine root production. Static measures of fine root mass at the end of a growing season probably underestimate the amount of carbon allocated by the plant to fine roots. An alternative, indirect measure is provided by the CO_2 efflux from the soil, which results from a combination of root respiration and soil microbial respiration. In the *Liriodendron system*, CO_2 efflux tended to be higher in elevated CO_2 (Norby et al., 1992), and this response was found to be a consistent and statistically significant one beneath the *Quercus* trees during the next year (R. J. Norby, unpublished data). Increased CO_2 efflux was interpreted as resulting from increased carbon allocation to fine roots in elevated CO_2 leading to an increase in carbon cycling through the soil. In the *Quercus* system, specific fine root respiration was lower in elevated CO_2, but the increase in fine root density more than compensated, and the proportionate increases in calculated total fine root respiration with increasing atmospheric CO_2 concentration were similar to the proportionate increases in CO_2 efflux from the soil (R. J. Norby, unpublished data).

Although it is easy to speculate that increased fine root production should be physiologically beneficial to the plant, the *Populus* study (Zak et al., 1993) has been the only one which attempted to quantify the linkages between CO_2 effects on root production and secondary responses. *Populus grandidentata* root sprouts were planted in nutrient-poor sand within open-top chambers and grown for 152 days in ambient or twice-ambient CO_2. Total plant biomass, of which about 80% was in roots, was significantly increased by CO_2 enrichment. The number of roots, root length, and root length production all were substantially greater at elevated CO_2. Associated with these root responses, labile carbon in the rhizosphere of CO_2-enriched plants was significantly greater than for plants in ambient air, and microbial biomass carbon was increased in both rhizosphere and bulk soil. The apparent increase in N mineralization was evidence for a positive feedback on soil C and N dynamics, effected through stimulation of belowground production. This experiment illustrates the importance of root production, mortality and metabolism, and their influence on soil microorganisms in understanding ecosystem responses to rising CO_2.

Root decomposition and soil organic matter

The mean residence time of live fine roots generally is about 1 year, although longer times have also been reported (Joslin and Henderson, 1987). Increased fine root production in elevated CO_2 cannot, therefore, represent an important mechanism for increasing long-term carbon sequestration by plants (except, perhaps, indirectly by improving nutrient or water status). However, the large annual production of fine roots coupled with their relatively slow decomposition rates after they are shed implies that a large portion of soil organic matter is derived from fine roots (McClaugherty et al., 1982). At least 25% of the soil organic matter in hardwood forests could be accounted for by fine roots (McClaugherty et al., 1984). If CO_2 enrichment causes a specific stimulation of fine root production or a decrease in fine root decomposition, ultimately there could be increased carbon sequestration by the ecosystem, even if not by the plant.

Root decomposition is in part a function of the chemical composition of the tissue in relation to the nutritional requirements of decomposer organisms, and is most commonly characterized as the C-to-N ratio. Carbon-to-N ratio (or N concentration) of live fine roots may be a reasonable predictor of root decomposition because N reportedly is not mobilized from fine roots prior to senescence (Nambiar, 1987). (In contrast, C-to-N ratio of green leaves is not a good predictor of leaf litter decomposition because N is withdrawn prior to leaf senescence; O'Neill, 1994.) Carbon-to-N ratio of roots of the marsh plant *Scirpus olneyi* was 22% higher in elevated CO_2 (Curtis et al., 1990). Nitrogen concentration in roots also tended to be lower in elevated CO_2 on the tallgrass prairie (Owensby et al., 1993a), in *Quercus alba* and dead *Liriodendron tulipifera* fine roots from field-grown saplings (R. J. Norby, unpublished data), and in fine roots from potted *Quercus* and *Liriodendron* seedlings (Norby et al., 1986; Norby and O'Neill, 1989).

Decomposition of fine roots produced in elevated CO_2 has not been measured. The possibility of changes in fine root production or decomposition leading to increased carbon sequestration as soil organic matter, while supported by ecosystem models (Hunt et al., 1991; Post et al., 1992; Thornley et al., 1991), is wholly a matter of speculation. The link between relatively short-term plant responses to CO_2 enrichment and any longer-term consequences at the ecosystem level will be very difficult to measure, but the possibility should not be ignored when attempting to extrapolate plant physiological data to predict ecosystem responses.

The need for better questions

The discussion above about the potential significance of CO_2 effects on fine roots illustrates an important need in global change research. It is imperative that the issues of interest be clearly defined so that the research questions are asked correctly and the data are interpreted in the right context. Three perspectives for analyzing the significance of root responses as a component of the response of the terrestrial biosphere to increasing atmospheric CO_2 are apparent:

(1) Roots as a platform for nutrient acquisition and a mediator of whole-plant response to CO_2.

(2) Carbon storage in roots as a component of whole-plant carbon storage.

(3) Effects of CO_2 enrichment on root turnover and the implications for carbon storage as soil organic matter.

The relative importance of these different perspectives will vary depending on whether the primary issue of interest is the role of vegetation in the global carbon cycle or rather plant and ecosystem responses for their own sake (i.e., environmental quality, economic benefit). For analyses of the global carbon cycle, carbon storage in roots and contributions of fine roots to soil organic matter are of paramount importance, but the role of roots in mediating whole-plant response is only important indirectly insofar as whole-plant carbon storage is increased. For analyses of plant and ecosystem responses, physiological responses of roots and associated microbial responses increase in importance (O'Neill et al., 1991). Implicit in all of these issues is the recognition that root responses cannot be considered apart from the rest of the system.

These issues will also vary in importance depending on the ecosystem of interest. In agricultural systems the only perspective of interest is the physiological role of roots. (Root crops are again an exception since carbon storage in roots has an economic attribute). In heavily fertilized and irrigated agrosystems, as distinct from dryland and less intensive systems, even CO_2 effects on nutrient acquisition might be of limited importance. In grasslands, both the physiological role of roots and the linkage between root production and soil organic matter could be of interest, but without woody roots, carbon storage in the living root system might have limited direct bearing on ecosystem carbon storage. In forest systems, all three perspectives could be critical, depending on the temporal and spatial scale of interest.

If the root response data from short-term experiments with potted seedlings are to be relevant to the larger-scale issues of interest, there must be some recognition that different perspectives are needed for different situations. Clearly, data describing CO_2 effects on root-to-shoot ratio are insufficient to address all of the critical questions concerning ecosystem responses to increasing atmospheric CO_2.

There is increasing recognition that responses below the ground to the CO_2 environment above the ground may be a critical missing link in our analyses of ecosystem responses to a changing atmosphere. Those belowground responses begin with changes in root growth and root processes. The extant data set that is relevant to the larger-scale issues of interest is partial, fragmentary, and sometimes ambiguous. Nevertheless, that data set has helped to provide us with sufficient understanding of the important and potentially responsive processes of root systems to study in CO_2 enrichment experiments, as well as providing the proper perspectives for analyzing those processes. It is no longer appropriate to state simply that "we need to know more about belowground responses." We can now better define our questions about roots and formulate specific testable hypotheses. A primary objective of the new generation of field-oriented experiments should now be to test and refine those hypotheses and evaluate the important role of roots and associated belowground responses in ecosystem responses to elevated atmospheric CO_2.

Acknowledgements

I thank Peter Curtis, Gerry O'Neill, and James Teeri for organizing the workshop "Belowground Responses to Rising Atmospheric CO_2: Implications for Plants, Soil Biota and Ecosystems Processes" for which this manuscript was prepared. I also thank Nelson Edwards and Stan Wullschleger for their critical reviews of the manuscript. Research sponsored by the Global Change Research Program, Office of Energy Research, U.S. Department of Energy, under Contract No. DE-AC05-840R21400 with Martin Marietta Energy Systems, Inc. Publication No. 4235, Environmental Sciences Division, Oak Ridge National Laboratory.

References

Bazzaz F A 1990 Response of natural ecosystems to the rising global CO_2 levels. Annu. Rev. Ecol. Syst. 21, 167–196.

Bazzaz F A and McConnaughay K D M 1992 Plant-plant interactions in elevated CO_2 environments. Aust. J. Bot. 40, 547–563.

Berntson G M and Woodward F 1192 The root system architecture and development of *Senecio vulgaris* L. in elevated CO_2 and drought. Funct. Ecol. 6, 324–333.

Berntson G M, McConnaughay K D M and Bazzaz F A 1993 Elevated CO_2 alters deployment of roots in "small" growth containers. Oecologia 94, 558–564.

Brown K and Higginbotham K O 1986 Effects of carbon dioxide enrichment and nitrogen supply on growth of boreal tree seedlings. Tree Physiol. 2, 223–232.

Chomba B M, Guy R D and Weger H G 1993 Carbohydrate reserve accumulation and depletion in Engelmann spruce (*Picea engelmannii* Parry): effects of cold storage and pre-storage CO_2 enrichment. Tree Physiol. 13, 351–364.

Cure J D 1985 Carbon dioxide doubling responses: a crop survey. *In* Direct Effects of Increasing Carbon Dioxide on Vegetation. Eds. B R Strain and J D Cure. pp 99–116. DOE/ER-0238. U.S. Department of Energy, Washington, DC.

Curtis P S, Balduman L M, Drake B G and Whigham D F 1990 Elevated atmospheric CO_2 effects on belowground processes in C3 and C4 estuarine marsh communities. Ecology 71, 2001–2006.

Del Castillo D, Acock B, Reddy V R and Acock M C 1989 Elongation and branching of roots on soybean plants in a carbon dioxide-enriched aerial environment. Agron. J. 81, 692–695.

Dyson F 1992 From Eros to Gaia. Pantheon Books, New York.

Eamus D and Jarvis P G 1989 The direct effects of increase in the global atmospheric CO_2 concentration on natural and commercial temperate trees and forests. Adv. Ecol. Res. 19, 1–55.

El Kohen A, Rouhier H and Mousseau M 1992 Changes in dry weight and nitrogen partitioning induced by elevated CO_2 depend on soil nutrient availability in sweet chestnut (*Castanea sativa* Mill). Ann. Sci. For. 49, 83–90.

Hunt H W, Trlica M J, Redente E F, Moore J C, Detling J K, Kittel T G F, Walter D E, Fowler M C, Klein D A and Elliott E T 1991 Simulation model for the effects of climate change on temperate grassland ecosystems. Ecol. Modelling 53, 205–246.

Idso S B, Kimball B A and Mauney J R 1988 Effects of atmospheric CO_2 enrichment on root:shoot ratios of carrot, radish, cotton and soybean. Agric. Ecosyst. Environ. 21, 293–299.

Idso S B and Kimball B A 1991 Effects of two and a half years of atmospheric CO_2 enrichment on the root density distribution of three-year-old sour orange trees. Agric. For. Meteorol. 55, 345–349.

Idso S B and Kimball B A 1992 Seasonal fine-root biomass development of sour orange trees grown in atmospheres of ambient and elevated CO_2 concentration. Plant Cell Environ. 15, 337–341.

Ingestad T 1982 Relative addition rate and external concentration; driving variables used in plant nutrition research. Plant Cell Environ. 5, 443–453.

Joslin J D and Henderson G S 1987 Organic matter and nutrients associated with fine root turnover in a white oak stand. For. Sci. 33, 330–346.

Kaushal P, Guehl J M and Aussenac 1989 Differential growth response to atmospheric carbon dioxide enrichment in seedlings of *Cedrus atlantic* and *Pinus nigra* ssp. Laricio var. *Corsicana*. Can. J. For. Res. 19, 1351–1358.

Körner C and Arnone J A III 1992 Responses to elevated carbon dioxide in artificial tropical ecosystems. Science 257, 1672–1675.

Lekkerkerk L J, Van de Geijn A S C and Van Veen J A 1990 Effects of elevated atmospheric CO_2-levels on the carbon economy of a soil planted with wheat. *In* Soils and the Greenhouse Effect. Ed. A F Bouwman. pp. 423–429. John Wiley and Sons, Chicester, UK.

Luxmoore R J, O'Neill E G, Ells J M and Rogers H H 1986 Nutrient uptake and growth response of Virginia pine to elevated atmospheric carbon dioxide. J. Environ. Qual. 15, 244–251.

McClaugherty C A, Aber J D and Melillo J M 1982 The role of fine roots in the organic matter and nitrogen budgets of two forested ecosystems. Ecology 63, 1481–1490.

McClaugherty C A, Aber J D and Melillo J M 1984 Decomposition dynamics of fine roots in forested ecosystems. Oikos 42, 378–386.

McConnaughay K D M, Berntson G M and Bazzaz F A 1993 Limitations to CO_2induced growth enhancement in pot studies. Oecologia 94, 550–557.

Mooney H A 1991 Biological response to climate change. Ecol. Appl. 1, 112–117.

Nambiar E K S 1987 Do nutrients retranslocate from fine roots? Can. J. For. Res. 17, 913–918.

Norby R J, O'Neill E G and Luxmoore R J 1986 Effects of atmospheric CO_2 enrichment on the growth and mineral nutrition of *Quercus alba* seedlings in nutrient-poor soil. Plant Physiol. 82, 83–89.

Norby R J, O'Neill E G, Hood W G and Luxmoore R J 1987 Carbon allocation, root exudation, and mycorrhizal colonization of *Pinus echinata* seedlings grown under CO_2 enrichment. Tree Physiol. 3, 203–210.

Norby R J 1987 Nodulation and nitrogenase activity in nitrogen-fixing woody plants stimulated by CO_2 enrichment of the atmosphere. Physiol. Plant. 71, 77–82.

Norby R J and O'Neill E G 1989 Growth dynamics and water use of seedlings of *Quercus alba* L. in CO_2-enriched atmospheres. New Phytol. 111, 491–500.

Norby R J and O'Neill E G 1991 Leaf area compensation and nutrient interactions in CO_2-enriched yellow-poplar (*Liriodendron tulipifera* L.) seedlings. New Phytol. 117, 515–528.

Norby R J, Gunderson C A ,Wullschleger S D, O'Neill E G and McCracken M K 1992 Productivity and compensatory responses of yellow-poplar trees in elevated CO_2. Nature 354, 322–324.

Norby R J, O'Neill E G, Wullschleger S D, Gunderson C A and Nietch C T 1993 Growth enhancement of *Quercus alba* saplings by CO_2 enrichment under field conditions. Bull. Ecol. Soc. Am. 74 (Suppl.), 375.

Norby R J, O'Neill E G and Wullschleger S D 1994 Belowground responses to atmospheric carbon dioxide in forests. *In* Carbon Forms and Functions in Forest Soils. Eds. W W McFee and J M Kelly. Soil Science Society of America, Madison, WI (*In press*).

Oechel W C and Strain B R 1985 Naive species responses to increased atmospheric carbon dioxide concentration. *In* Direct Effects of Increasing Carbon Dioxide on Vegetation. Eds. B R Strain and J D Cure. pp 117–154. DOE/ER-0238. U.S. Department of Energy, Washington, DC.

O'Neill E G 1994 Responses of soil biota to elevated atmospheric carbon dioxide. Plant and Soil 165, 55–65.

O'Neill E G, O'Neill R V and R J Norby 1991 Hierarchy theory as a guide to mycorrhizal research on large-scale problems. Environ. Pollut. 73, 271–284.

Owensby C E, Coyne P I and Auen L M 1993a Nitrogen and phosphorus dynamics of a tallgrass prairie ecosystem exposed to elevated carbon dioxide. Plant Cell Environ. 16, 843–850.

Owensby C E, Coyne P I, Ham J M, Auen L M and Knapp A K 1993b Biomass production in a tallgrass prairie ecosystem exposed to ambient and elevated CO_2. Ecol. Appl. 3, 644–653.

Pettersson R and McDonald A J S 1992 Effects of elevated carbon dioxide concentration on photosynthesis and growth of small birch plants (*Betula pendula* Roth.) at optimal nutrition. Plant Cell Environ. 15, 911–919.

Poorter H, Pot S and Lambers H 1988 The effect of an elevated atmospheric CO_2 concentration on growth, photosynthesis and respiration of *Plantago major*. Physiol. Plant. 73, 553–559.

Porter M A and Grodzinski B 1985 CO_2 enrichment of protected crops. Hortic. Rev. 7, 345–398.

Post W M, Peng T-H, Emanuel W R, King A W, Dale V H and DeAngelis D L 1990 The global carbon cycle. Am. Sci. 78, 310–326.

Post W M, Pastor J, King A W and Emanuel W R 1992 Aspects of the interaction between vegetation and soil under global change. Water Air Soil Pollut. 64, 345–363.

Rogers H H, Prior S A and O'Neill E G 1992 Cotton root and rhizosphere responses to free-air CO_2 enrichment. Crit. Rev. Plant Sci. 11, 251–263.

Rogers H H, Runion G B and Krupa S V 1994 Plant responses to atmospheric CO_2 enrichment with emphasis on roots and the rhizosphere. Environ. Pollut. 83, 155–189.

Sasek T W and Strain B R 1988 Effects of carbon dioxide enrichment on the growth and morphology of kudzu (*Pueraria lobata*). Weed Sci. 36, 28–36.

Siegenthaler U and Sarmiento J L 1993 Atmospheric carbon dioxide and the ocean. Nature 365, 119–125.

Stulen I and den Hertog J 1993 Root growth and functioning under atmospheric CO_2 enrichment. Vegetatio 104/105, 99–115.

Tans P P, Fung I Y and Takahashi T 1990 Observational constraints on the global atmospheric budget. Science 247, 1431–1438.

Thibaud J-B, Toulon V, Sentenac H, Davidian J-C and Grignon C 1990 CO_2 high partial pressure in the rhizosphere affect mineral nutrition of plants: an approach to the mechanisms involved. Symbiosis 9, 69–70.

Thomas R B and Strain B R 1991 Root restriction as a factor in photosynthetic acclimation of cotton seedling grown in elevated carbon dioxide. Plant Physiol. 96, 627–634.

Thornley J H M, Fowler D and Cannell M G R 1991 Terrestrial carbon storage resulting from CO_2 and nitrogen fertilization in temperate grasslands. Plant Cell Environ. 14, 1007–1011.

Tschaplinski T J, Norby R J and Wullschleger S D 1993 Responses of loblolly pine seedlings to elevated CO_2 and fluctuating water supply. Tree Physiol. 13, 283–296.

Wong S-C 1990 Elevated atmospheric partial pressure of CO_2 and plant growth. II. Non-structural carbohydrate content in cotton plants and its effect on growth parameters. Photosyn. Res. 23, 171–180.

Whipps J M 1985 Effect of CO_2 concentration on growth, carbon distribution and loss of carbon from the roots of maize. J. Exp. Bot. 36, 644–651.

Wullschleger S D, Post W M and King A W 1994 On the potential for a CO_2 fertilization effect in forests — Estimates of the biotic growth factor based on 58 controlled-exposure studies. *In* Biospheric Feedbacks in the Global Climate System: Will Warming Speed the Warming? Ed. G M Woodwell. Oxford University Press, Oxford, UK (*In press*).

Zak D R, Pregitzer K S, Curtis P S, Teeri J A, Fogel R and Randlett D L 1993 Elevated atmospheric CO_2 and feedback between carbon and nitrogen cycles in forested ecosystems. Plant and Soil 151, 105–117.

Plant and Soil **165**: 21–32, 1994.

Root response to CO_2 enrichment and nitrogen supply in loblolly pine

A. Larigauderie, J. F. Reynolds and B. R. Strain
Department of Botany, Duke University, Phytotron Building, Durham, NC 27708-0340, USA. Present address: Botanisches Institut der Universität, Schönbeinstrasse 6, CH-4056 Basel, Switzerland

Key words: allocation, CO_2, image analysis, loblolly pine, nitrogen, root morphology

Abstract

This paper examines how elevated CO_2 and nitrogen (N) supply affect plant characteristics of loblolly pine (*Pinus taeda L.*) with an emphasis on root morphology. Seedlings were grown in greenhouses from seeds during one growing season at two atmospheric CO_2 concentrations (375 and 710 μL L^{-1}) and two N levels (High and Low). Root morphological characteristics were determined using a scanner and an image analysis program on a Macintosh computer. In the high N treatment, elevated CO_2 increased total plant dry weight by 80% and did not modify root to shoot (R/S) dry weight ratio, and leaf and plant N concentration at the end of the growing season. In the low N treatment, elevated CO_2 increased total dry weight by 60%. Plant and leaf N concentration declined and R/S ratio tended to increase. Nitrogen uptake rate on both a root length and a root dry weight basis was greater at elevated CO_2 in the high N treatment and lower in the low N treatment. We argue that N stress resulting from short exposures to nutrients might help explain the lower N concentrations observed at high CO_2 in other experiments; Nitrogen and CO_2 levels modified root morphology. High N increased the number of secondary lateral roots per length of first order lateral root and high CO_2 increased the length of secondary lateral roots per length of first order lateral root. Number and length of first order lateral roots were not modified by either treatment. Specific root length of main axis, and to a lower degree, of first order laterals, declined at high CO_2, especially at high N. Basal stem diameter and first order root diameters increased at high CO_2, especially at high N. Elevated CO_2 increased the proportion of upper lateral roots within the root system.

Introduction

Understanding the interaction between elevated atmospheric CO_2 and limiting resources such as nitrogen is a key element in predicting the response of plants to CO_2 enrichment in natural communities (Bazzaz, 1990). Greater root dry weights and higher root to shoot ratios (R/S) are commonly observed in plants grown at elevated atmospheric CO_2 concentrations (Rogers et al., 1993). It is often assumed that increased root dry weight and R/S ratio will be advantageous to plants for acquiring nutrients and water in a CO_2 enriched atmosphere, especially under limiting soil conditions (Stulen and den Hertog, 1993). However, before this conclusion can be drawn, root systems developed under elevated CO_2 need to be better described in terms of their morphology and architecture. The increased root biomass at elevated CO_2 could be the result of increased root storage in which case no improvement in nutrient acquisition might be expected.

There is some evidence that root systems may become less efficient at nutrient uptake on a dry weight basis at elevated CO_2. In some studies, elevated CO_2 increased the size of the root system and R/S ratio but did not stimulate N uptake, resulting in lower plant N concentrations (e.g. Hocking and Meyer, 1985; Norby et al., 1986; O'Neill et al., 1987; Larigauderie et al., 1988; Rogers et al., 1993). This suggests that the amount of nitrogen taken up per unit root dry weight was lower at elevated CO_2 as, for example, was observed by Hocking and Meyer (1985) on *Xanthium occidentale*. Understanding this lack of concomitant increase in N uptake is crucial for predicting long term effects of elevated CO_2 on trees. Norby et al. (1986) observed that less translocatable N was present in perennial woody tissues of CO_2-enriched white oak seedlings showing no increase in N uptake at elevated CO_2. These authors speculated that no further response to CO_2 was to be expected during the second year of CO_2 enrichment in a nutrient poor environment if N uptake was not stimulated.

We conducted a greenhouse experiment on loblolly pine (*Pinus taeda* L.) seedlings to study the following questions: Does elevated CO_2 modify root architecture and morphology? If so, how do these modifications relate to plant nutritional status? The ultimate purpose of our project is to predict how loblolly pine, a widespread forest species in the south-eastern portion of the United States, will be affected by future levels of atmospheric CO_2. Since this species is often N limited in its natural habitat, we conducted our CO_2 enrichment study at low and high levels of nitrogen supply.

Methods

Materials

Seeds of loblolly pine (*Pinus taeda* L.) were obtained from a single wild type seed source in piedmont North Carolina, from the North Carolina Forest Service (Lot# LB-NC-P-84-27). Seeds were planted on March 30, 1992 in four greenhouses, two maintained at 375 $\mu L\ L^{-1}$ and two maintained at 710 $\mu L\ L^{-1}$, under natural light conditions, at the Duke phytotron. Temperature in the greenhouses was 26°C day/20°C night. Seeds germinated on April 18, and on June 8 seedlings were transplanted from Styrofoam cups into 24-L, 30-cm-deep plastic pots containing a mixture of Turface, vermiculite, gravel and soil (4:2:2:1) and one tablespoon of an inoculate of *Pisolithus tinctorius*, an ectomycorrhizal fungi that associates with conifers (Marx, 1977). Plants were randomly placed in the greenhouse they had germinated in. Nitrogen treatment (Low or High N) was randomly assigned to each pot. Nitrogen is believed to be the primary limiting factor for growth in the Piedmont region of North Carolina where our seeds came from. Plants were watered to saturation every Monday, Wednesday and Friday morning with a modified Hoagland's solution containing 1 m*M* KH_2P0_4, 2 m*M* K_2S0_4, 1.5 m*M* $MgS0_4.7H_20$, 3 m*M* $CaCl_2$, .167 NaOH, .14 m*M* FeSeq., and micronutrients and either 1.75 m*M* or 5.5 m*M* NH_4N0_3 for the Low N and the High N treatment, respectively. Soil in the pot mixture provided an extra 4.2 mg L^{-1} $N0_3$ and had a low cation exchange capacity of 3.8 meq/100 cm^3. Plants were watered every Sunday with de-ionized water to prevent salt build up.

Harvests

Plants were harvested on June 8 (Harvest 1), July 13 (Harvest 2), August 20 (Harvest 3), September 15 (Harvest 4), and October 7 (Harvest 5), 51, 86, 124, 150 and 172 days after germination, respectively. Five plants per treatments were harvested at harvests 1 to 4, and 10 plants were harvested at harvest 5. At each harvest, the main stem was excised and needles were separated. Branches were counted and measured and projected needle area was determined with a LI-COR 3100 Leaf Area Meter. Needle area was calculated using the following regression between projected area and geometric measurements: $y = 0.177 + 3.793x$ where y is the area of fascicle needles and x is the projected area (Thomas et al., 1993). Stem and needles were oven-dried at 80°C. At harvests 3 to 5, needle area was determined on a subsample of approximately 3 g dry weight. Specific needle weight was calculated as the ratio of needle dry weight to needle area. At harvest 5, basal stem diameter was measured with digital calipers 1 cm above the soil surface.

Roots were washed carefully to preserve the structure of the root system. First order laterals were cut off the main tap root and counted. The length of the main root was measured. The root system was divided in four parts: 1) the upper 10 cm portion, and 2) the remaining portion of the main root axis (upper and lower main axis), and 3) the first order lateral and sublateral roots removed from the upper 10 cm of root main axis (upper laterals), and 4) the remaining first order lateral and sublateral roots (lower laterals). These root fractions were oven-dried separately. Since most lateral roots branch from the upper main axis, this portion was studied in more detail in this experiment. The main tap root is considered of order 0, and the laterals branching from it are called first order laterals. Sublaterals refer to second and higher order laterals. Loblolly pine roots were not branched beyond the third order as observed by Kozlowski and Scholtes (1948).

Photocopies of lateral roots were made for detailed morphological measurements using image analysis techniques adapted from Zoon and Van Tienderen (1990). Four upper and four lower first order lateral roots with their attached sublateral roots were photocopied at harvest 2, and three upper and four lower lateral roots were photocopied at harvest 3. No root morphological measurements were made at harvest 4.

At harvest 5, sub-samples were used instead of complete first order laterals. Four upper first order lateral roots were randomly selected and cut in 10 cm segments, and three of these 10-cm segments were randomly selected and copied. Photocopies were made by spreading wet roots with a needle between transparent plastic sheets. Oven dry weight of photocopied roots was determined. Numbers of second order laterals were counted by hand directly on the photocopies and diameter of first order laterals was measured 1 cm below the excision point using digital calipers.

Photocopies of roots were scanned on a Macintosh computer with a HP Scan Jet llc scanner and analyzed using NIH Image software. Scanned root pictures were first edited to remove extraneous images. The image was then made binary and skeletonized to a thickness of one pixel. The program counted the number of black pixels which was transformed into a length using the following calibration curve equation: y= -0.2068 + 0.0393 x (n=10; r^2=0.993), where y is a length in cm and x is the number of black pixels. The calibration curve was established by placing wires of various shapes (e.g. lines, circles) and of known lengths on the scanner. Lengths of selected sub components of the image, e.g. laterals, were obtained by manually selecting portions of the image for analysis. This technique allowed us to calculate the following root morphological characteristics: length of sublaterals per length of first order lateral root, number of secondary roots per length of first order lateral root, mean length of a first order lateral root, mean length of a secondary root with its sublaterals. Specific root lengths (SRL) of lateral roots were calculated as the ratio of root length to root dry weight.

Nitrogen analysis

Dried needles, stems and roots from harvest 5 were ground in a Wiley mill for Kjeldahl N analysis. Sub samples from each seedling (about 50 mg) were digested in a Technicon block digestor using a sulfuric acid-hydrogen peroxid flux (Lowther, 1980). Total N was analyzed using a TRACCS 800 Continuous Flow Analytical System.

Photosynthesis

Photosynthesis was measured on 5 plants per treatment 170 days after germination, immediately before the last harvest (harvest 5). Measurements were made between 10 and 12 a.m. in the greenhouse in which plants were growing, by placing 6 needles (2 fascicles) in the chamber of a LICOR 6200 portable photosynthesis system. All measurements were made at the growth CO_2 treatment of each plant and were expressed on a total leaf area basis.

Data analyses

All measurements were analyzed as randomized block designs, the greenhouse being treated as a block, with ANOVAs. All data were checked for normality. Root to shoot ratios were analyzed after square-root transformation. No transformation of the proportions presented in Figure 2 was necessary. Since block effect was significant in only a few cases, it is not shown in this paper. Error bars on graphs are standard errors. Significance levels of ANOVAs are indicated as follows: $**$: $p<.01$; $*$: p .01<<.05; m: .05 <p<.1.

Nitrogen uptake per unit root weight over the course of the experiment was calculated with the Runge Kutta method of numerical integration using the following equation:

$$\int \text{N Uptake} / \int \text{Root dry weight} \qquad (1)$$

where $\int$ is equal to the finite integral from time t_0=55 days (harvest 1) to time t=172 days (harvest 5) of N uptake and root dry weight. Mean root dry weights obtained at each of the five harvests for each treatment were used in the integration of root dry weight. For the integration of N uptake, N measurements obtained at harvest 5 were used for each treatment as well as a common initial plant N concentration of 1.5 mg g^{-1} for the four treatments (Strain and Thomas, unpublished data).

Nitrogen uptake per unit root length was computed between harvest 2 (86 days) and harvest 5 (172 days), since root length data were not available for harvest 1. Total length of the upper 10 cm of the root system calculated for harvest 2, 3 and 5 were used to integrate root length, since root length was not measured on the lower remaining part of the root system. The upper 10 cm of the root system represented between 79 and 90% of the total root system dry weight at harvest 5. Total length of lateral roots was calculated from the number of first order lateral roots (A; Fig. 3A), the mean length of a first order lateral root (B; Fig. 3B), and the length of sublateral per length of lateral root (C; Fig. 3E), using this formula:

$$\text{Length of lateral roots} = (\text{A} \times \text{B}) + ((\text{A} \times \text{B}) \times \text{C}), \qquad (2)$$

where A × B represents the total length of first order lateral roots, and ((A × B)×C) represents the total length of sublaterals. No statistical analysis was done on N uptake rates since calculations could only be done with mean values.

Results

Biomass

Total dry weight per plant was significantly increased by high CO_2 at harvests 2 to 5 in both nitrogen (N) treatments (Fig. 1A; Table 1). The effect of CO_2 on total dry weight was more pronounced than the effect of N; the N effect was significant only at harvest 4, and the interaction between CO_2 and N was significant only at harvest 3 (Table 1). At the final harvest, total dry weight at high CO_2 was increased by 60% at low N and by 80% at high N.

Root to shoot ratio (R/S) varied during the growing season, first declining until mid-August and then increasing later in the growing season (Fig. 1B). There was no significant effect of either CO_2 or N on R/S ratio except at harvest 4 where high CO_2 tended to increase R/S ratio in both N treatments (Table 1). Root to shoot ratio tended to be higher at high CO_2 at the low N, but the CO_2 × N interaction was significant only at harvest 2.

Shoot characteristics were modified by elevated CO_2 but not by N. At harvest 5 plants grown at elevated CO_2 had a greater needle dry weight per plant and a greater total needle area, but showed little change in specific needle weight (Table 2). Plants were taller, had more first and second order branches, and a greater total branch length. Photosynthetic rate measured at the CO_2 concentration received during growth was about doubled by high CO_2 at both N treatments (Table 2). Leaf nitrogen concentration was lowered by high CO_2 only at low N.

Root dry weight was significantly increased by elevated CO_2 at all sampling dates (Fig. 1C; Table 1). Nitrogen had a significant effect only at harvest 4. Within the root system, all parts were not equally stimulated by CO_2. The upper lateral root fraction was the biggest root compartment and represented between 54 and 73% of total root dry weight at harvest 5, depending on the treatment (Fig. 2A). This root fraction was significantly increased by elevated CO_2 but did not significantly vary with N supply (Table 1). The increase due to CO_2 appeared greater at low compared to high

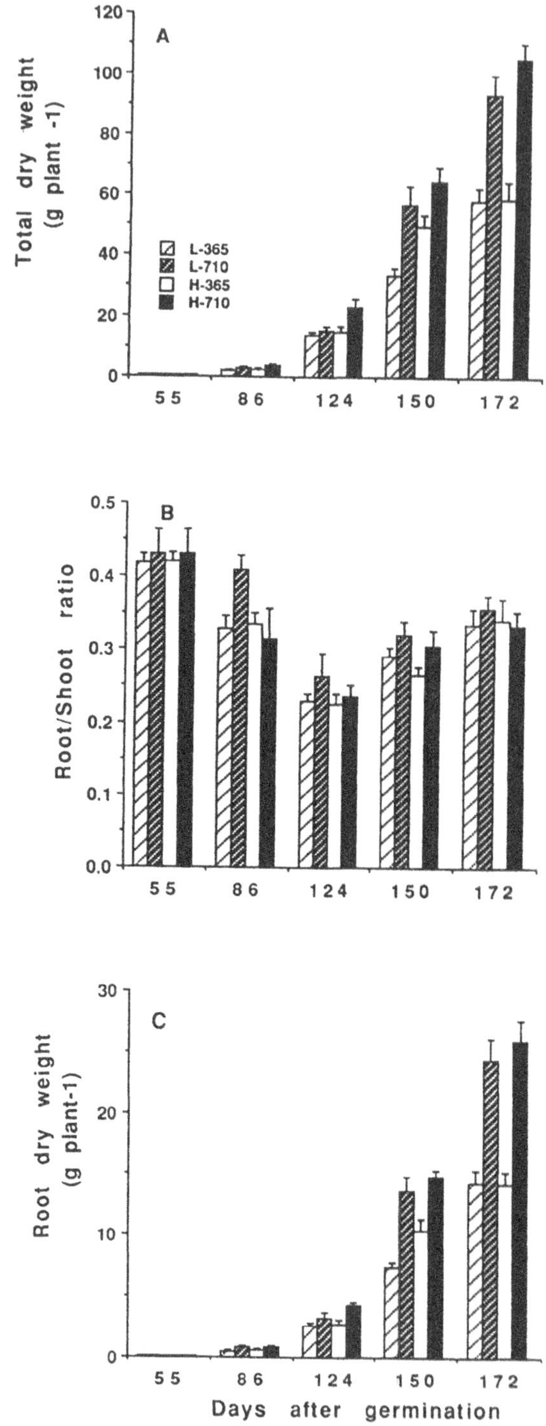

Fig. 1. Biomass characteristics of loblolly pine seedlings grown at either high (H) or low (L) nitrogen level and at either 375 or 710 $\mu L\ L^{-1}$ CO_2 with time after germination; total dry weight per plant (**A**), root to shoot dry weight ratio (**B**), and root dry weight per plant (**C**). Error bars indicate one standard error.

Table 1. Significant levels of ANOVAs on total dry weight, Root/Shoot dry weight ratio, and root morphological characteristics at harvests 2, 3, 4 and 5. Root morphology was not measured at harvest 4. The four divisions of root dry weight were analyzed as proportions of total root dry weight. **:p<.01;*: .01<p<.05; m: .05<p<.1

Variable	H2 (86 Days)			H3 (124)			H4 (150)			H5 (172)		
	CO_2	N	$CO_2 \times N$	CO_2	N	$CO_2 \times N$	CO_2	N	$CO_2 \times N$	CO_2	N	$CO_2 \times N$
Total dry weight	0.016*	0.449	0.481	0.014*	0.729	0.081^{m}	0.000**	0.026*	0.373	0.000**	0.948	0.328
Root/Shoot	0.234	0.864	0.061^{m}	0.297	0.922	0.557	0.052^{m}	0.291	0.737	0.772	0.847	0.497
Root dwt	0.006**	0.317	0.645	0.007**	0.782	0.208	0.000**	0.02*	0.338	0.000**	0.957	0.552
Proportion root dwt:												
Main axis (upper)	0.048*	0.143	0.167	0.129	0.557	0.669	0.004**	0.658	0.232	0.006**	0 .65	0.269
Main axis (lower)	0.484	0.828	0.408	0.028*	0.03*	0.181	0.04*	0.009**	0.012*	0.01**	0.166	0.161
Laterals (upper)	0.985	0.225	0.912	0.164	0.53	0.687	0.002**	0.128	0.117	0.001**	0.501	0.244
Laterals (lower)	0.550	0.217	0.710	0.008**	0.813	0.794	0.226	0.200	0.853	0.007**	0.401	0.229^{m}
m sublat./m first lat.	0.003**	0.987	0.685	0.419	0.265	0.069^{m}				0.052^{m}	0.665	0.907
No sec. roots/m first lat.	0.042*	0.009**	0.622	0.382^{m}	0.087^{m}	0.164				0.484	0.001**	0 .438
No upper first lat.	0.214	0.754	0.698	0.493	0.391	0.07^{m}	0.572	0.809	0.65	0.87	0.625	0.956
No lower first lat.	0.105	0.407	0.376	0.511	0.312	0.741	0.028*	0.011*	0.154	0.010**	0.212	0.019*
Mean length one first lat.	0.048*	0.682	0.526	0.477	0.968	0.116						
Mean length one sec.lat. with its sublat.	0.015*	0.005**	0.496	0.939	0.666	0.583				0.063^{m}	0.112	0.423
SRL upper main axis	0.509	0.046*	0.590	0.053^{m}	0.451	0.242^{m}				0.012*	0.088^{m}	0.133
SRL lower main axis	0.192	0.027*	0.853	0.134	0.861	0.069^{m}						
SRL upper laterals	0.058^{m}	0.042*	0.977	0.000**	0.790	0.100^{m}				0.985	0.745	0.971
SRL first order										0.306	0.897	0.734
SRL >first order										0.207	0.977	0.652
Diameter first order lat.	0.944	0.014*	0.944	0.000**	0.460	0.196						

Table 2. Aboveground characteristics of plants at harvest 5 and significance level obtained for ANOVAs **: p< .01; *: .01<p<.05; m: .05<p<.1. Photosynthesis was measured at growth CO_2 concentration

	Treatment				anova		
	Low N		High N		CO_2	N	$CO_2 \times N$
Variable	375 μL L^{-1}	710 μL L^{-1}	375 μL L^{-1}	710 μL L^{-1}			
Height (cm)	57.2	65.6	53.4	70.8	.000**	.453	.213
No 1° branches	12.8	18.6	1.2	14.4	.000**	.619	.141
No 2° branches	3.5	5.8	3.4	5.3	.250	.969	.912
Total branch lgth (m)	1.9	2.8	1.7	2.6	.001**	.491	.876
Needle dwt (g)	28.6	45.1	30.1	51.5	.000**	.73	.427
Stem dwt (g)	15.4	24.0	14.5	27.7	.000**	.744	.238
Needle area (m^2)	.78	1.15	.92	1.53	.000**	.384	.296
Specific needle weight (g m^{-2})	32.9	35.4	34.4	37.0	.018*	.270	.958
Stem diameter (cm)	13.9	15.6	13.1	16.6	.000**	.284	.083^{m}
Photosynthetic rate (μmol m^{-2} s^{-1})	4.2	8.9	4.6	9.3	.000**	.400	.968
Leaf [N] mg g^{-1}	19.9	15.9	20.9	20.8	.053^{m}	.492	.048*

N. For example, at high N and at harvest 5, the upper lateral fraction increased by 9% at high compared to ambient CO_2 and comprised 67% of total root dry weight. At low N, this fraction increased by 19% and comprised 73% of total root dry weight. The interaction between N and CO_2 was, however, not significant (Table 1). The proportion of the remaining root compartments generally declined at elevated CO_2 and

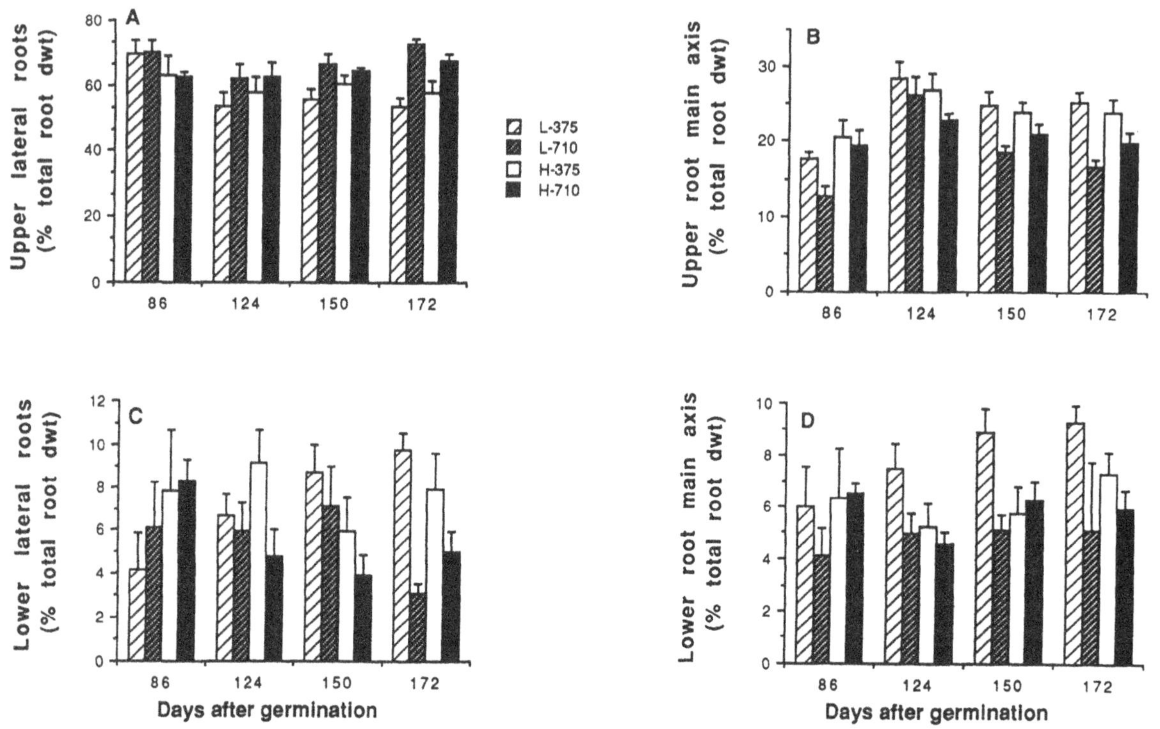

Fig. 2. Root biomass characteristics (percent of total root dry weight) of loblolly pine seedlings grown at either high (H) or low (L) nitrogen level and at either 375 or 710μL L^{-1} CO_2 with time after germination. Root systems were divided into four parts: upper lateral roots attached on the upper main axis with their sublaterals (**A**), upper 10 cm main root axis (**B**), lower laterals attached on the lower root main axis (**C**), and remaining root axis (**D**). Error bars indicate one standard error.

especially in the low N treatment, but did not change with N supply (Fig. 2B, C and D; Table 1). The lower root main axis and lower lateral roots fractions represented a small amount of total root dry weight and variability among these fractions was higher than in the upper root parts.

Root morphology

Root morphology is presented only for the upper portion of the root system (upper lateral roots and upper root main axis), which represented 80 to 90% of total root dry weight. The total number and the average length of first order laterals was not affected by either CO_2 or N (Table 1; Fig. 3A, B). The number of second order laterals per length of first order lateral, which is a measure of second order branching, was significantly greater at high N treatment throughout the harvests and was not significantly modified by elevated CO_2 (Fig. 3C). For example, at harvest 5 (day 172), high N plants grown at 710 μL L^{-1} CO_2 had twice as many second order laterals per cm of first lateral than low N plants grown at 710 μL L^{-1} CO_2. Secondary lateral roots were on average longer at elevated CO_2 and shorter at high N (Fig. 3D). The mean length of sublateral roots (i.e. secondary laterals with their sublaterals) per length of first order lateral was overall stimulated by elevated CO_2 and not modified by N (Fig. 3E). For example, at harvest 5, this value increased by 62% at elevated compared to ambient CO_2 for both N treatments.

The mean specific root length (SRL), the ratio between length and dry weight, of the main axis was significantly lowered by elevated CO_2 at harvest 3 and harvest 5, and especially so in the high N treatment (Fig. 4A; Table 1). For example, at harvest 5, elevated CO_2 decreased SRL by 10% at low N and 40% at high N. The interaction between CO_2 and N was, however, not significant (Table 1). The same trend was observed at harvest 2, but the CO_2 effect was not significant. Specific root length was significantly lowered by high compared to low N at harvest 2 and 5. At harvest 5, high N lowered SRL by only 2% at 375 μL L^{-1} and 50% at 710 μL L^{-1}. The same trend was observed at harvest 3, but the N effect was not significant. Specific

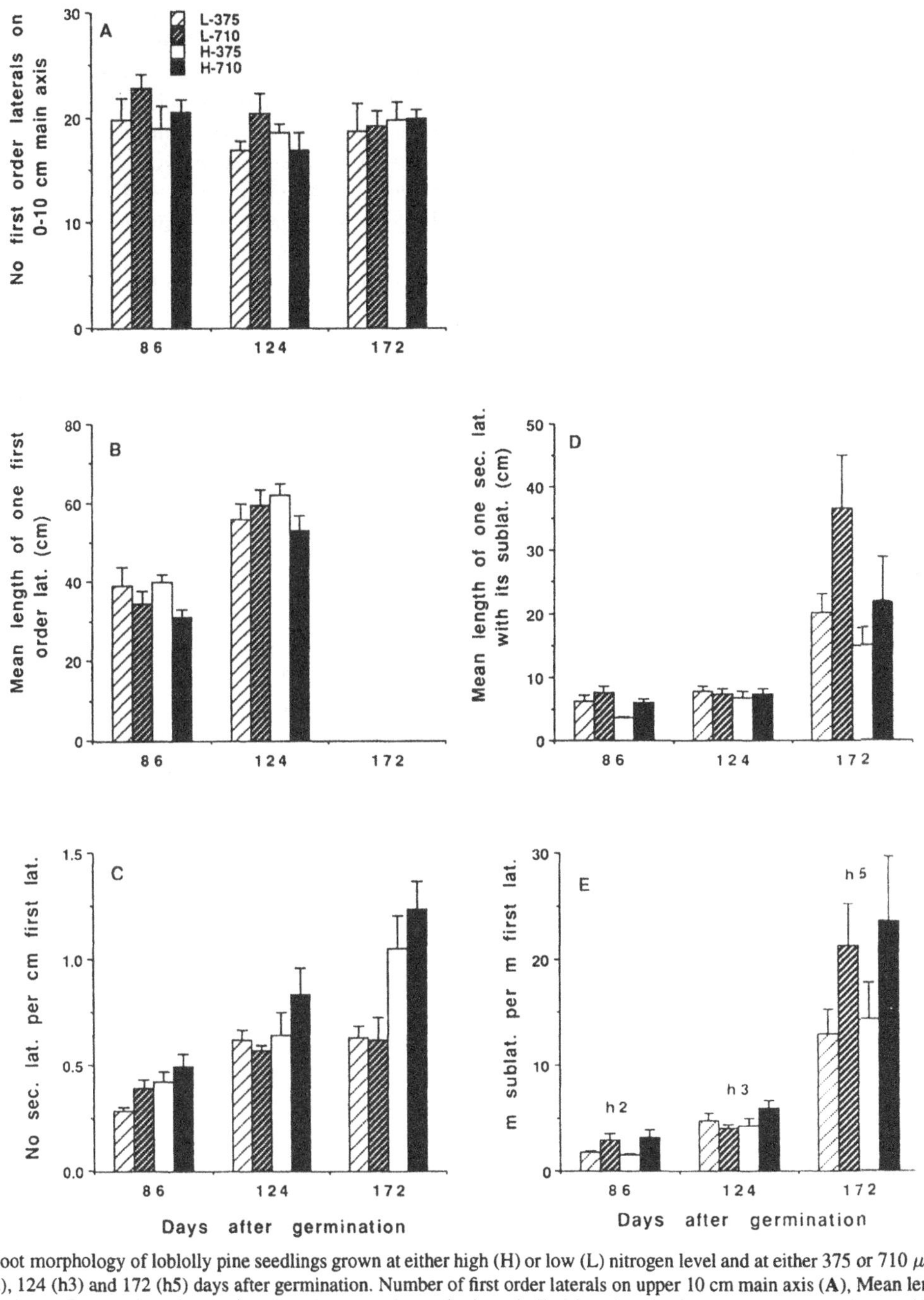

Fig. 3. Root morphology of loblolly pine seedlings grown at either high (H) or low (L) nitrogen level and at either 375 or 710 $\mu L\ L^{-1}\ CO_2$, 86 (harvest 2), 124 (h3) and 172 (h5) days after germination. Number of first order laterals on upper 10 cm main axis (**A**), Mean length of one first order lateral root (**B**), Number of secondary laterals per cm first order lateral (**C**), Mean length of one secondary lateral root with its sublaterals (**D**), and Length of sublateral roots per length of first lateral (**E**). Data are presented for upper lateral roots only. Error bars indicate one standard error.

root length of the lower main axis showed a similar response at harvest 2 but not at harvest 3 and was not measured at harvest 5 (Data not shown; Table 1). SRL of the upper lateral roots showed a response similar to that of the upper main axis but less pronounced (Fig. 4B; Table 1). It declined at high CO_2 at harvest 2 and 3 and declined with increased N at harvest 2. The trend toward an interaction between CO_2 and N was also

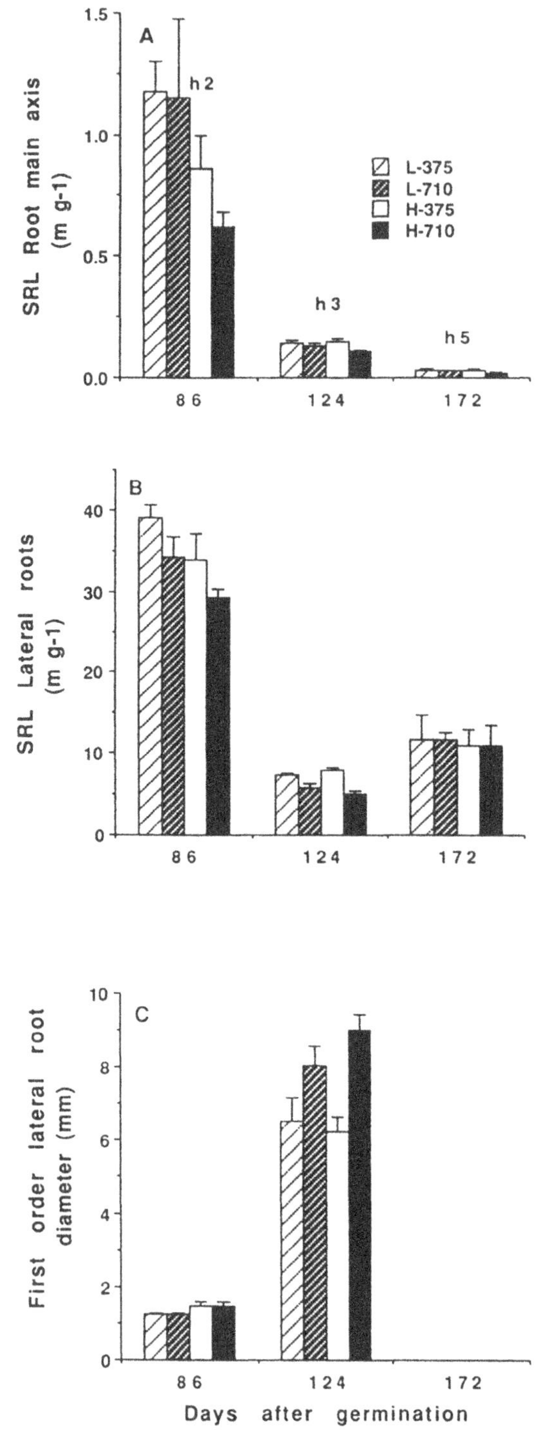

Fig. 4. Root morphology of loblolly pine seedlings grown at either high (H) or low (L) nitrogen level and at either 375 or 710 μL L^{-1} CO_2, 86 (harvest 2), 124 (h3) and 172 (h5) days after germination. Specific root length (SRL; m g^{-1}) of root main axis (**A**) and lateral roots (**B**) and mean diameter of one first order lateral root (**C**; mm). Data are presented for upper roots only. Error bars indicate one standard error.

less pronounced. No differences were observed at harvest 5, but when laterals were divided into first and sublateral roots, a similar trend was observed for first laterals, but not for sublaterals (data not shown) .

In agreement with these measurements, main stem diameter, measured at harvest 5 only, significantly increased at elevated CO_2, and more so at high N (Table 2). The interaction between CO_2 and N on main stem diameter was marginally significant.

The diameter of first order laterals also significantly increased at high CO_2, especially at high N at harvest 3 (44% at high N vs. 2% at low N; Fig. 4C; Table 1).

Nutrient analyses

Total plant N content (g N per plant) at the final harvest was significantly increased by CO_2, especially at high N (92 vs. 32% increase; Fig. 5A). There was no significant effect of N, but there was a significant interaction between CO_2 and N. Plant N concentration (mg N per g of plant) was not affected by either CO_2 or N (Fig. 5B). Leaf and stem nitrogen concentrations declined by about 20% at low N (significant interaction in both cases), but were not modified by elevated CO_2 at high N. Root nitrogen concentration was not affected by elevated CO_2 at either N treatment (data not shown).

Nitrogen uptake rate per unit root dry weight calculated for the duration of the experiment increased at elevated CO_2 by about 20% in the high N but decreased by the same percentage at low N (Fig. 5C; see Methods). The greater N supply did not modify N uptake rate at 375 μL L^{-1} CO_2 but increased it by 67% at elevated CO_2. Similar data were obtained for N uptake rate expressed on a root length basis (Fig. 5C). When expressed per unit root length, N uptake rate was 50% greater at elevated compared to ambient CO_2 in the high N treatment.

Discussion

Response to elevated CO_2 (at high N)

In the high N treatment, after one growing season at elevated CO_2, dry weight of loblolly pine seedlings was 80% greater than in control plants grown at 375 μL L^{-1} CO_2. Sionit et al. (1985) and Thomas et al. (1993) reported a comparable stimulation of growth in the same species after one growing season in pots (ca. 60%). In our study plant nitrogen concentrations were similar in both CO_2 treatments and so were leaf

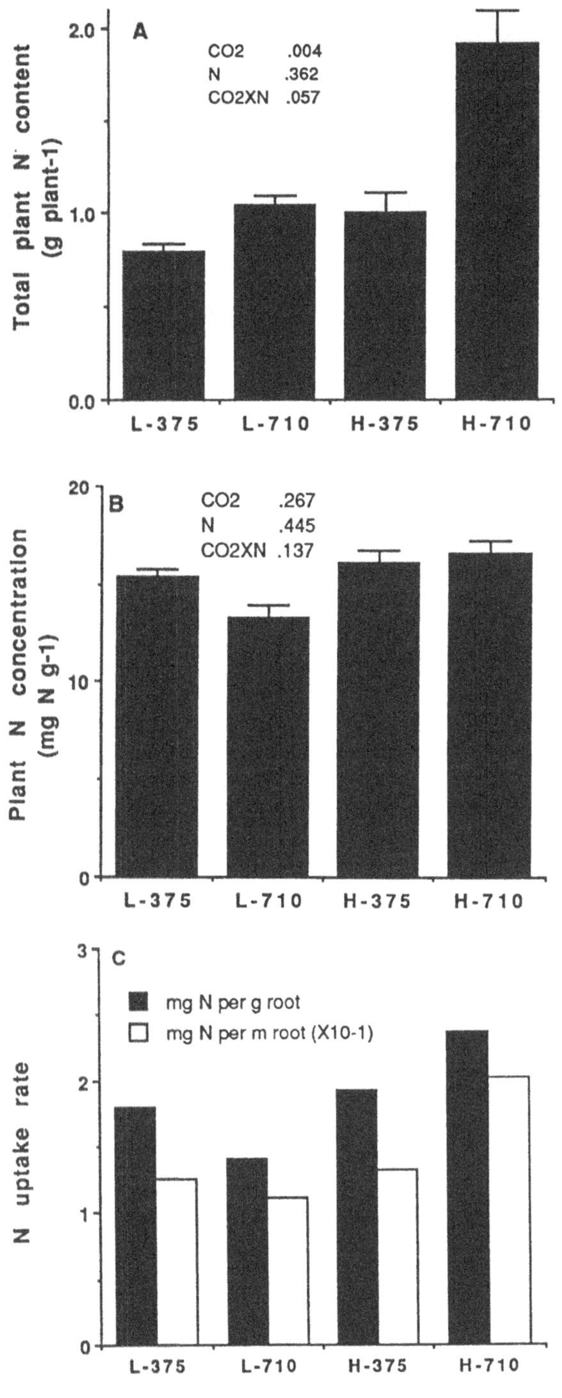

Fig. 5. Total plant N content (**A**; g plant $^{-1}$), plant N concentration (**B**; mg N g dry weight^{-1}) of loblolly pine seedlings grown at either high (H) or low (L) nitrogen level and at either 375 or 710 μL L^{-1} CO_2, 172 days after germination (harvest 5), with results of ANOVAs. N uptake rate (**C**; mg N per g root or per m root $\times$ 10^{-1}) was calculated for the full duration of the experiment.

N concentrations on both dry weight and area bases (data not shown). In a similar experiment on loblolly pine, Thomas et al. (1993) found needle N concentrations similar to those found in this experiment (18 mg g^{-1}) after one growing season at 375 μL L^{-1} CO_2 with a slightly higher N supply (7 m*M* NH_4N0_3). They, however, observed a 50% decline in leaf nitrogen concentration in the elevated CO_2 treatment.

Elevated CO_2 did not modify R/S ratio in this experiment, an observation also made by Sionit et al. (1985) and Thomas et al. (1993). Root to shoot ratio in loblolly pine seedlings has been shown to remain constant under a broad spectrum of environmental conditions (Ledig et al., 1970).

The structure and function of the root system, however, was modified by elevated CO_2. Roots built at elevated CO_2 had greater N uptake rates on both a dry weight and a length basis. Biomass allocation within the root system also was modified; in particular, the proportion of upper laterals significantly increased. A greater proportion of lateral roots might at least partially explain the greater N uptake rates obtained at elevated CO_2. Similarly, seedlings of *Pinus echinata* grown at 695 μL L^{-1} CO_2 allocated proportionately more ^{14}C-labeled photosynthate to fine roots than seedlings grown at 368 μL L^{-1} CO_2 (Norby et al., 1987). In our study, high CO_2 did not modify the number of first and second order lateral roots, but increased the length of second and higher order laterals. Similar findings were reported by Rogers et al. (1992b) for soybean. Del Castillo et al. (1989), however, obtained opposite data on soybean; high CO_2 increased the number of roots, i.e. branching, but did not modify root elongation rate. Cotton roots as in Del Castillo et al. (1989) also appeared to have a greater number of lateral roots per length of tap root at high CO_2 (Rogers et al., 1992a; 1989 experiment).

Lower root SRL and greater root diameter at elevated CO_2 have been reported in other studies. Sasek and Strain (1988) measured a greater effect of CO_2 on total root dry weight than on total root length in kudzu. For example, 14 days after planting, root dry weight was 100% greater at 675 μL L^{-1} CO_2 and root length only 30% greater. Consistent with these measurements, they reported a 20% greater maximum root diameter at elevated CO_2. Rogers et al. (1992b) found that root dry weight in soybean is significantly more stimulated by elevated CO_2 than root length, and root diameter in the root hair zone was 27% greater, due to both an increase in stele diameter and cortex width. In cotton, lateral root length and dry weight were increased by

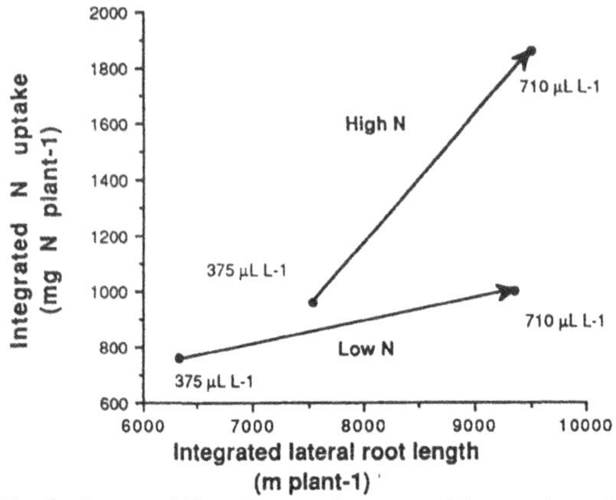

Fig. 6. Integrated N uptake over the course of the experiment (mg N per plant) as a function of integrated upper lateral root length (m plant^{-1}) for loblolly pine seedlings grown at either high (H) or low (L) nitrogen level and at either 375 or 710 μL L^{-1} CO_2, 172 days after germination (harvest 5). Graph shows how an increase in root length at elevated CO_2 resulted in a much greater N uptake at high N but not at low N over the course of the experiment.

100 and 157%, respectively, and roots diameters were greater at elevated CO_2 (Rogers et al., 1992a). After 4 months of growth at elevated CO_2, mean SRL of ponderosa pine roots was on average 57% lower at elevated compared to ambient CO_2 under conditions of high N supply (T. Ball and D. Johnson, pers. com.). Pettersson and McDonald (1992), however, found an opposite response in *Betula pendula* seedlings grown hydroponically in Ingestad-type units. The decline in SRL in the study presented here was strongest for the main axis and declined with higher order laterals. This effect of elevated CO_2 on SRL and root diameter might only be a consequence of the effect of CO_2 on root system size, i.e. CO_2 does not directly modify root characteristics but only accelerates plant growth. As the root system becomes bigger, the proportion of transport and support tissues increases, resulting in lower SRL and greater diameters (Boot, 1990), especially in lower branching orders. On the other hand, CO_2 could directly modify these root characteristics. Our measurements do not allow us to interpret these differences in root characteristics.

Our data show that plants grown at elevated CO_2 have the same proportion of roots, but that these roots have a greater proportion of lateral roots, due to a greater elongation of sublateral roots and not to increased branching, and lower SRL. Over the course of our experiment, integrated root length was 26% greater at elevated CO_2, and N uptake 94% greater, which illustrates the greater efficiency of elevated CO_2 roots at N uptake (Fig. 6).

Response to N availability (at 375 μL L^{-1} CO_2)

We found no difference between low N and high N plants grown at 375 μL L^{-1} CO_2 in terms of total dry weight, R/S ratio and biomass allocation within the root system. Low N plants had mean N content 20% lower than high N plants, but this difference was not significant. There were however some differences in root morphology between the two treatments. The first order lateral roots of the low N plants were less branched, as indicated by the lower number of secondary lateral roots per length of first order lateral root at the last harvest. A higher degree of branching is a common response of roots to a greater nutrient supply (e.g. Drew and Saker, 1975). Secondary lateral roots were, however, on average longer at low N, resulting in similar length of secondary roots per length of first order. Calculations based on morphological measurements indicate that integrated length of upper laterals was about 20% lower in the low N compared to the high N plants (Fig. 6). This lower value was not due to a single lower parameter, but to the fact that all the measurements used in the calculations are slightly lower at low N. The fact that roots had the same diameter and that a 20% lower length corresponded to 20% lower integrated N uptake shows that low and high N plants had similar N uptake rate and were functionally similar at 375 μL L^{-1} CO_2. Plants, however, differed in their response to elevated CO_2.

Interaction between CO_2 and N availability

Low N plants weighed on average only 10% less than high N plants grown at high CO_2 and were able to reach this comparable dry weight with 50% less N in their tissues. These data differ greatly from those of Thomas et al. (1993) who observed no stimulation of growth by elevated CO_2 in their low N treatment. Plants in their low N treatment were in average four times smaller than ours. The fact that we obtained large plants in our low N treatment and that leaf and plant N concentrations did not decline in our high N treatment at elevated compared to low CO_2 level (see first paragraph discussion) suggest that plants had more N available to them than was the case in Thomas et al. (1993) experiment. Concentrations of N applied in the nutrient solution, however, did not differ much between the two experiments (1 and 7 m*M* in Thomas et al., vs 1.75

and 5.5 m*M* here). The time of exposure to nutrients, however, varied greatly between the two experiments. Seedlings in the experiment of Thomas et al. (1993) were exposed three half-days per week to the nutrients, since it is a common phytotron practice to rinse with deionized water in the afternoon when nutrient solution is applied in the morning. In our experiment, seedlings were exposed one week at a time to the nutrient solution. We propose that time of exposure to nutrients was as important as the concentration of nutrients in controlling plant growth. Lower N concentrations in Thomas et al. (1993) and maybe in other experiments may be due to episodic N limitations, in spite of the high N concentrations supplied. The low N concentration supplied in our low N treatment may have been compensated for by a longer time of exposure. Furthermore, our growing medium contained some soil which did not contain much extra nitrogen, but may have enhanced exposure to nutrients by adsorption on soil particles. Experiments using hydroponic culture with automatic replenishment of nutrients lead to a stimulation of growth at high CO_2 but no change in plant N concentration (Knecht and O'Leary, 1983; Vessey et al., 1990). Furthermore, increasing NO_3^- concentration in the solution did not further stimulate plant growth (Vessey et al., 1990). Vessey et al. (1990) argue that lower N concentrations obtained at elevated CO_2 in experiments using porous media as opposed to hydroponic culture may only reflect nutrient stress. The question of reduced tissue N concentration under elevated CO_2 has been the object of much debate. Coleman et al. (1993) showed that reduced tissue N concentration is an indirect and size-dependent effect of CO_2, and that N concentration does not vary with CO_2 when plants of similar size are compared. Other workers (e.g. Wong, 1990) have suggested that starch accumulation dilutes N and lowers its concentration in the tissues. Here we suggest that the way to supply nutrients to plants growing in pots might create this pattern of reduced tissue N concentration observed at elevated CO_2 (e.g. short exposure to high N level) and should be included in this debate.

High CO_2 produced changes in root morphology at low N comparable to those observed at high N. Calculations of upper lateral root length based on morphological measurements revealed an interesting interaction between CO_2 and N (Fig. 6). Integrated upper lateral root length over the course of the experiment was stimulated by CO_2 by 48% at low N and only 26% at high N. Integrated root lengths of low and high N plants were equal at high CO_2. However, integrated N uptake increased 31% in the low N compared to 94% in the high N treatment at elevated CO_2. These calculations suggest that at elevated CO_2 low N plants compensated for their lower root length observed at 375 μL L^{-1}, but that they were unable to absorb the same quantity of nitrogen due to the lower N supply.

These data give a rather complex picture of the effects of elevated CO_2 on roots in this experiment and suggest that no conclusion about nutrient status should be drawn from measurements of total root dry weight and root to shoot ratio alone. This experiment shows that CO_2 can affect a number of within-organ processes, such as within-root biomass allocation and root morphological characteristics which can potentially modify plant nutrient status, without shifts in whole plant allocation patterns. This experiment finally suggests that time of exposure may be as important as nutrient concentration in interpreting phytotron experiments. Characterizing and quantifying the response of roots to elevated CO_2 remains one of the major tasks in our effort to better understand how plants grow at elevated CO_2.

Acknowledgements

We thank Mary Brunelle, Paul Kemp, Paul Leadley, Jeff Pippen, Richard Thomas, David Tremmel, the staff of the loblolly pine CO_2 field site, the staff of the Duke phytotron, and the two anonymous reviewers of this manuscript. This research was supported by Department of Energy, CO_2 Research Division contract DE-FG0587ER60575, Electric Power Research Institute Forest Response to CO_2 Program and by NSF Grant BSR87–06429 for support of the Duke University Phytotron.

References

Bazzaz F A 1990 The response of natural ecosystems to the rising global CO_2 levels. Annu. Rev. Ecol. Syst. 21, 167–196.

Boot R G A 1990 The significance of size and morphology of root systems for nutrient acquisition and competition. *In* Causes and Consequences of Variation in Growth Rate and Productivity of Higher Plants. Eds. H Lambers, M L Cambridge, H Konings and T L Pons. pp 299–311. SPB Academic Publishing, The Hague, The Netherlands.

Coleman J S, McConnaughay K D M and Bazzaz F A 1993 Elevated CO_2 and plant nitrogen-use: is reduced tissue nitrogen concentration size-dependent? Oecologia 93, 195–200.

Del Castillo D, Acock B, Reddy V R and Acock M C 1989 Elongation and branching of roots on soybean plants in a carbon-dioxide-enriched aerial environment. Agron. J. 81, 692–695.

Drew M C and Saker L R 1975 Nutrient supply and the growth of the seminal root system in barley. J. Exp. Bot. 26, 79–90.

Hocking P J and Meyer C P 1985 Responses of Noogoora burr (*Xanthium occidentale* Bertol.) to nitrogen supply and carbon dioxide enrichment. Ann. Bot. 55, 835–844.

Knecht GN and O'Leary J W 1983 The influence of carbon dioxide on the growth, pigment, protein, carbohydrate, and mineral status of lettuce. J. Plant Nutr. 6, 301–312.

Kozlowski T T and Scholtes W H 1948 Growth of roots and root hairs of pine and hardwood seedlings in the Piedmont. J. For. 46, 750–754.

Larigauderie A, Hilbert D W and Oechel W C 1988 Effect of CO_2 enrichment and nitrogen availability on resource acquisition and resource allocation in a grass, *Bromus mollis*. Oecologia 77, 544–549.

Ledig F T, Bormann F H and Wenger K F 1970 The distribution of dry matter growth between shoot and roots in Loblolly pine. Bot. Gaz. 131, 349–359.

Lowther J R 1980 Use of a single sulfuric acid-hydrogen peroxide digest for the analysis of *Pinus radiata* needles. Comm. Soil Sci. Plant Anal. 11, 175–188.

Marx D H 1977 Tree host range and world distribution of the ectomycorrhizal fungus *Pisolithus tinctorius*. Can. J. Microbiol. 23, 217–223.

Norby R J, Pastor J and Melillo J M 1986 Carbon-nitrogen interactions in CO_2 enriched white oak: physiological and long-term perspectives. Tree Physiol. 2, 233–241.

Norby R J, O'Neill E G, Hood W G and Luxmoore R J 1987 Carbon allocation, root exudation and mycorrhizal colonization of *Pinus echinata* seedlings grown under CO_2 enrichment. Tree Physiol. 3, 203–210.

O'Neill E G, Luxmoore R J and Norby R J 1987 Elevated atmospheric CO_2 effects on seedling growth, nutrient uptake, and rhizosphere bacterial populations of *Liriodendron tulipifera* L. Plant and Soil 104, 3–11.

Pettersson R and McDonald A J S 1992 Effects of elevated carbon dioxide concentration on photosynthesis and growth of small birch plants (*Betula pendula* Roth.) at optimal nutrition. Plant Cell Environ. 15, 911–919.

Rogers H H, Peterson C M, McCrimmon J N and Cure J D 1992a Response of plant roots to elevated atmospheric carbon dioxide. Plant Cell and Environ. 15, 749–752.

Rogers H H, Prior S A and O'Neill E G 1992b Cotton root and rhizosphere responses to free-air CO_2 enrichment. Crit. Rev. Plant Sci. 11, 251–263.

Rogers H H, Runion G B and Krupa S V 1993 Plant responses to atmospheric CO_2 enrichment with emphasis on roots and the rhizosphere. Environ. Pollut. (*In press*).

Sasek T W and Strain B R 1988 Effects of carbon dioxide enrichment on the growth and morphology of kudzu (*Pueraria lobata*). Weed Sci. 36, 28–36.

Sionit N, Strain B R, Hellmers H, Riechers G H and Jaeger C H 1985 Long-term atmospheric CO_2 enrichment affects the growth and development of *Liquidambar styraciflua* and *Pinus taeda* seedlings. Can. J. For. Res. 15, 468–471.

Stulen I and den Hertog J 1993 Root growth and functioning under atmospheric CO_2 enrichment. Vegetatio 104/105, 99–115.

Thomas R B and Strain B R 1991 Root restriction as a factor in photosynthetic acclimation of cotton seedlings grown in elevated carbon dioxide. Plant Phys. 96, 627–634.

Thomas R B, Lewis D and Strain B R 1993 Effects of leaf nutrient status on photosynthetic capacity in loblolly pine (*Pinus taeda* L.) seedlings grown in elevated atmospheric CO_2. Tree Physiol. (submitted).

Tissue D T, Thomas R B and Strain B R 1993 Long-term effects of elevated CO_2 and nutrients on photosynthesis and Rubisco in loblolly pine seedlings. Plant Cell Environ. (*In press*).

Vessey J K, Henry L T and Raper C D 1990 Nitrogen nutrition and temporal effects of enhanced carbon dioxide on soybean growth. Crop Sci. 30, 287–294.

Wong S C 1990 Elevated atmospheric partial pressure of CO_2 and plant growth. II Nonstructural carbohydrate content in cotton plants and its effect on growth parameters. Photosyn. Res. 23, 171–180.

Zoon F C and Van Tienderen P H 1990 A rapid quantitative measurement of root length and root branching by microcomputer image analysis. Plant and Soil 126, 301–308.

Plant and Soil **165:** 33–44, 1994.

Free-air CO_2 enrichment of cotton: vertical and lateral root distribution patterns

S. A. Prior[1], H. H. Rogers[1], G. B. Runion[2] and G. R. Hendrey[3]
[1]*National Soil Dynamics Laboratory, Agricultural Research Service, U.S. Department of Agriculture, P.O. Box 3439, Auburn, Alabama, 36831–3439, USA,* [2]*School of Forestry, Auburn University, 108 M. White Smith Hall, Auburn University, Alabama 36849–5418, USA and* [3]*Department of Applied Science, Brookhaven National Laboratory, Upton, NY 11973, USA*

Key words: Gossypium hirsutum, rising CO_2, root dry weight density, root length density, root lineal density

Abstract

The objective of this investigation was to determine how free-air carbon dioxide enrichment (FACE) of cotton (*Gossypium hirsutum* L.) affects root distribution in a natural soil environment. For two years cotton was grown on a Trix clay loam under two atmospheric CO_2 concentrations (370 and 550 μmol mol^{-1}) and two water treatments [wet, 100% of evapotranspiration (ET) replaced and dry, 75% (1990) and 67% (1991) of ET replaced] at Maricopa, AZ. At early vegetative and mid-reproductive growth, 90 cm soil cores were taken at 0, 0.25, and 0.5 m perpendicular to row center; root variables were ascertained at three 30 cm depth increments. The effect of water stress alone or its interaction with CO_2 on measured variables during both samplings were rare and showed no consistent pattern. There was a significant $CO_2 \times$ position interaction for root length density at the vegetative stage (both years) and reproductive stage (1990 only); the positive effects of extra CO_2 were more evident at interrow positions (0.25 and 0.5 m). A $CO_2 \times$ depth $\times$ position interaction at the vegetative phase (1990) indicated that FACE increased root dry weight densities for the top soil depth increment at all positions and at the middle increment at the 0.5 m position. Similar trends were seen at the reproductive sampling for this measure as well as for root length density at both sample dates in 1990. In 1991, a $CO_2 \times$ depth interaction was noted at both periods; CO_2 enhancement of root densities (i.e., both length and dry weight) were observed within the upper and middle depths. Although variable in response, increases for root lineal density under high CO_2 were also seen. In general, results also revealed that the ambient CO_2 treatment had a higher proportion of its root system growing closer to the row center, both on a root length and dry wight basis. On the other hand, the FACE treatment had proportionately more of its roots allocated away from row center (root length basis only). Results from this field experiment clearly suggest that increased atmospheric CO_2 concentration will alter root distribution patterns in cotton.

Introduction

Increased atmospheric CO_2 concentration (Keeling et al., 1989) arising from anthropogenic sources (e.g., fossil fuel use, deforestation; Pearman, 1980; Woodwell, 1978) are predicted to double in the next century (Gribbon, 1981). Although CO_2-induced shifts in climate are debatable (Idso, 1989; Rosenberg et al., 1990; Smit et al., 1988), increased CO_2 levels will directly affect vegetation in natural and agro-ecosystems. There is solid evidence showing that increases in atmospheric CO_2 level will increase biomass production, growth, photosynthesis (Allen et al., 1987; Rogers et al., 1983; Sionit et al., 1984; Strain and Cure, 1985; Warrick, 1988) and water use efficiency (Carlson and Bazzaz, 1980; Dahlman et al., 1985; Kimball and Idso, 1983). There have been several recent reviews of plant responses to CO_2 (Allen, 1990; Kimball et al., 1990; Krupa and Kickert, 1989; Rogers and Dahlman, 1993). Most reported investigations have used plants grown in containers within controlled environments which may not entirely represent the variable and complex environments which exist under field conditions. Furthermore, it has recently been reported that the results of studies using pot-grown plants where roots are confined may not always be indicative of field responses, i.e., source-sink relationships may be affected (Arp, 1991; Thomas and Strain, 1991). These findings not only emphasize the need for field studies, but also highlight the importance of investigating belowground responses, an area that has often been neglected.

There are far fewer belowground than aboveground investigations in the CO_2 response literature. The most commonly reported root descriptor, showing CO_2-induced enhancement, is root dry weight (Rogers et al., 1994). This has been observed in some important crop species such as sorghum [*Sorghum bicolor* (L.) Moech.], (Chaudhuri et al., 1986), soybean [*Glycine max* (L.) Merr.], (Del Castillo et al., 1989; Rogers et al., 1992b), and winter wheat [*Triticum aestivum* (L.)], (Chaudhuri et al., 1990). A few crop experiments have indicated that root increases (i.e., dry weight and/or length) were concentrated in the upper portions of the soil profile (Chaudhuri et al., 1990; Del Castillo et al., 1989) suggesting a more thorough exploration of a given volume of soil under elevated CO_2. Others have observed enhancements at all soil depths (Chaudhuri et al., 1986; Rogers et al., 1992b), thus implying that the volume of soil explored may be greater. Roots of CO_2 enriched plants also have been shown to reach deeper (Rogers et al., 1992b) or to attain maximum rooting depth ahead (in time) of plants grown under ambient CO_2 levels (Chaudhuri et al., 1990). Detailed morphological studies have also demonstrated that high CO_2 levels increase stele and cortex diameter, root diameter in the root hair zone, length of unbranched first order laterals, and total root length and volume (Rogers et al., 1992b). Collectively, results from the above studies suggest that more plant rooting may occur in a CO_2 enriched world; however, further work is warranted on root system response in natural soil profiles in the field.

The present study utilized cotton plants grown in the field under free-air carbon dioxide enrichment (FACE). This exposure system ensures continuous CO_2 treatment with minimal alteration of plant microclimate while allowing plants to grow in their natural environments. The objective was to investigate how CO_2 enrichment of a cotton canopy affects root distribution (vertical and horizontal) during two growth phases (vegetative and reproductive) under two water regimes.

Materials and methods

Experimental site

This study was conducted for two years (1990–1991) on a Trix clay loam [fine, loamy, mixed (calcareous), hyperthermic Typic Torrifluvents] at the Maricopa Agricultural Center for Resources and Extension (MAC) of the University of Arizona at Maricopa located 25 miles south of Phoenix, AZ. Cotton seeds [*Gossypium hirsutum* (L.) 'Delta Pine 77'] were sown into dry raised beds (1 meter row spacing) on 23 April 1990 and 16 April 1991 with plants being thinned to 10 per meter after 50% emergence. Recommended farming practices of the region as reported by Mauney et al. (1994) were used in managing both the soil and crop.

Experimental design

The experimental design was a split-plot arranged as a randomized complete block with four replications. The main plot factor was CO_2 concentration at two levels (ambient=370 μmol mol^{-1} and FACE=550 μmol mol^{-1}). The split plot treatments were soil water content (wet =100% of evapotranspiration (ET) replaced and dry = 75 and 67% of ET replaced in 1990 and 1991, respectively).

CO_2 exposure system

A technique called free-air carbon dioxide enrichment (FACE) was utilized to create large-scale test atmospheres of CO_2 in the field with minimal alteration of microclimate (Fig. 1). Each circular array (i.e., exposure unit) was constructed of a 22 m diameter PVC torus (plenum chamber) with 32 vertical ventpipes (2 m height) spaced evenly around its diameter. Each ventpipe was individually valved and had gas exit ports drilled at vertical intervals along its length. A computer program (based on an algorithm keyed to wind velocity and CO_2 measured at the center of the experimental plot; see Fig. 1) controlled the release of CO_2 into the array. Carbon dioxide was dumped upwind from open sectors of the vertical pipes in amounts equivalent to windspeed such that a circular plot was uniformly fumigated (Hendrey et al., 1993). There were eight experimental plots, four at nominal CO_2 concentrations of 550 μmol mol^{-1} (surrounded by the 22 m FACE array) and four controls at ambient, 370 μmol mol^{-1} (surrounded by a dummy array). Plots were spaced so as to maximize distance between them

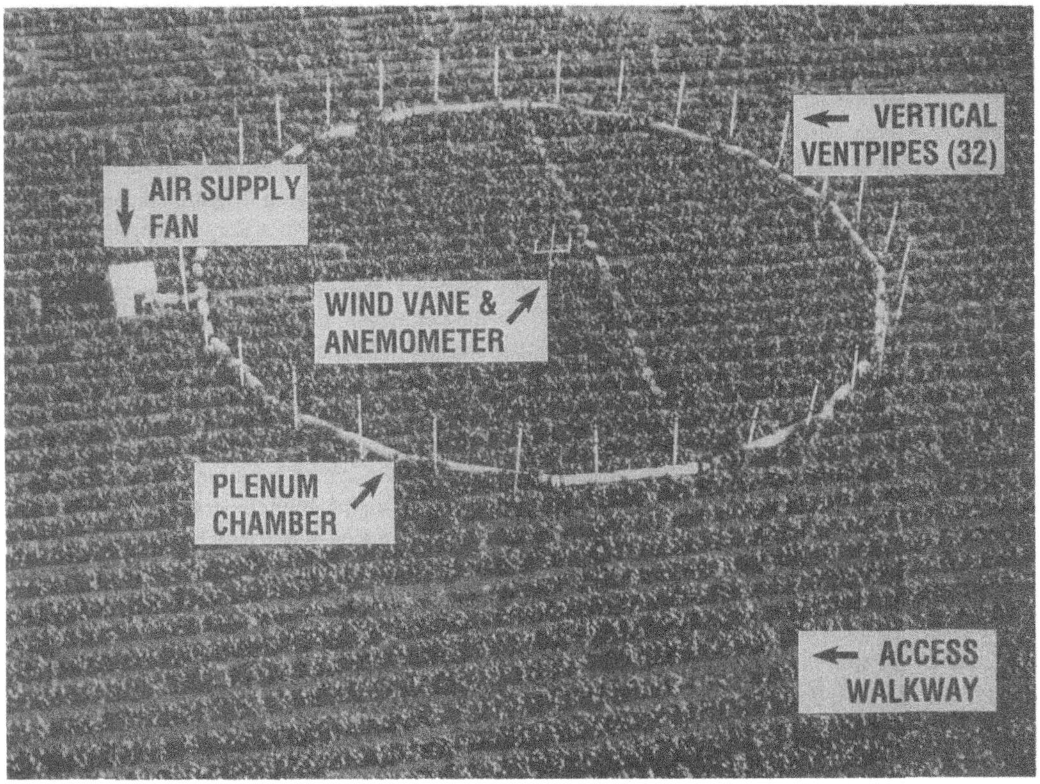

Fig. 1. Photograph of a FACE array in the cotton experiment at Maricopa, AZ, in 1990. Shown are the air supply fan housing, 22 m diameter plenum chamber, the 32 vertical ventpipes, wind measurement mast, and access walkway. (Hendrey et al., 1993).

(at least 100 m), thus reducing the possibility of CO_2 enriched air blowing into controls. Arrays were put in place immediately after planting. Exposure was initiated at 50% seedling emergence and continued until plants reached physiological maturity. One quarter of each experimental plot was used for plant root sampling.

Irrigation

The cotton crop was irrigated with a subsurface drip tube system. A single drip tube was installed 25 cm below each crop row. Initial irrigations of 254 and 272 mm of water were applied immediately after planting in 1990 and 1991, respectively. Main plots were split and each half differentially irrigated on a biweekly basis starting on 3 July in 1990 and 20 May in 1991. The irrigation system was also used for fertilizer N and micronutrient applications during the season as described by Mauney et al. (1994).

Root samples

Soil cores (38 mm diameter) measuring 90 cm in length were collected to determine root length density, root dry weight density, and root lineal density on 18 June and 17 September in 1990, and 12 June and 2 August in 1991. These dates corresponded to aboveground harvests occurring at early vegetative growth and at middle or late reproductive growth as reported elsewhere (Prior et al., 1994). At each sampling date respective measures of leaf area index were about 1.0 and 4.0 in 1990 and 0.7 and 3.0 in 1991. During each sampling, soil cores were taken in-row (0 m) and at distances of 0.25 and 0.50 m perpendicular to the row. In 1990, 192 soil cores were taken at each sampling, representing four cores per position within each subplot. In 1991, the number of cores per position was increased to eight, increasing the total number of cores per sampling to 384. Pneumatic hammers for driving core tubes and electric core tube extraction devices were used to collect root-soil cores (Prior and Rogers, 1992; Prior et al., 1994). A large custom-built cooler was used to transport samples by truck to Auburn, AL, where they were placed in cold storage until processing. A hand-held electric band saw was used to cut each 90 cm core into 15 cm segments. After processing, core segments were combined to form three 30 cm depth increments (i.e., 0–30, 30–60, and 60–90 cm) for statistical analyses. Roots were washed from each core segment with a hydropneumatic elutriation system (Gillison's 1990; Smucker et al., 1982) and stored in 20% ethanol (Bohm, 1979) at 4°C. After organic debris had been removed with tweezers and spring-loaded suction pipettes, root length was measured with a Comair Root Length Scanner (Hawker de Havilland, 1985). Root weight determinations were made after drying samples at 55°C.

Data analyses

All analyses were performed using the general linear models procedure of the Statistical Analysis System (SAS, 1985). Core position and depth increments were treated as additional split-plot treatments within the overall study design and an average of the replicate core samples (four in 1990 and eight in 1991) was used for statistical analyses. Proper error terms were specified for each split-plot treatment. Contrast statements were used to determine the significance between interacting main effect variables. Differences were considered significant at the $p < 0.10$ level. Values which differed at the $0.10 < p < 0.20$ level were considered trends.

Results and discussion

At both growth stages, the effects of water stress alone or its interaction with CO_2 were rare and, when detected, showed no consistent pattern. Previous findings (Prior et al., 1993) suggest that such variable results may be related to an interaction occurring in certain root diameter size classes and not in others. In that study, first order lateral roots were partitioned into different size classes based on root diameter. During vegetative growth, significant CO_2 × water interactions were detected for root variables (length, dry weight, and volume) in the smallest class size (< 1.00 mm); as growth progressed, interactions were detected (length, dry weight, volume, number, and tissue density) for the third largest class size (1.5–2.0 mm). This indicated that some smaller roots continue to exhibit CO_2 × water interactions later in the cotton crop life cycle. If patterns such as these do occur in the field, then trying to detect interactions becomes very arduous. This topic, without a doubt, requires further investigation.

A few factors which may have contributed to these variable root responses (i.e., between years) under CO_2 enrichment bear mentioning. Some differences could be related to changing irrigation patterns. More specifically, less irrigation water was applied between planting and sampling in 1991 than in 1990 due to the use of different methods of estimating evapotranspiration for the two years. A detailed description of these methods and the rationale for their use, including a chronological record of soil water content and seasonal rainfall, have been reported elsewhere (Hunsaker et al., 1994; Mauney et al., 1994). In addition, some of the variability may be ascribed in part to the fact that in 1991, early season night-time temperatures were lower than the previous season which may have slowed crop development resulting in smaller plants. Furthermore, spatial variability of the soil profile often encountered in the field can cause roots to be non-uniformly distributed and in itself may have hindered statistical separation of treatment effects. Nevertheless, some important differences in root response due to CO_2 enrichment were detected and are discussed below.

Effect of increased CO_2 on root length density

Positive shifts in rooting patterns due to extra atmospheric CO_2 during vegetative growth may have important implications in terms of seedling vigor. In the present study, there were significant CO_2 × position interactions for root length density (p = 0.04–0.08) at the vegetative sampling in both years (Fig. 2). At the in-row position (0 m) FACE tended to increase root length density in 1990 ($p = 0.13$; Fig. 2A) and significantly increased this measure (15%) in 1991 (Fig. 2B). FACE enhanced root length density more at the 0.5 m position in 1990, whereas in 1991, increases in this measure were most apparent at the 0.25 m position. For example, at the 0.25 and 0.5 m positions this measure was increased by 33 and 64% in 1990 and by 38 and 27% in 1991, respectively. Results show that elevated CO_2 resulted in more prolific rooting, especially at the interrow positions. This may well confer an advantage to CO_2-enriched plants for more readily acquiring soil water and nutrients needed for development, especially when these factors limit seedling growth. Consequently, establishment of crop stands would be enhanced. In the present study, FACE significantly increased corresponding measures of aboveground dry matter by 56

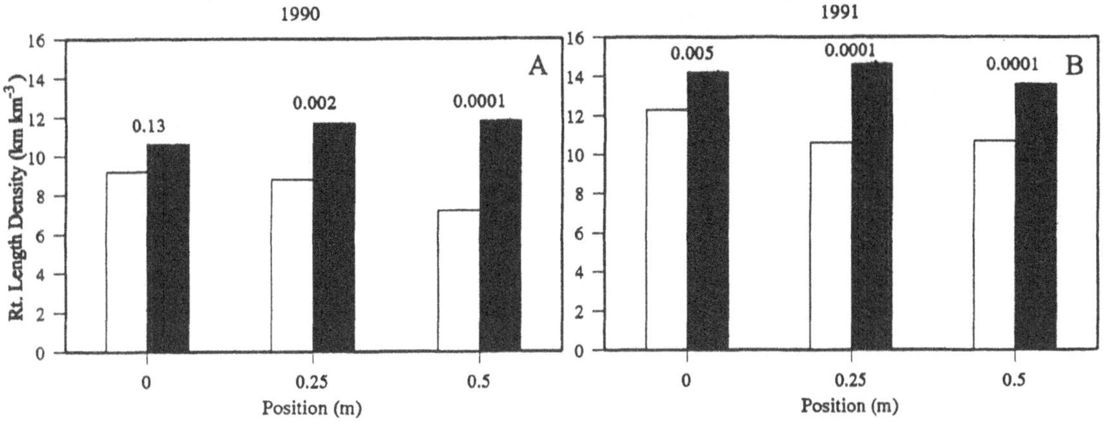

Fig. 2. The effect of CO_2 concentration (□ = 370 μmol mol^{-1}; ■ = 550 μmol mol^{-1}) on cotton root length density at three positions (0, 0.25, and 0.5 m away from the crop row center) during vegetative growth in 1990 (**A**) and 1991 (**B**). Means and probabilities are shown.

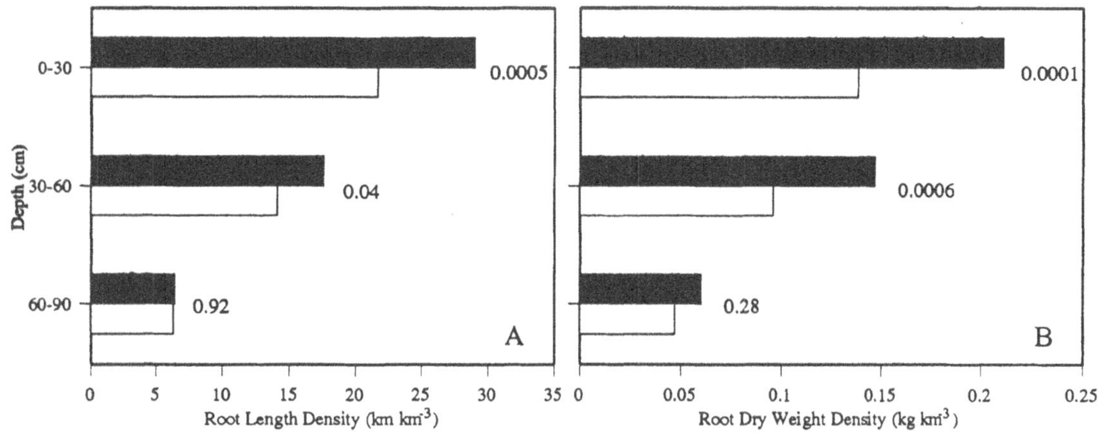

Fig. 3. The effect of CO_2 concentration (□ = 370 μmol mol^{-1}; ■ = 550 μmol mol^{-1}) on cotton root length density and root dry weight density at three depth increments during reproductive growth in 1990 (**A**) and 1991 (**B**). Means and probabilities are shown.

and 43% relative to ambient CO_2 levels in 1990 and 1991, respectively (Prior et al., 1994).

Changes in rooting patterns under high CO_2 environments may also have important implications for crop performance during critical periods of reproductive growth, such as boll filling, when plants have high demands for water and nutrients. Positive shifts in root length density due to FACE were seen at the reproductive samplings while corresponding measures of aboveground responses also showed that FACE increased boll biomass values by 65 and 44% in 1990 and 1991, respectively (Prior et al., 1994). Observations for the root samples collected during reproductive growth (data not shown) were similar to those reported for the vegetative growth samples in 1990 (Fig. 2A); for example, significant increases in root length densities due to elevated CO_2 were observed at both interrow positions (p = 0.002–0.02). A trend toward a CO_2 × depth × position interaction observed at the initial sampling (p = 0.14) was also seen during the reproductive stage (p = 0.19; data not shown); root length densities under FACE were higher at the two upper most depths (i.e., 0–30 cm and 30–60 cm) at the interrow positions. In the following year, a significant CO_2 × depth interaction was detected at the vegetative sampling (p = 0.02; data not shown) and was also observed during the reproductive sampling (Fig. 3A; p = 0.05). At both times, increased root length densities due to elevated CO_2 were found at the upper and middle depth increments; the 0–30 cm depth showed increases of 30 and 34% while the 30–60 cm depth had increases of 21 and 25%. Chaudhuri et al. (1986) observed that sorghum grown in glass-sided boxes filled with topsoil under CO_2 enrichment also exhibited root length increases in the upper parts of the soil profile. Likewise, enhancements in root length density due to CO_2 enrichment in our study are similar to those reported by others (Rogers et al., 1992a).

Effect of increased CO_2 on root dry weight density

Although root dry weight often exhibit substantial increases due to elevated atmospheric CO_2 (Rogers et al., 1994), little data exist on how extra CO_2 alters horizontal and vertical distribution patterns within actual soil profiles. In our study, the effect of increased CO_2 on root dry weight density varied between samplings and years. In 1990, there were significant CO_2 × depth × position interactions for root dry weight density at the vegetative sampling ($p = 0.02$; Fig. 4), but only a trend could be seen at the reproductive sampling ($p = 0.19$; data not shown). In comparison, CO_2 × depth interactions were detected at both sampling dates in 1991 ($p = 0.04$–0.10; Fig. 3B, only reproductive data shown). In general, the root dry weight distribution patterns were similar to those observed for root length density.

Representative results for root dry weight density are shown in Figures 3B and 4. At the initial sampling in 1990, FACE increased root dry weight densities at the upper depth increment, 0–30 cm, at all sample positions (0, 0.25, and 0.5 m) by 40, 33, and 111%, respectively (Fig. 4). Although values were numerically higher for some depth increments at the 0 m position (i.e., 30–60 cm; Fig. 4A) and at the 0.25 m position (i.e., 60–90 cm; Fig. 4B), these differences were only marginally non-significant ($p = 0.11$). However, the middle depth increment (30–60 cm) at the 0.5 m position exhibited a large increase in root dry weight density (200%) due to FACE conditions (Fig. 4C). Rogers et al. (1992b) reported that in soybean (18 day old) grown under CO_2 enriched conditions in growth chambers, rooting increased over a depth of 45 cm which also resulted in a substantial rise in root dry weight (143%). In the current study, by the reproductive sampling there was only a tendency for a CO_2 × depth × position interaction ($p = 0.19$; data not shown); rooting patterns were somewhat like those observed at the vegetative sampling (Fig. 4). However, the extra aerial CO_2 did significantly increase root dry weight density by 38% when averaged over all horizontal positions and depth increments at this sampling ($p = 0.04$; data not shown). In 1991, the significant CO_2 × depth interaction was related to a CO_2 induced increase in rooting at the two upper most depth increments (50% each) at both sample periods (Fig. 3B). By comparison, Chaudhuri et al. (1986) reported that sorghum grown in glass-sided boxes filled with topsoil under CO_2 enrichment exhibited increases in root length in the upper parts of the soil profile while increases in root dry weight were observed at all depths. Other containerized studies utilizing outdoor sunlit chambers for CO_2 exposure of soybean (Del Castillo et al., 1989) and winter wheat (Chaudhuri et al., 1990) have shown that CO_2 enrichment had little affect on root dry weight at depth increments below about 45 cm. Utilization of different CO_2 exposure systems, measurement methods, growth media, and cultivar studied may account for these varying responses.

Our root dry weight density data (as well as root length density data) resemble the distribution pattern for cotton roots described by Klepper et al. (1973) where rooting density generally declines exponentially with depth under optimum soil water conditions. A common pattern noted in our data was that CO_2 enrichment exerted most of its affect at the upper and middle portions of the profile while having no apparent affect on the distribution of root dry matter (or length) to the lowest profile depth (Figs. 3 and 4). Rogers et al. (1992b) reported that roots of high CO_2 grown soybean penetrated deeper into the profile and Chaudhuri et al. (1990) that maximum rooting depth of wheat was reached quicker. We are unable to conclusively determine whether elevated CO_2 had no effect on deep rooting since core samples were not collected beyond 90 cm. However, it is possible that other factors (e.g., oxygen diffusion rate, high soil strength, etc.) at lower depths were limiting root development even for the CO_2-enriched cotton plants. Furthermore, it is important to note that the influence of irrigation line placement within the profile as a reason for more root proliferation at the upper to middle depths was unlikely since results by Kamara et al. (1991) demonstrated that the depth of trickle irrigation lines had no significant affect on the depth or horizontal distribution of cotton roots during the course of a growing season.

Effect of increased CO_2 on spatial distribution of roots

In addition to examining changes in absolute root length density, data were expressed as a percent of total root length to ascertain whether proportionately more rooting occurred at a given position under one treatment than another. There were significant CO_2 ×

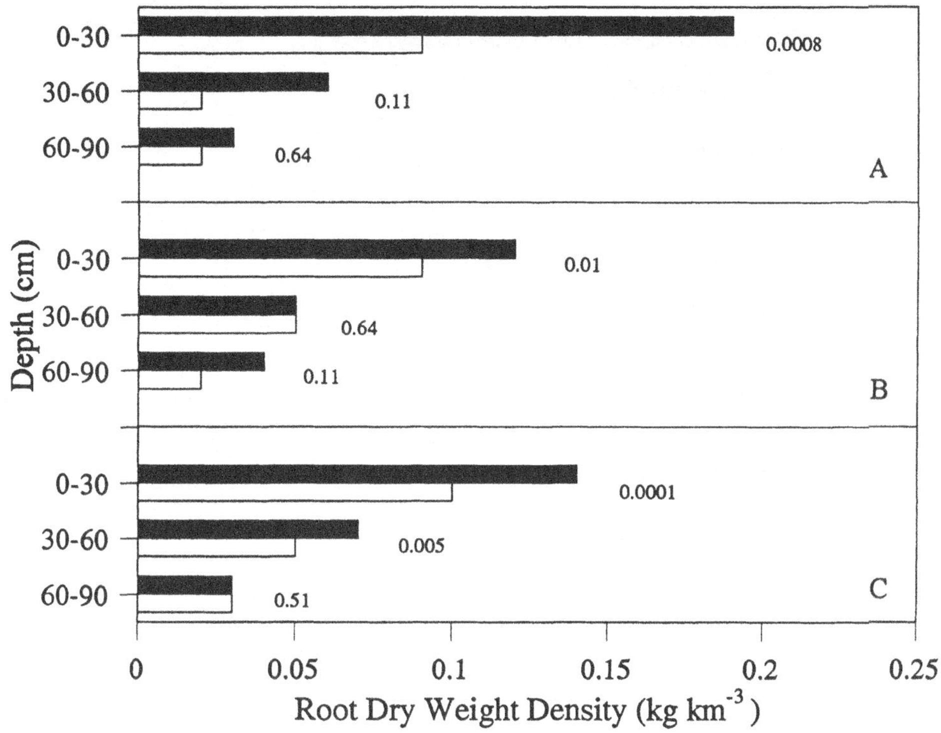

Fig. 4. The effect of CO_2 concentration (□ = 370 μmol mol^{-1}; ■ = 550 μmol mol^{-1}) on cotton root dry weight density at the three depth increments and positions (**A** = 0 m, **B** = 0.25 m, and **C** = 0.5 m) away from crop row center during vegetative growth in 1990. Means and probabilities are shown.

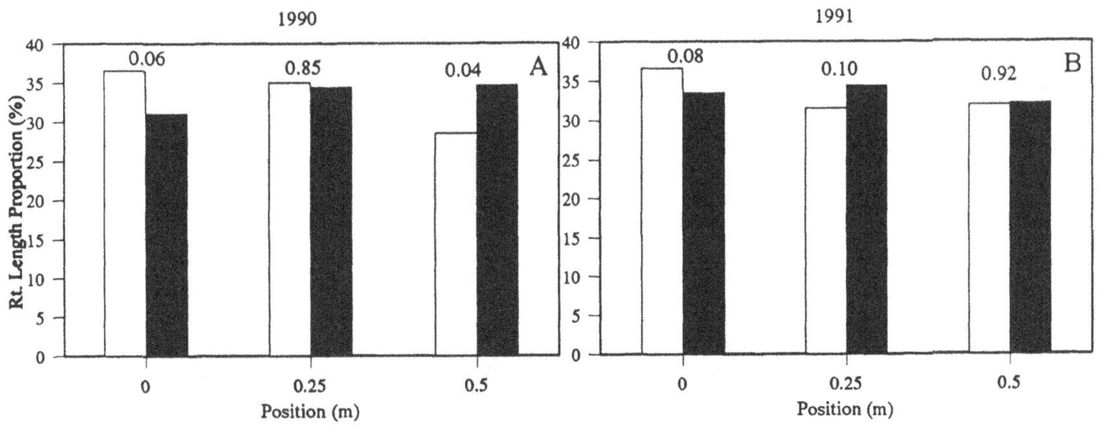

Fig. 5. The effect of CO_2 concentration (□ = 370 μmol mol^{-1}; ■ 550 μmol mol^{-1}) on root length proportion at three positions (0, 0.25, and 0.5 m away from crop row center) during vegetative growth in 1990 (**A**) and 1991 (**B**). Means and probabilities are shown.

position interactions for root length proportion (p = 0.02) at both samplings in 1990, but only for the vegetative sampling in 1991. At the vegetative samplings, the ambient treatment had a higher percentage of its total root length distributed at the in-row (0 m) position compared to the elevated CO_2 treatment (Fig. 5). However, at the interrow positions, the opposite was observed; in 1990 significantly more root length was found at the 0.5 m position under FACE (Fig. 5A) while in 1991 a similar pattern was observed at the

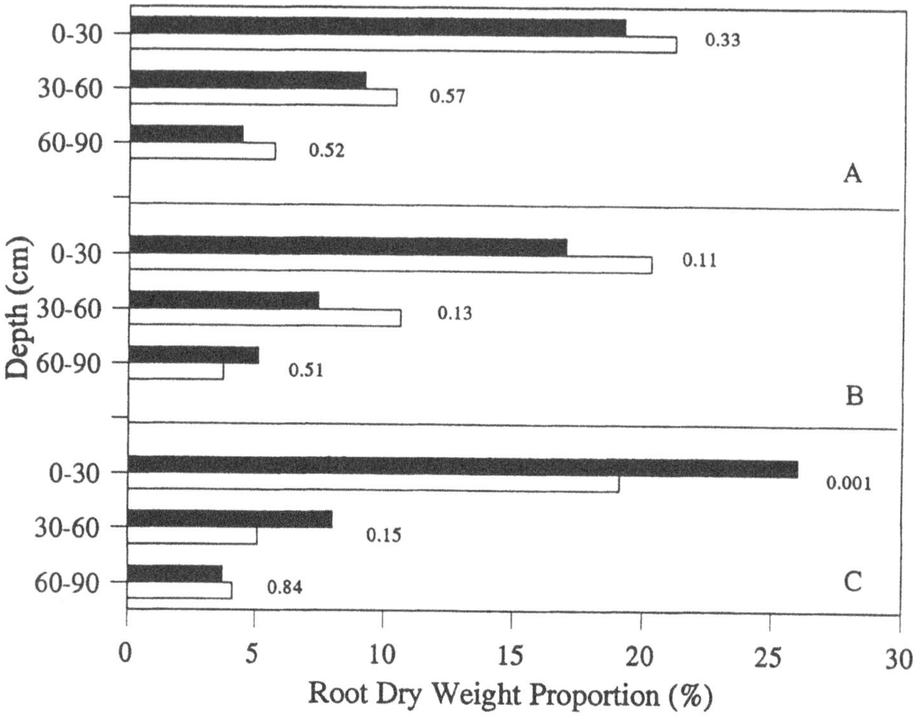

Fig. 6. The effect of CO_2 concentration (□ = 370 μmol mol^{-1}; ■ 550 μmol mol^{-1}) on root dry weight proportion at three depth increments and positions (**A** = 0 m, **B** = 0.25 m, and **C** = 0.5 m away from crop row center) during vegetative growth in 1990. Means and probabilities are shown.

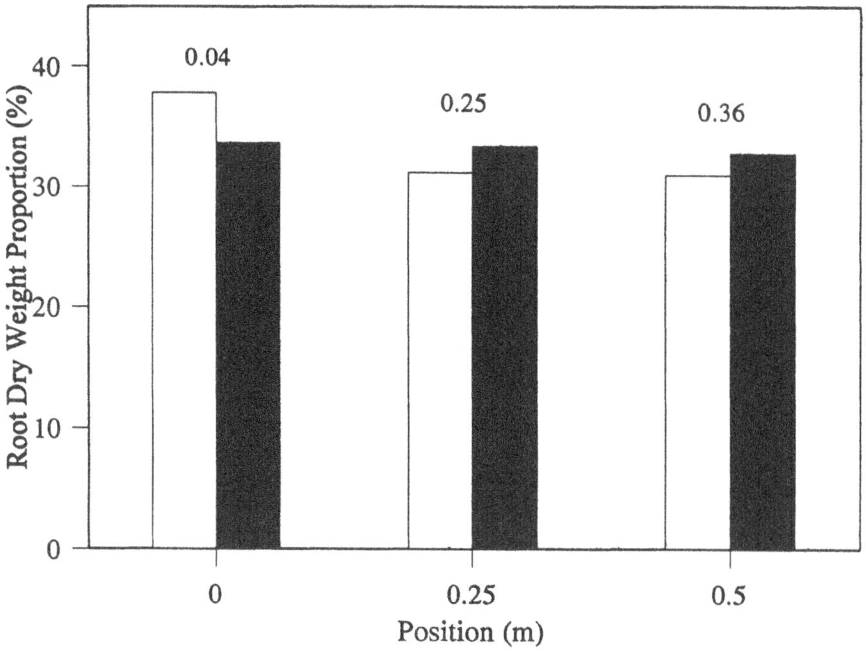

Fig. 7. The effect of CO_2 concentration (□ = 370 μmol mol^{-1}; ■ = 550 μmol mol^{-1}) on root dry weight proportion at three positions (0, 0.25, and 0.5 m away from crop row center) during vegetative growth in 1991. Means and probabilities are shown.

0.25 m position (Fig. 5B). Observations at the reproductive sampling in 1990 (data not shown) were like those observed at the vegetative sampling (Fig. 5A). A trend toward a CO_2 × depth × position interaction for this variable was also seen at both samplings in 1990 (p = 0.14–0.17; data not shown) where patterns were similar to data described above, but limited to the upper depth (0–30 cm) increment.

Root proportion was also calculated on a dry weight basis (Figs. 6 and 7). There were CO_2 × depth × position interactions at both samplings in 1990 (p = 0.06–0.09). At the vegetative sampling, there was a trend for a higher root dry weight proportion at the two upper depths (0–30 cm and 30–60 cm) at the 0.25 m position under ambient CO_2 conditions (p = 0.11 and 0.13; Fig. 6B). Conversely, at the 0.5 m position, values from FACE were significantly higher at 0–30 cm depth while this same measure at the next depth level (30–60 cm) tended to be higher (p = 0.15; Fig 6C). At the reproductive sampling, (this measure at the 0 and 0.25 m positions) was significantly higher or tended to be higher under ambient CO_2 levels at the 0–30 cm depth (data not shown; p = 0.0002 and p = 0.14, respectively). However, under FACE more of the root system's dry weight, relative to the ambient CO_2 treatment, tended to be allocated to the lowest depth level at the 0 m position (p = 0.10) and 0.5 m position (p = 0.15). By comparison, in 1991 there were CO_2 × position interactions with the same pattern being observed at both dates (p = 0.04–0.10); the ambient CO_2 treatment had a higher proportion of its root dry weight allocated to the in-row position (0 m) compared to FACE (Fig. 7). No differences were seen at the remaining two positions.

The changes in root proportion reported here may be related to different rates of extension of first-order lateral roots and further root branching. The higher proportion of root length and weight observed at the in-row (0 m) position under ambient CO_2 may be due to first-order lateral roots having slower elongation rates (relative to high CO_2 plants) which may have resulted in more higher-order branching occurring close to the taproot or center row position. On the other hand, increases in lateral root growth (and further root branching) possibly explain the greater root length proportion at the interrow positions under FACE. It is, however, unclear why root length proportion differed away from row center while corresponding measures of root dry weight did not, unless it was somehow related to root fineness.

We did not make detailed measurements of root branching patterns or root elongation rates. However, results from a few other studies conducted in growth chambers should be mentioned. Rogers et al. (1992b) reported that elevating CO_2 had no significant effect on soybean root branching but did increase the length of first-ordered laterals (110%). The use of perlite (a media with large pore size distribution) may have allowed for unimpeded root elongation, leading to large increases in root length without the added expense of extensive branching. Conversely, Del Castillo et al. (1989) observed that CO_2 enrichment enhanced root branching rather than elongation rate in soybean. In this case, it appears that elongation rates were measured on higher-order lateral roots rather than on first-order laterals. In addition, a soil mixture (i.e., equal volume mix of sand and vermiculite) was used which probably represented a media having an overall smaller pore size distribution which may have impeded root extension enough to induce more branching (Russell, 1977). We suspect that the findings of both investigations may occur in natural soil profiles, depending upon edaphic conditions. Our data for the larger taproot systems (total volume, dry weight, and length) under FACE (Prior et al., 1994) revealed larger diameter taproots which supported greater number of first-order lateral roots, collectively representing more root length (Prior et al., 1993). These changes may well have been accompanied by more branching which could explain the higher proportion of root length found within interrow positions under FACE. These observations could have important implications for both agricultural and natural ecosystems where such change might alter competitive effectiveness for belowground resources, especially if some plant species are affected more or less, relative to neighboring species.

Effect of increased CO_2 on root lineal density

Calculations of root lineal density (i.e., root weight per unit root length) were made to make inferences about possible changes in root structure and composition due to CO_2 enrichment. The CO_2-induced shifts in root lineal density were highly variable between both samplings and year. In the first year a significant CO_2 × depth × position interaction (p = 0.008) was seen at the vegetative sampling (Fig. 8). FACE increased lineal root density at the 0 m position (i.e., 0–30 cm depth; Fig. 8A) and 0.5 m position (i.e., 0–30 cm and 30–60 cm depths; Fig. 8C). An increase in root lineal density under ambient CO_2 conditions was only seen

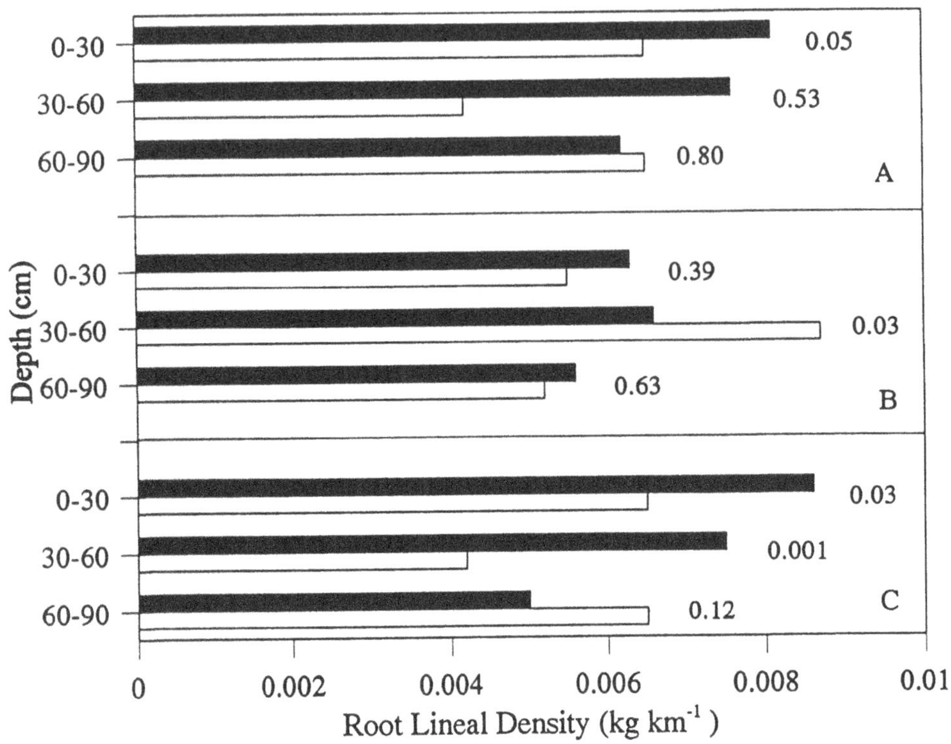

Fig. 8. The effect of CO_2 concentration (□ = 370 μmol mol^{-1}; ■ = 550 μmol mol^{-1}) on root lineal density at three depth increments and positions (**A** = 0 m, **B** = 0.25 m, and **C** = 0.5 m away from crop row center) during vegetative growth in 1990. Means and probabilities are shown.

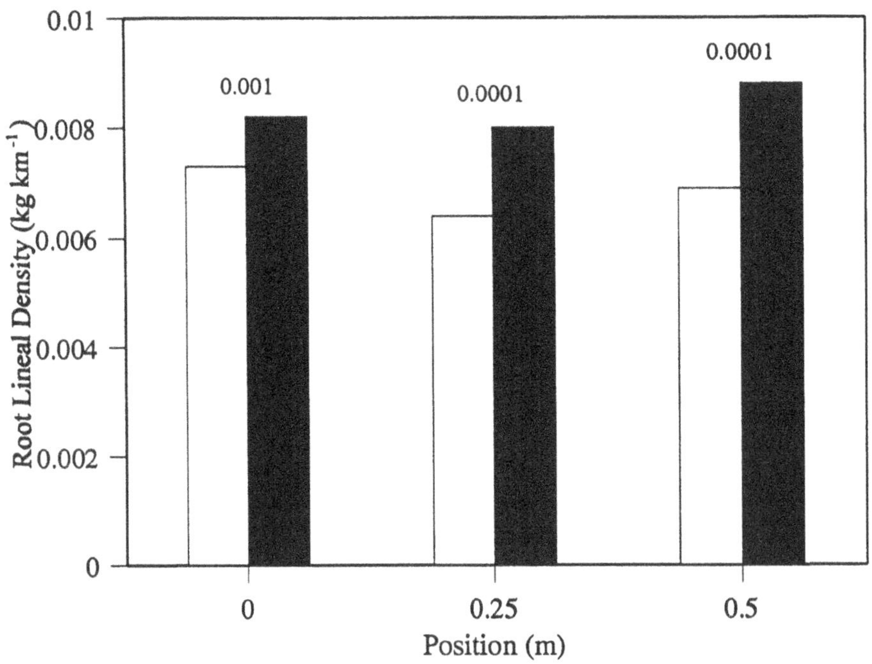

Fig. 9. The effect of CO_2 concentration (□ = 370 μmol mol^{-1}; ■ = 550 μmol mol^{-1}) on root lineal density at three positions (0, 0.25, and 0.5 m away from crop row center) during reproductive growth in 1991. Means and probabilities are shown.

at the 0.25 m position (30–60 cm depth; Fig. 8B). A similar trend was observed at the 0.5 m position (p = 0.12; 60–90 cm depth). We have no clear explanation for this reversal which only occurred once. By comparison, data from 1991 indicated a trend for an overall effect of CO_2 on root lineal density ($p = 0.11$; data not shown). At the reproductive sampling, the $CO_2 \times$ position interaction significance level was $p = 0.11$ in 1990 and $p = 0.03$ in 1991. However, in 1990 additional CO_2 resulted in a 27% increase in root lineal density when data were averaged across all depths and positions (p = 0.05; data not shown). In 1991, root lineal density was higher under elevated CO_2 at all three positions (i.e., 0, 0.25, and 0.5 m; Fig. 9).

Changes in root lineal density most likely occurred in smaller lateral roots and/or successive branching orders of first-order lateral roots since large root segments (i.e., woody first-order lateral roots) were rarely found in individual samples. Increased lineal root density has been related to roots having larger diameters (Klepper, 1992). It is also possible that changes in root lineal density may be related to more compact or denser tissue or to alteration in carbohydrate storage, cell number, cell size, and other structural modifications, or combinations thereof. A growth chamber study showed that roots of young soybean plants grown at high CO_2 concentrations exhibited larger diameters (in the root hair zone) (Rogers et al., 1992b). A significant enhancement of both stele diameter and cortex width was also seen. In the present study, taproots and their associated first-ordered lateral roots often showed increases in root tissue density (g/cm^3) under FACE (Prior et al., 1993). Although such measurements were not made on smaller higher-order branch roots, the increases in root lineal density reported here possibly coincide with changes in root tissue density. If such is the case, root function may be altered along with the quality of carbon inputs into the soil. This, in turn, could affect both microbial activity and soil physical properties as well as soil carbon storage.

Conclusions

Despite variation between years, results from this field study suggest that rising atmospheric CO_2 concentration could be beneficial to a cotton crop with respect to enhanced root proliferation. Stimulation of crop rooting may help ensure exploitation of soil resources. In general, vertical changes in absolute root density (both length and dry weight) due to CO_2 enrichment are most likely to occur in the uppermost soil layers (0–30 cm and 30–60 cm depths in this study). Results gathered here also indicate that additional CO_2 can alter horizontal rooting patterns by increasing root proliferation into interrow positions (0.25 m and/or 0.50 m). It was also noted that under ambient CO_2 conditions a higher proportion of the root system (length and/or dry weight basis) was often distributed closer to center row whereas the FACE treatment usually had proportionately more of its root system allocated away from row center. If some plant species are affected in this manner while others are not, this finding could have implications for both agricultural and natural ecosystems associated with the strategy by which plants compete for belowground resources. However, before final conclusions are made, further work with other important plant species is needed. Functional aspects of root growth as affected by elevated atmospheric CO_2 clearly need detailed study. Future work must consider not only how CO_2 enrichment affects crop rooting, but also how it impacts other belowground processes, both physical and biological.

Acknowledgements

The authors are indebted to Bernard E McLean, Tammy K Counts, Trina L Cagle, John H McFarlin, John H Walden, J Crayton Walker and Billy J Vice for technical assistance. Support from the Environmental Sciences Research Division, U.S. Dept. of Energy, and the Alabama Agricultural Experiment Station, Auburn University, is gratefully acknowledged.

References

Allen L H Jr 1990 Plant responses to rising carbon dioxide and potential interactions with air pollutants. J. Environ. Qual. 19, 15–34.

Allen L H Jr, Boote K J, Jones J W, Jones P H, Valle R R, Acock B, Rogers H H and Dahlman R C 1987 Response of vegetation to rising carbon dioxide: photosynthesis, biomass, and seed yield of soybean. Global Biogeochem. Cycles 1, 1–14.

Arp W J 1991 Effects of source-sink relations on photosynthetic acclimation to elevated CO_2. Plant, Cell Environ. 14, 869–975.

Bohm W 1979 Methods of studying root systems. Spring-Verlag, NY. 188 p.

Carlson R W and Bazzaz F A 1980 The effects of elevated CO_2 concentration on growth, photosynthesis, transpiration, and water use efficiency of plants. *In* Environmental and Climatic Impact of Coal Utilization. Eds. J Singh and A Deepak. pp 609–612. Academic Press, NY.

Chaudhuri U N, Kirkham M B and Kanemasu E T 1990 Root growth of winter wheat under elevated carbon dioxide and drought. Crop Sci. 30, 853–857.

Chaudhuri U N, Burnett R B, Kirkham M B and Kanemasu E T 1986 Effect of carbon dioxide on sorghum yield, root growth, and water use. Agric. For. Meteorol. 37, 109–122.

Dahlman R C, Strain B R and Rogers H H 1985 Research on the response of vegetation to elevated atmospheric carbon dioxide. J. Environ. Qual. 14, 1–8.

Del Castillo D, Acock B, Reddy V R and Acock M C 1989 Elongation and branching of roots on soybean plants in a carbon dioxide-enriched aerial environment. Agron. J. 81, 692–695.

Gillison's Variety Fabrication, Inc 1990 Basic operation of the hydropneumatic root washer. 3033 Benzie Hwy., Benzonia, MI 49616.

Gribbon J 1981 The politics of carbon dioxide. New Sci. 90, 82–84.

Hawker de Havilland 1985 User's Manual. 304 Lorimer Street, Port Melbourne, Victoria, 3207, Australia.

Hendrey G R, Lewin K F and Nagy J 1993 Free-air carbon dioxide enrichment: development, progress, results. Vegetatio 104/105, 17–31.

Hunsaker D J, Hendrey G R Kimball B A, Lewin K F, Mauney J R and Nagy J 1994 Cotton evapotranspiration under field conditions with CO_2 enrichment and variable soil moisture regimes. Agric. For. Meteorol. (*In press*).

Idso S B 1989 Carbon Dioxide and Global Change: Earth in Transition. IBR Press, Tempe, AZ. 292 p.

Kamara L, Zartman R and Ramsey R H 1991 Cotton-root distribution as a function of trickle irrigation emitter depth. Irrig. Sci. 12, 141–144.

Keeling C D, Bacastow R B, Carter A F, Piper S C, Whorf T P, Heimann M, Mook W G and Roeloffzen H 1989 A three dimensional model of atmospheric CO_2 transport based on observed winds: Observational data and preliminary analysis. *In* Aspects of Climate Variability in the Pacific and the Western Americas. Vol. 55. American Geophysical Union.

Kimball B A, Rosenberg N J and Allen L H Jr 1990 Impact of CO_2, trace gases, and climate change on global agriculture. ASA Special Publication No. 53, Am. Soc. Agron., Madison, WI. 133 p.

Kimball B A and Idso S B 1983 Increasing CO_2: Effects on crop yield, water use and climate. Agric. Water Manage. 7, 55–72.

Klepper B, Taylor H M, Huck M G and Fiscus E L 1973 Water relations and growth of cotton in drying soil. Agron. J. 65, 307–310.

Klepper B 1992 Development and growth of crop root systems. *In* Limitations to plant root growth. Eds. J L Hatfield and B A Stewart. Advances in Soil Science Volume 19. pp 1–25. Springer-Verlag, New York.

Krupa S V and Kickert R N 1989 The greenhouse effect: Impacts of ultraviolet-B (UV-B) radiation, carbon dioxide (CO_2), and ozone (O3) on vegetation. Environ. Pollut. 61, 263–396.

Mauney J R, Kimball B A, Pinter Jr P J, LaMorte R L, Lewin K, Nagy J and Hendrey, G R 1994 Growth and yield of cotton in response to a free-air carbon dioxide enrichment (FACE) environment. Agric. For. Meteorol. (*In press*).

Pearman G I 1980 The global carbon cycle and increasing levels of atmospheric carbon dioxide. *In* Carbon Dioxide and Climate: Australian research. Ed. G I Pearman. pp 11–20. Australian Academy of Science, Canaberra.

Prior S A and Rogers H H 1992 A portable soil coring system that minimizes plot disturbance. Agron. J. 84, 1073–1077.

Prior S A, Rogers H H and Runion G B 1993 Effects of free-air CO_2 enrichment on cotton root morphology. Supplement to Plant Physiol. 102 (1), 173 (Abstract).

Prior S A, Rogers H H, Runion G B and Mauney J R 1994 Effects of free-air CO_2 enrichment on cotton root growth. Agric. For. Meteorol. (*In press*).

Rogers H H and Dahlman R C 1993 Crop responses to CO_2 enrichment. Vegetatio 104/105, 117–131.

Rogers H H, Runion G B and Krupa S V 1994 Plant responses to atmospheric CO_2 with emphasis on roots and the rhizosphere. Environ. Pollut. 83, 155–189.

Rogers H H, Thomas J F and Bingham G E 1983 Response of agronomic and forest species to elevated atmospheric carbon dioxide. Science 220, 428–429.

Rogers H H, Prior S A and O'Neil E G 1992a Cotton root and rhizosphere responses to free-air CO_2 enrichment. Crit. Rev. Plant Sci. 11, 251–263.

Rogers H H, Peterson C M, McCrimmon J M and Cure J D 1992b Response of soybean roots to elevated atmospheric carbon dioxide. Plant, Cell Environ. 15, 749–752.

Rosenberg N J, Kimball B A, Martin P and Cooper C F 1990 From climate and CO_2 enrichment to evapotranspiration. *In* Climate Change and U.S. Water Resources. Ed. P E Waggoner. pp 151–175. John Wiley and Sons, NY.

Russell R S 1977 Plant Root Systems: Their Function and Interaction with the Soil. McGraw-Hill, UK. 298 p.

Sionit N, Rogers H H, Bingham G E and Strain B R 1984 Photosynthesis and stomatal conductance with CO_2-enrichment of container and field-grown soybeans. Agron. J. 65, 207–211.

SAS Institute, Inc 1985 SAS User's guide: statistics. Statistical Analysis System (SAS) Institute, Inc., Cary, NC. 956 p.

Smit, B, Ludlow L and Brklacich M 1988 Implications of a global climatic warming for agriculture: A review and appraisal. J. Environ. Qual. 17, 519–527.

Smucker A J M, McBurney S L and Srivastava A K 1982 Quantitative separation of roots from compacted soil profiles by the hydropneumatic elutriation system. Agron. J. 74, 500–503.

Strain B R and Cure J D 1985 Direct effects of increasing carbon dioxide on vegetation. DOE/ER-0238. The Office of Energy Research. U. S. Dept. of Energy, Washington, DC. 286 p.

Thomas R R and Strain B R 1991 Root restriction as a factor in photosynthetic acclimation of cotton seedlings grown in elevated carbon dioxide. Plant Physiol. 96, 627–634.

Warrick R A 1988 Carbon dioxide, climatic change and agriculture. Geographical J. 154, 221–233.

Woodwell G M 1978 The carbon dioxide question. Sci. Am. 238, 34–43.

Plant and Soil **165**: 45–51, 1994.

Above- and belowground response of *Populus grandidentata* to elevated atmospheric CO_2 and soil N availability

Peter S. Curtis[1], Donald R. Zak[2], Kurt S. Pregitzer[3] and James A. Teeri[4]
[1]*Department of Plant Biology, The Ohio State University, 1735 Neil Avenue, Columbus, OH 43210, USA,* [2]*School of Natural Resources and Environment, University of Michigan, Ann Arbor, MI 48109, USA,* [3]*Department of Forestry, Michigan State University, East Lansing, MI 48824, USA and* [4]*Department of Biological Sciences, University of Michigan, Ann Arbor, MI 48109, USA*

Key words: CO_2, gas exchange, nitrogen, *Populus*

Abstract

Soil N availability may play an important role in regulating the long-term responses of plants to rising atmospheric CO_2 partial pressure. To further examine the linkage between above- and belowground C and N cycles at elevated CO_2, we grew clonally propagated cuttings of *Populus grandidentata* in the field at ambient and twice ambient CO_2 in open bottom root boxes filled with organic matter poor native soil. Nitrogen was added to all root boxes at a rate equivalent to net N mineralization in local dry oak forests. Nitrogen added during August was enriched with ^{15}N to trace the flux of N within the plant-soil system. Above– and belowground growth, CO_2 assimilation, and leaf N content were measured non-destructively over 142 d. After final destructive harvest, roots, stems, and leaves were analyzed for total N and ^{15}N.

There was no CO_2 treatment effect on leaf area, root length, or net assimilation prior to the completion of N addition. Following the N addition, leaf N content increased in both CO_2 treatments, but net assimilation showed a sustained increase only in elevated CO_2 grown plants. Root relative extension rate was greater at elevated CO_2, both before and after the N addition. Although final root biomass was greater at elevated CO_2, there was no CO_2 effect on plant N uptake or allocation. While low soil N availability severely inhibited CO_2 responses, high CO_2 grown plants were more responsive to N. This differential behavior must be considered in light of the temporal and spatial heterogeneity of soil resources, particularly N which often limits plant growth in temperate forests.

Introduction

The role of soil fertility in regulating responses of plants to rising atmospheric CO_2 is important in understanding the consequences of global climate change for terrestrial ecosystems. In controlled environment studies, as mineral nutrients limit growth, the relative magnitude of CO_2 responses also decline (Cure, 1985). Plant growth can be stimulated on very nutrient-poor soils by elevated CO_2 (Norby et al., 1986, 1992), although the increase is often largely belowground. Increased soil exploration by roots and mycorrhizae of high CO_2 grown plants may in turn have a positive effect on long-term net nutrient uptake (O'Neill et al., 1987).

Although plant growth in many temperate ecosystems is considered N limited (Pastor and Post, 1988), soil N availability varies temporally with net N mineralization rates and the timing of exogenous inputs. In a nutrient-poor, oak-dominated forest in northern Lower Michigan net N mineralization rates varied by a factor of two over the growing season with the highest rates occurring in mid-summer (Zak and Pregitzer, 1990). Leaf N also varies temporally, particularly in deciduous forest species. Reich et al. (1991) found a close correlation between seasonal changes in leaf N and photosynthetic rate in mature *Quercus ellipsoidalis*, *Acer rubrum*, and *A. saccharum* growing in Wisconsin. This general relationship between photosynthesis and leaf N has been supported by many other studies (Field and Mooney, 1986).

Temporal changes in soil and leaf N under natural conditions may have important consequences for plant growth at elevated CO_2. Variation in either could stimulate or limit CO_2 assimilation. The potential for N availability to control plant response to elevated CO_2 has been largely ignored in field studies. We have proposed and tested components of a model linking above- and belowground C and N cycles at elevated CO_2 (Zak et al., 1993). The present study is part of an ongoing investigation into plant–soil feedbacks with rising atmospheric CO_2. We report here the response of *Populus grandidentata* to an increase in soil N availability under conditions of elevated CO_2.

Materials and methods

Plant growth

In November 1990, 12 open bottom root boxes (0.7m × 0.7m × 1.3m) were installed at the University of Michigan Biological Station (UMBS) in northern Lower Michigan, USA. The boxes were lined with plastic and each was fitted with two minirhizotron tubes extending from the top to the bottom of the box at a 30° angle. The surface horizon of the excavated soil (sandy, mixed, frigid Entic Haplorthod) was discarded in order to achieve a rooting medium low in organic C and available N. The Bs and C subsoil horizons were sieved (4mm), homogenized, and used to fill the root boxes which were left fallow until mid-May 1991. Initial organic C (1977 $\mu g\ C\ g^{-1}$) and total N (70 $\mu g\ N\ g^{-1}$) were determined with a Carlo Erba NA 1500.

Root segments from a single clone of *Populus grandidentata* were collected in November 1990 at UMBS and used for the propagation of all plant material. Shoots growing from adventitious buds along root segments were excised, treated with indole butyric acid, and rooted in peat moss pots (Barry and Sachs, 1968). Nine individual rooted sprouts were planted into each of the 12 root boxes on 20 May 1991 (Julian Day (JD) 140).

Open top chamber (0.5m × 0.5m × 1.0m), identical to those described by Curtis and Teeri (1992), were used to used to elevate CO_2 around the trees. Experimental treatments were: chambered trees at ambient atmospheric CO_2 (chambered ambient), chambered trees at twice-ambient CO_2 (chambered elevated), and unchambered boxes with trees (unchambered ambient). Each treatment was replicated 4 times in a randomized block design. Plants from the unchambered ambient treatment were not measured for physiological or belowground growth responses and are not discussed further here.

Carbox dioxide partial pressure inside elevated chambers was increased by dispensing 100% CO_2 into an input blower; concentrations were controlled with manual flowmeters. The atmosphere inside chambers was continuously monitored by an infra-red gas analyser that logged data to a personal computer (Curtis and Teeri, 1992). Temperatures inside and outside chambers were monitored every 15 min using shaded thermocouples connected to an LI 1000 datalogger (LICOR Inc, Lincoln NB). Mean seasonal daytime CO_2 partial pressures were 692 ± 19 μbar (mean ± s.d.) inside elevated chambers and 342 ± 6 μbar inside ambient chambers. Daytime temperatures were 3.0 °C higher inside chambers than outside, but there was no temperature difference between elevated and ambient chambers.

Nitrogen was added to each root box in two stages. In the first stage, intended to simulate natural rates of N mineralization, a total of 745 mg N was applied as aqueous NH_4Cl to each plot over a 29 d period beginning 12 July (JD 193). In the second stage, intended to stimulate natural increases in N availability during mid-summer, a total of 1440 mg N was applied as ^{15}N enriched NH_4Cl (5% atom excess) to each plot over an 8 d period beginning 20 August (JD 232). Total N added (4.5 $g\ m^{-2}$) was similar to that mineralized on an annual basis in dry oak forests (Zak and Pregitzer, 1990). Plants were kept well watered by adding 3 L water daily to each root box except during periods of rainfall.

Plant growth was evaluated on 6 dates during the course of the experiment. Prior to planting (JD 140), 36 individuals were randomly selected to determine leaf area and above- and belowground dry weights by destructive harvest. On JD 169, 206, 231, and 266, stem height and leaf area were measured non–destructively. Leaf area was calculated from the equation $LA = -0.2032 + 0.007372(LW)$ ($r^2 = 0.99$, n=83), obtained from destructive harvests of leaves from plants of the same clone, where LA = leaf area (cm^2), L = leaf length (cm) along the midvein, and W = leaf width (cm) at the widest point. Complete above- and belowground harvests were conducted on 19 October 1991 (JD 291). Individual trees were cut at the soil surface, leaves excised, and leaf area measured. Large roots were recovered directly from the soil by hand excavation, and fine roots were separated from soil by sieving (3 mm) the entire contents of each root box. All tissue was dried at 70 °C to constant weight.

Leaf gas exchange

Leaf gas exchange was measured on individual attached leaves using an LCA3 portable photosynthesis system (ADC Ltd, Hoddesdon, UK) fitted with a water-jacketed cuvette. Bottled air enriched with CO_2 (ca. 1000 μbar) was supplied to the LCA3, allowing net assimilation and stomatal conductance measurements at growth CO_2 partial pressure. A saturating light intensity (1800 μmol m^{-2} s^{-1}) was supplied by a quartz halogen lamp. Previous work (Curtis and Teeri, 1992) had shown this PAR level to saturate net assimilation (A) in ambient and elevated CO_2 grown plants. Leaf temperature was maintained between 25° and 27° C, and cuvette relative humidity was always greater than 50%. Gas exchange was measured on JD 176 (n=3), 190 (n=2), 207 (n=3), 227 (n=4), 245 (n=l), 263 (n=6), 273 (n=6) and 288 (n=6). All gas exchange measurements were conducted on the youngest fully expanded leaf.

Tissue N analysis

Tissue N was analysed non-destructively during the growing season and from tissue collected at the final harvest. Immediately following all gas exchange measurements, light absorbance at 650 nm was measured at 4, 0.06 cm^2 areas on the same leaf with a Minolta SPAD meter (Minolta Corp., Ramsey NJ). Leaf N content (mg cm^{-2}) was related to absorbance by the equation N = -0.037 + 0.00616S (r^2=0.89), where S is absorbance in SPAD units. This relationship was determined by destructive leaf harvests from a separate set of plants. Leaf, stem, and root tissue collected at the final harvest was dried, ground, and composited by chamber. This tissue was then analysed for total N and total ^{15}N using a Europa Scientific Tracermass mass spectometer.

Root dynamics

Minirhizotron images were recorded on VHS tape using a Circon Microvideo 9011 Color Agricultural Camera (Circon Corp., Santa Barbara CA). Video images were recorded at 2-3 week intervals throughout the growing season following the procedure of Hendrick and Pregitzer (1992a). ROOTS, an interactive PC-based software program, was used to digitize images from the VHS tape, calculate length and diameter of each root and store these data in a data base. Data from the minirhizotron tubes in each box were pooled, and root box means were used to calculate the average number of roots, their combined length, and production by treatment (Hendrick and Pregitzer, 1992a, b). Root relative extension rate (RER) was calculated by linear regression of ln transformed root lengths against time. For this analysis, growth was divided into two stages; an early stage (JD 189-233) prior to the increase in N availability, and a later stage (JD 247-274) following the increase. Root length at JD 189 in the Ambient treatment was not included in this calculation because of 0 values in two replicates.

Statistical analysis

Treatment effects on leaf area on a given sampling date, net assimilation at growth CO_2, and tissue nutrient content were analysed by analysis of variance (ANOVA) for a randomized block design, where the unit of replication was the chamber (n=4). Treatment effects on RER were analysed by F-tests for the equality of regression coefficients (Sokal and Rohlf, 1981).

Results

Early in the season, when N availability was low, A in chambered elevated CO_2 grown plants (Elevated plants) was only marginally greater than that in chambered ambient CO_2 grown plants (Ambient plants), and this difference was only significant on JD 190 (Fig. 1A). Late in the season, when N availability was high, A increased significantly in both Elevated and Ambient plants ($p < 0.05$, t-test). In Ambient plants, A declined rapidly after JD 245. Mean A during 28 days following the increase in N availability (10.6 ± 1.7 umol m^{-2} s^{-1}, JD 245-273) was not significantly different than mean A during 37 days preceeding the increase in N availability (11.8 ± 1.0 umol m^{-2} s^{-1}, JD 190-227).

Plants grown at elevated CO_2 showed a greater, and more sustained, response to changing N availability than did Ambient plants (Fig. 1A). While A also declined after JD 245, it remained higher than early season levels until after JD 273. Consequently, late season A in Elevated plants was significantly greater than in Ambient plants, ranging from 58% to 115% higher. Mean A following the increase in N availability (22.8 ± 2.2 umol m^{-2} s^{-1}, JD 245-273) was significantly greater than mean A preceeding the increase in N availability (14.3 ± 1.6 umol m^{-2} s^{-1}, JD 190-227). Stomatal conductance increased with increasing

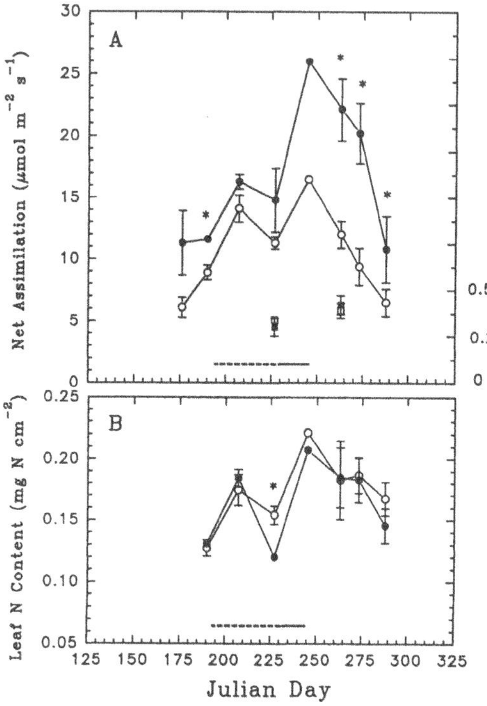

Fig. 1. Net assimilation (circles), and stomatal conductance, g_s, (squares) (**A**) and leaf N content (**B**) over time in plants grown at ambient (open symbols) and elevated (closed symbols) CO_2. Net assimilation was measured at growth CO_2 partial pressure. Nitrogen was added between JD 193 and 240 to produce conditions of low (dashed horizontal line) or high (solid horizontal line) availability. Vertical bars indicate one standard error, * indicates a significant difference, $p < 0.05$.

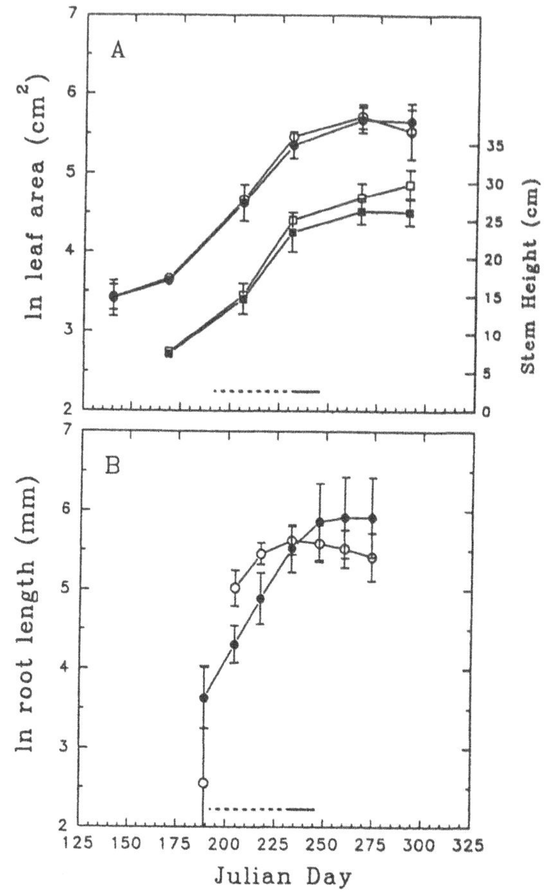

Fig. 2. Changes in leaf area (circles) and stem height (squares) (**A**), and root length (**B**) over time in plants grown at ambient (open symbols) and elevated (closed symbols) CO_2. Vertical bars indicate one standard error. Note that ambient root length on JD 189 was not connected to subsequent points and was not included in RER calculations.

N availability but was not affected by CO_2 treatment (Fig. 1A).

The CO_2 treatment had no significant effect on growth in leaf area or stem height (Fig. 2A). In both treatments, exponential growth in leaf area began ca. 30 d after planting and continued until JD 230, after which growth rates declined. By JD 233, 44% of Elevated plants and 26% of Ambient plants had set bud. However, following the increase in N availability, 35% of Elevated and 49% of Ambient plants that had previously set bud broke bud although with minimal subsequent leaf growth. All of these plants had reset bud by the final harvest.

Root length increased concomitantly with leaf area but showed evidence for CO_2 treatment effects in the timing and rate of growth (Fig. 2B). Early season RER was significantly greater at elevated CO_2 (0.0437 ± 0.0010 d^{-1}) than ambient CO_2 (0.0205 ± 0.0065 d^{-1}, $p < 0.05$), but growth appeared delayed about 15 days in Elevated relative to Ambient plants. The period of high RER was significantly longer at elevated CO_2, extending until approximately JD 247 versus JD 217 (t-test, $p < 0.05$) at ambient CO_2. Late season RER was also greater at elevated CO_2 (0.0020 ± 0.0012 d^{-1}) than at ambient CO_2 (-0.0063 ± 0.0007 d^{-1}, $p < 0.05$). There were no treatment effects on absolute root length measured against the mini-rhizotron tubes at any date, however, because of large within treatment variability. Final root biomass, determined by destructive harvest, was significantly greater in Elevated plants than in Ambient plants (Zak et al., 1993). Final root length was not determined.

Nitrogen content per unit leaf area in the youngest fully expanded leaf varied during the growing season

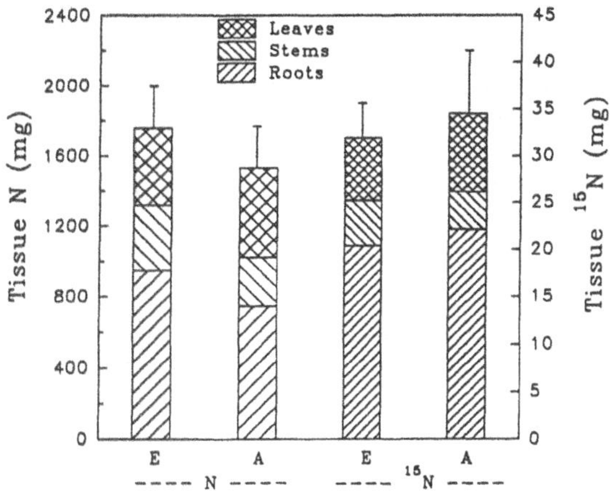

Fig. 3. Total N and ^{15}N distribution within plants grown at ambient (A), and elevated (E) CO_2. Tissue was collected at the final harvest, JD 291. Vertical bars indicate one standard error for whole plant N or ^{15}N content.

and was similar in the two CO_2 treatments (Fig. 1B). Leaf N increased significantly in both CO_2 treatments with increased N availability ($p < 0.05$, t-test) and then declined. At the final harvest, total N and ^{15}N in leaf, stem, and root did not vary significantly between CO_2 treatments (Fig. 3). Approximately 75% of total N applied was accounted for in harvested plant parts, with ca. 50% in roots, 20% in stems, and 30% in leaves. Percentage ^{15}N recovered was somewhat less (ca. 50%), with relatively more found in roots (65%) and less in stems (13%) and leaves (22%).

Discussion

There are important interactions between mineral nutrients and CO_2 as they affect plant growth. Numerous studies have shown that in pot culture, when N concentrations are low, positive growth responses to elevated CO_2 are diminished and may be eliminated entirely (Patterson and Flint, 1982; Israel et al., 1990; Thomas et al., 1991). In field grown plants, although nutrient or water stress clearly influences the magnitude of the CO_2 response, stimulation of photosynthesis by elevated CO_2 is rarely inhibited entirely, even with reductions in leaf N content (Radin et al., 1987; Ziska et al., 1990). Under low nutrient conditions, root growth in high CO_2 grown plants often increases more than shoot growth (Curtis et al., 1990, 1992; Wong et al., 1992; Stulen and den Hertog, 1993) and in some instances may be the only positive growth response (Norby et al., 1992). Soil nutrient availability in natural systems is rarely static, however, and previous studies have not considered the ability of plants to respond to changing soil conditions and how this might be affected by exposure to elevated CO_2. Our results provide a dynamic picture of above- and belowground responses to changing soil N availability in field grown plants at ambient and elevated CO_2.

Soil N availability at the beginning of this experiment was very low and was augmented over a 29 d period at a rate equivalent to N mineralization rates of native, N-poor soil (Zak and Pregitzer, 1990). At the end of this period (JD 233) N availability severely limited photosynthetic and growth responses to CO_2: there were no measurable differences in photosynthetic rate or above- or belowground growth between CO_2 treatments. Interestingly, early season RER was greater at high CO_2, although there was no difference in minirhizotron observable root length between CO_2 treatments. This could reflect greater frequency of fine root branching, and a functional shift in the pattern of soil exploration by the root system at high CO_2 (Berntson and Woodward, 1992).

Between JD 232 and 240, each root box received 1440 mg N to simulate in situ increases in N availability that occur during mid to late summer (Zak and Pregitzer, 1990). Acquistion and allocation of this N was smilar in both CO_2 treatments: young leaf N content increased to the same extent and both absolute and relative allocation of absorbed N to roots, stems and leaves was the same. Net assimilation, however, responded quite differently in the two CO_2 treatments. In the chambered ambient treatment, while there was an initial increase in A, mean A in the month following increased N availability was no different than in the preceeding month. In the chambered elevated treatment, there was a sustained increase in A, resulting in a 59% increase in mean A in the month following the increased N availability. During this period, Elevated plants had a more than two-fold greater A than Ambient plants.

Our observation of greater late season fine root RER in Elevated plants is consistent with final harvest results suggesting an increased C flux belowground at high CO_2. Final root biomass increased 60%, microbial C increased 44%, and rhizosphere labile C increased 17% in Elevated relative to Ambient plants (Zak et al., 1993). The activity of C sinks, whether above- or belowground may be an important determinant of photosynthetic CO_2 responses (Arp, 1991). Low sink activity and consequent carbohydrate accumulation in source leaves under elevated CO_2 can reduce A either directly via feedback inhibition or indirectly through effects on protein synthesis (Stitt, 1992). In *Populus*, roots are major C sinks following bud set (Nelson and Isebrands, 1983). The differential C allocation belowground we observed in Elevated plants could therefore have contributed to the prolonged late season stimulation of A.

Phenological events also influenced late season photosynthetic capacity in both CO_2 treatments. Day length was decreasing and bud set had occurred, although there were no hard frosts before the final harvest. Leaf N and photosynthetic capacity ordinarily decline in tandem following bud set in temperate trees (Reich et al., 1991) as part of normal autumnal senescence. Curtis and Teeri (1992), however, found that growth at elevated CO_2 delayed photosynthetic senescence in *P. grandidentata*. In the present study, changing N availability caused late season photosynthetic activity to differ substantially from that reported by Curtis and Teeri (1992), making direct comparisons difficult. However, effects of CO_2 on the timing of senescence could have influenced both the responsiveness of our plants to soil and leaf N and the magnitude of C fluxes belowground.

Although root biomass was greater at the final harvest in Elevated plants (Zak et al., 1993), they did not recover any more of the applied N than did Ambient plants. Rather, root growth was more N use efficient at high CO_2 (i.e., less N g^{-1} dry weight). That this increased root growth did not result in greater N uptake was probably due to the extremely low N content of the rooting medium and the short-term nature of the experiment. Based on recovery of ^{15}N, at least 50% of the N pulse was taken up and incorporated into plant biomass. There was little ^{15}N present either in the bulk soil or microbial biomass at the final harvest (D. Zak, unpublished data). This suggests that N not taken up soon after application was quickly leached from the open bottom root box filled with sandy soil. Differential water use or leaching between treatments was unlikely since neither leaf area nor stomatal conductance was affected by CO_2. Given the low soil N content and minimal potential for N mineralization, new root growth or more branching would not enable the plants to acquire additional N, at least in the short term. In higher fertility soil, or with further inputs of N either exogenously or via mineralization, plants grown at elevated CO_2, with more extensive roots, would likely acquire more net N than ambient CO_2 grown plants.

Our results provide further evidence of the importance of soil fertility in the CO_2 response of field grown plants. With N available at a rate equivalent to that mineralized in native forests on the same soil series, positive physiological and growth responses to CO_2 were largely inhibited. With increased N availability, such as could occur during periods of accelerated mineralization in mid summer or via atmospheric deposition (Pregitzer et al., 1992), a positive CO_2 response in net assimilation and RER was present. Spatial and temporal heterogeneity of soil resources suggests this responsiveness may be an important factor for forest growth on infertile sites under future atmospheric CO_2 levels.

Acknowledgements

Support for this research was provided by a USDA competitive grant (90-37290-5668), and the University of Michigan Biological Station. We thank Robert Fogel and John Lussenhop for critical input throughout the course of this experiment and Diana Randlett, Ronald Hendick, Melissa Moss, Tony Sutterley, and Richard Spray for assistance in the field and laboratory.

References

Arp W J 1991 Effects of source-sink relations on photosynthetic acclimation to elevated CO_2. Plant Cell Environ. 14, 869–875.

Barry W G and Sachs R M 1968 Vegetative propagation of quaking aspen. Calif. Agric. 22, 14–16.

Berntson G M and Woodward F I 1992 The root system architecture and development of *Senecio vulgaris* in elevated CO_2 and drought. Funct. Ecol. 6, 324–333.

Cure J D 1985 Carbon dioxide doubling responses: a crop survey. *In* Direct Effects of Increasing Carbon Dioxide on Vegetation. Eds. B R Strain and J D Cure. pp 99-116. U.S. Dept. of Energy, Washington, D.C.

Curtis P S, Balduman L S, Drake B G and Whigham D F 1990 Elevated atmospheric CO_2 effects on belowground processes in C_3 and C_4 estuarine marsh communities. Ecology 71, 2001–2006.

Curtis P S and Teeri J A 1992 Seasonal responses of leaf gas exchange to elevated carbon dioxide in *Populus grandidentata*. Can. J. For. Res. 22, 1320–1325.

Field C and Mooney H A 1986 The photosynthesis-nitrogen relationship in wild plants. *In* On the Economy of Plant Form and Function. Ed. T J Givinish. pp 25–55. Cambridge Univ. Press., UK.

Hendrick R L and Pregitzer K S 1992a Spatial variation in tree root distribution and growth associated with minirhizotrons. Plant and Soil 143, 283–288.

Hendrick R L and Pregitzer K S 1992b The demography of fine roots in a northern hardwood forest. Ecology 73, 1094–1104.

Israel D W, Rufty T W J and Cure J D 1990 Nitrogen and phosphorus nutritional interactions in a CO_2 enriched environment. J. Plant Nutr. 13, 1419–1433.

Nelson N D and Isebrands J G 1983 Late-season photosynthesis and photosynthate distribution in an intensively-cultured *Populus nigra* × *laurifolia* clone. Photosynthetica 17, 537–549.

Norby R J, O'Neill E G and Luxmoore R G 1986 Effects of atmospheric CO_2 enrichment on the growth and mineral nutrition of *Quercus alba* seedlings in nutrient-poor soil. Plant Physiol. 82, 83–89.

Norby R J, Gunderson C A, Wullschleger S D, O'Neill E G and McCracken M K 1992 Productivity and compensatory responses of yellow-popular trees in elevated CO_2. Nature 357, 322–324.

O'Neill E G, Luxmoore R J and Norby R J 1987 Increases in mycorrhizal colonization and seedling growth in *Pinus echinata* and *Quercus alba* in an enriched CO_2 atmosphere. Can. J. For. Res. 17, 878–883.

Pastor J and Post W M 1988 Response of northern forests to CO_2 induced climate change. Nature 334, 55–58.

Patterson D T and Flint E P 1982 Interacting effects of CO_2 and nutrient concentration. Weed Science 30, 389–394.

Pregitzer K S, Burton A J, Mroz G D, Liedity H O and MacDonald N W 1992 Foliar sulfur and nitrogen along an 800-km pollution gradient. Can. J. For. Res. 22, 1761–1769.

Radin J W, Kimball B A, Hendrix D A and Mauney J R 1987 Photosynthesis of cotton plants exposed to enhanced levels of CO_2 in the field. Photosyn. Res. 12, 191–203.

Reich P B, Walters M B and Ellsworth D S 1991 Leaf age and season influence the relationships between leaf nitrogen, leaf mass per area and photosynthesis in maple and oak trees. Plant Cell Environ. 14, 251–259.

Sokal R R and Rohlf F J 1981 Biometry. WH Freeman and Co., San Francisco. 859 p.

Stitt M 1991 Rising CO_2 levels and their potential significance for carbon flow in photosynthetic cells. Plant Cell Environ. 14, 741–762.

Stulen I and den Hertog J 1993 Root growth and functioning under atmospheric CO_2 enrichment. Vegetatio 104/105, 99–115.

Thomas R B, Richter D D, Ye H, Heine P R and Strain B R 1991 Nitrogen dynamics and growth of seedlings of an N-fixing tree (*Gliricidia sepium* (Jacq. Walp.) exposed to elevated atmospheric carbon dioxide. Oecologia 88, 415–421.

Wong S C, Kriedemann P E and Farquhar G D 1992 CO_2 × nitrogen interaction on seedling growth of four species of eucalypt. Aust. J. Bot. 40, 457–472.

Zak D R and Pregitzer K S 1990 Spatial and temporal variability of nitrogen cycling in northern lower Michigan. For. Sci. 36, 367–380.

Zak D R, Pregitzer K S, Curtis P S, Teeri J A, Fogel R and Randlett D L 1993 Elevated atmospheric CO_2 and feedback between carbon and nitrogen cycles. Plant and Soil 151, 105–117.

Ziska L H, Drake B G and Chamberlain S 1990 Long-term photosynthetic response in single leaves of a C_3 and C_4 salt marsh species grown at elevated atmospheric CO_2 *in situ*. Oecologia 83, 469–472.

SOIL BIOTA

Plant and Soil **165:** 55–65, 1994.

Responses of soil biota to elevated atmospheric carbon dioxide

Elizabeth G. O'Neill
Environmental Sciences Division, Oak Ridge National Laboratory, Oak Ridge, TN, USA

Key words: CO_2 enrichment, decomposition, mycorrhizae, nitrogen fixation, rhizosphere, soil biota, soil fauna

Abstract

Increasing concentrations of atmospheric CO_2 could have dramatic effects upon terrestrial ecosystems including changes in ecosystem structure, nutrient cycling rates, net primary production, C source-sink relationships and successional patterns. All of these potential changes will be constrained to some degree by below ground processes and mediated by responses of soil biota to indirect effects of CO_2 enrichment. A review of our current state of knowledge regarding responses of soil biota is presented, covering responses of mycorrhizae, N-fixing bacteria and actinomycetes, soil microbiota, plant pathogens, and soil fauna. Emphasis will be placed on consequences to biota of increasing C input through the rhizosphere and resulting feedbacks to above ground systems. Rising CO_2 may also result in altered nutrient concentrations of plant litter, potentially changing decomposition rates through indirect effects upon decomposer communities. Thus, this review will also cover current information on decomposition of litter produced at elevated CO_2.

Introduction

Ecosystems are largely constrained by the rates at which soil processes occur. These processes are in turn constrained by rates of carbon input from the above-ground portion of the system. An increasing number of studies on effects of rising elevated CO_2 point to belowground systems as critical sources or sinks in global C cycling and as critical determinants of ecosystem response to climate change (Norby et al., 1994). Changes in C input to the soil, whether in the form of increased or chemically altered exudation, through increased litter production, or through increased production and turnover of fine roots could have a dramatic effect on rates of belowground processes (Norby, 1994). Changes in chemical composition of either plant litter or fine roots as a result of atmospheric CO_2 enrichment could also affect process rates, especially nutrient cycling rates (Strain and Bazzaz, 1983). All of these changes can be manifested through their effects on the soil biota. As early as 1981 it was recognized that certain soil biota could play a vital role in the responses of unmanaged vegetation to CO_2 enrichment (Luxmoore, 1981). In response to Kramer's (1981) assertion that nutrient limitations in unmanaged ecosystems would preclude biomass increases in forest ecosystems, Luxmoore hypothesized that the existence of strong positive feedbacks between plants and symbiotic or root-associated microorganisms would result in increased acquisition of nutrients by plants and subsequent biomass increase. Luxmoore considered the roles of mycorrhizae, symbiotic nitrogen-fixing bacteria and mineralizing rhizosphere bacteria in his hypothesis, but did not include responses of other groups of soil biota. In 1982, a workshop was held in Athens, Georgia to assess the state-of-the-science on potential effects of rising CO_2 on vegetation and to make recommendations for future research (Lemon, 1983). The working group on microbial effects identified the following as critical areas of concern: symbiotic organisms, microbial mineralization-immobilization processes (especially nitrification and denitrification), soil organic matter changes and structural changes, and possible effects on soil-borne plant disease and decomposition (Lamborg et al., 1983). The working group did not deem direct effects of CO_2 enrichment on soil microbial populations to be likely, due to the order of magnitude difference between soil CO_2 concentrations and atmospheric CO_2 concentrations. However, the group did consider the possibility of indirect effects of changes in soil pCO_2 and pO_2. Specific effects of

CO_2 enrichment on soil fauna or on the potential for changes in food webs were not evaluated.

The objective of this paper is to assess the current state-of-the-science with regard to effects of CO_2 enrichment on the major classes of soil biota, beginning at the root with mycorrhizae and symbiotic N-fixers, and proceeding in turn through microbial responses, pathogens, and soil fauna. Finally, decomposition processes will be examined, in light of possible changes in litter quality that could determine rates at which soil microorganisms are able to break down plant material and release nutrients for eventual reentry into active cycles.

Mycorrhizal responses

The concept of mycorrhizal benefit to host plant nutrient uptake is now generally accepted. Both Luxmoore (1981) and the Athens working group (Lamborg et al., 1983) considered this symbiosis to be one key to nutrient acquisition in an enriched CO_2 world. However, mycorrhizae should be considered beyond their ability to enhance plant growth. They are frequently overlooked in questions of terrestrial C cycling, although mycorrhizae are the largest single consumer of NPP in terrestrial biomes (Allen, 1991). Although they may comprise only a small fraction of the standing organic matter pool at any one time, they are responsible for a major proportion of the annual flux of C to the soil. For example, while mycorrhizae and their associated fruiting structures comprised only 12 to 16.7% of the total soil organic matter in a second-growth Douglas fir forest, they were responsible for 53.5 to 89.5% of C flux to the soil. Litterfall only accounted for 10 to 10.4% of total C flux (Fogel and Hunt, 1979, 1983). Under elevated CO_2 conditions, increased translocation of carbon to roots could stimulate growth, activity or longevity of mycorrhizae and impact both plant nutrient uptake and C cycling.

Most plant species are colonized by either ectomycorrhizal (ECM) or vesicular-arbuscular mycorrhizal (VAM) fungi. These two groups differ in their morphology, physiology, and host associations, and these differences may dictate the extent of their response to elevated CO_2. ECM fungi invest considerable plant-produced C in specialized structures (mantles, fruiting bodies, rhizomorphs, and extensive extraradical mycelium). This demand for C may equate to heightened responsiveness (compared to VAM) to increased C supply. CO_2 supplied at twice-ambient levels was shown to result in both higher numbers of mycorrhizal short root tips (proportion of total fine root tips) and a more rapid rate of colonization in white oak, *Quercus alba* L. (O'Neill et al., 1987; Fig. 1a). In shortleaf pine, (*Pinus echinata* L.), exposure to elevated CO_2 appeared to offset the temporary growth lag frequently seen in tree seedlings during the early stages of mycorrhizal establishment (O'Neill et al., 1987). In this study, when seedlings were grown at ambient CO_2 for 6 weeks, the total biomass of plants inoculated with the mycorrhizal fungus *Pisolithus tinctorius* (Pt) was reduced compared to uninoculated seedlings. However, Pt-inoculated seedlings grown at 700 μmol mol^{-1} CO_2 exhibited a 27% biomass *increase* over uninoculated seedlings. In another experiment, root colonization by *P. tinctorius* was also higher under elevated CO_2 on one population of loblolly pine (*Pinus taeda* L.), but not on another (Lewis et al., 1992, 1994). Biomass of these seedlings in a low P treatment was greater at elevated CO_2 only in the population of seedlings that experienced enhanced colonization.

In contrast to the results with ECM, VA-mycorrhizal yellow-poplar (*Liriodendron tulipifera* L.) grown at three atmospheric CO_2 concentrations (ambient, 550 μmol mol^{-1}, 700 μmol mol^{-1} CO_2) showed no difference in percent root colonized (O'Neill et al., 1991; Fig. 1b). It would be incorrect to say there was no response to elevated CO_2, however. In this experiment, as in the ECM series, total fine root length increased dramatically in plants grown at elevated CO_2, resulting in more mycorrhizal tissue (and presumably greater nutrient acquisition) per plant. Nutrient concentrations were not determined in all experiments, so there was no way to determine how the differences in response related to seedling nutrient uptake. Responses of both VAM and ECM were confirmed in the field, when mycorrhizal colonization late in the growing season was assessed on white oaks and yellow-poplars grown in the ground in open-top chambers at three concentrations of atmospheric CO_2 (Fig. 1a,b). Again, proportional infection of root length was increased by CO_2 enrichment in white oak but unchanged in yellow-poplar and both species had increased fine root biomass at their respective final harvests (Norby, 1994).

Although VAM have not been examined under CO_2 enrichment in other woody species, several researchers report results from exposures of grass species to elevated CO_2. In a serpentine grassland exposed to elevated CO_2 in open-top chambers and sampled at monthly intervals over a two-year period, VAM colonization was not increased (Whitbeck, 1993). However,

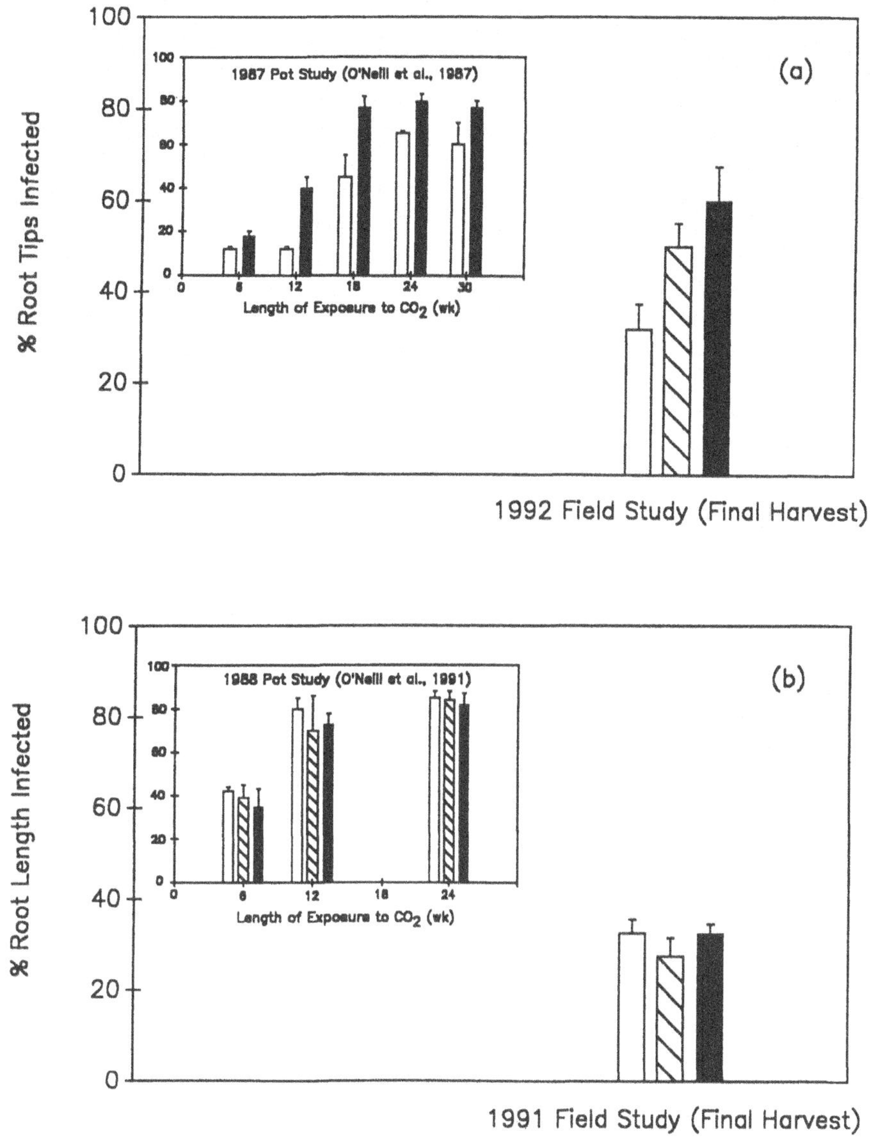

Fig. 1. Comparison of CO_2 enrichment effects on ECM vs. VAM infection in field and environmental chamber studies. Primary figures are results obtained at final harvest of field grown saplings exposed to three levels of atmospheric CO_2 for four years in open-top chambers. Inserts are data obtained from one-year-old seedlings in environmental chamber pot studies (O'Neill et al., 1987; 1991). Legend: **Open bars**, Ambient air + 0 μmol mol^{-1} CO_2; **Cross-hatched bars**, ambient + 150 μmol mol^{-1} CO_2; **Solid bars**, ambient + 300 μmol mol^{-1} CO_2. **a)** Ectomycorrhizal white oaks. Data represent percentage of total root tips that were mycorrhizal. **b)** VA-mycorrhizal yellow-poplar. Data represent proportion of fine root length that was mycorrhizal.

Monz et al. (1994) provide evidence that the responses of VAM to CO_2 enrichment are not uniform. In their study, VAM colonization in *Bouteloua gracilis* (a C4 grass) increased at elevated CO_2 but there was no increase in *Pascopyrum smithii* (a C3 grass). The work of Monz and co-workers is also notable in that interactive effects of CO_2 enrichment, temperature and water availability on VAM were also demonstrated. The combination of elevated CO_2, elevated temperature, and decreased precipitation reduced VAM colonization in *P. smithii* but increased colonization on *B. gracilis*; the latter effect was primarily due to elevated CO_2

Effects of CO_2 enrichment on the mycorrhizal symbiosis may not simply be related to colonization rates. In fact, undisturbed ecosystems may be "saturated" with regard to mycorrhizae. Decades of plant growth under steadily increasing CO_2 levels may result in differences in fungal species composition, rather than changes in percent infection or colonization rates. Little is known about interactions among differing species of mycorrhiza, although mycorrhizal diversity is extremely high. Up to 2000 species of ECM are potentially capable of colonizing Douglas-fir; several hundred may colonize a single tree at one time (Trappe, 1977). Other host/fungus combinations can be highly specific. Although mycorrhizae have been long regarded as functionally redundant in ecosystems, evidence from single species plant-fungus tests in controlled studies seems to indicate otherwise. The different host-fungus combinations are morphologically and physiologically different. They respond differently to experimental treatment, and colonization may result in varied benefit to the host plant. Environmental factors may influence the diversity of species upon a single host and this could in turn feed back to affect host plant response. Evidence for environmental effects on fungal community composition is sketchy. One ECM species, *Cenococcum graniforme*, while abundant on white oak seedlings grown at ambient CO_2, was consistently under-represented (on many seedlings absent altogether) on seedlings grown at elevated CO_2 (Norby et al., 1986a; O'Neill et al., 1987). A highly diverse mycoflora may be more important to the ability of a plant to adjust to elevated CO_2 and climate change than simply a threshold biomass. There is an urgent need to examine interactive effects of varied fungal species, in tests with simultaneous colonization by multiple fungi, on different hosts and in different soils (Bledsoe, 1992).

An additional area requiring study includes the relationship between mycorrhizal propagule dispersion rates and their roles in ecosystem migration and species composition (Allen, 1987; Allen et al., 1989; O'Neill et al., 1991). Mycorrhizae can structure communities by making resources more available to some plants than others (Allen, 1991; Caldwell et al., 1985). Ecosystem structure may change under elevated CO_2 and climate change if fungal species dominance changes or if rates of propagule dispersal cannot keep pace with host species migration.

Nitrogen fixers

The second symbiosis of importance in vegetation response to elevated CO_2 is the symbiotic association of nodulating bacteria or actinomycetes with plants which results in fixation of atmospheric nitrogen. Like mycorrhizae, symbiotic N-fixers receive all of their C from photosynthate, and furthermore, provide a source of available N for their host plant and the ecosystem of which the host is a member. Whether the host of interest is an agricultural species, a nonagricultural herbaceous plant, or a woody plant, the questions regarding CO_2 effects are the same. Given increased availability of photosynthate, will specific activity of the N-fixing nodules increase? Or will nodule mass per plant increase? Both will result in increased total N-fixation per plant. And finally, if total fixation per plant increases, can we expect plant N concentrations to increase?

Results to date suggest common patterns of response (Table 1). In general, N-fixation per plant increased whether plants were herbaceous or woody, fertilized or unfertilized. Generally nodule mass or number was increased. However, effects on specific nitrogenase activity (or fixation per g nodule) were variable. In one case, specific activity was initially stimulated by 1200 μmol mol^{-1} CO_2, but this effect declined through time (Phillips, 1976). Plant N content was increased in all herbaceous species, in contrast to woody species, where N content increased only when fertilizer N was supplied.

Stimulation of nitrogen fixation in woody species has been demonstrated in *Robinia pseudoacacia*, *Alnus glutinosa* and *Eleagnus angustifolia* (Norby, 1987), *Alnus rubra* (Arnone and Gordon, 1990) and *Gliricidia sepium* (Thomas et al., 1991). With one exception, specific nitrogenase activity (SNA) was not increased in these studies, rather, nodule mass and/or nodule number was stimulated by elevated CO_2 so total nitrogenase activity (TNA) per plant increased. Arnone and Gordon (1990) showed no increase in nodule mass, but a small increase in SNA, thus total activity was once again enhanced. Plant N content was not increased in these studies unless additional nitrogen was supplied to plants (Arnone and Gordon, 1990; Thomas et al., 1991), and leaf N concentrations were generally reduced at elevated CO_2, as has been commonly observed in plant species that are not N-fixing. Norby (1987) points out that P limitation may influence results; nitrogen fixation is strongly dependent upon phosphorus availability. In all nonagricultural N-fixing plants, the presence of mycorrhizae is critical to P acquisition. As mycorrhizal function may also be influenced by CO_2 enrichment, more studies are needed to examine N fixation in native soils with mycorrhizal inoculum present. More information is also needed on responses of tropical host species, where P limitation

Table 1. Comparison of effects of CO_2 enrichment on nitrogen fixation. SNA, specific nitrogenase activity; TNA, total nitrogenase activity (per plant)

Source	CO_2 enrichment effect on:				
Species/Symbiont	CO_2 (μmol mol^{-1})	Nodule Mass/Number	SNA	TNA	Plant N Content
Herbaceous plants					
Masterson and Sherwood, 1978					
Trifolium repens/Rhizobium	1200	No	Yes	Yes	Yes
Pisum sativum/Rhizobium	1200	Yes	No	Yes	Yes
Hardy and Havelka, 1976					
Glycine max/Rhizobium	1200	Yes	Yes	Yes	Yes
Finn and Brun, 1982					
Glycine max/Rhizobium	1020	Yes	No	Yes	—
Phillips et al., 1976					
Pisum sativum/Rhizobium	1200	Yes	No	Yes	Yes
Woody plants					
Norby, 1987					
Robinia pseudoacacia/Rhizobium	700	Yes	No	Yes	No
Alnus glutinosa/Frankia	700	Yes	No	Yes	No
Eleagnus angustifolia/Frankia	700	Yes	No	No	No
Arnone and Gordon, 1990					
Alnus rubra/Frankia	650	No	Slight	Yes	Yes (+N)
Thomas et al., 1991					
Gliricidia sepium/Rhizobium	650	Yes	—	—	Yes (+N)

is particularly critical and N-fixing woody species are of increased importance.

No work to date has been conducted on effects of rising CO_2 on associative nitrogen fixation. In some ecosystems, free-living soil microbes are responsible for the bulk of atmospheric nitrogen fixed. If C input to soil is increased through exudation by roots or mycorrhizal hyphae, this source of N could also be enhanced.

Microbial responses

Stimulation of microbial nutrient transformations in the rhizosphere as a result of increased C input is another mechanism hypothesized by Luxmoore (1981) to increase nutrient availability. The rhizosphere is a qualitatively defined area of the soil, influenced by the presence of the root, that contains higher populations of some soil organisms than the soil at large. The rhizosphere exists primarily because of the exudation of easily assimilated organic materials from the plant root. If competition for C is the primary factor limiting soil microbial populations (Lockwood, 1981), environmental factors that influence root exudation, either through changes in exudation rate or chemical composition, could affect rhizosphere microbial populations. This has been demonstrated in the case of light and temperature changes (Fenwick, 1973; Rouatt and Katznelson, 1960; Rouatt et al., 1963). It is reasonable to expect CO_2 effects as well. For a discussion of current information on exudation and fine root growth responses to CO_2 enrichment, see Norby (1994).

In mycorrhizal plants, CO_2 effects would not necessarily be limited to the rhizosphere. Mycorrhizal hyphae are very efficient at redistributing recently-fixed C away from roots (Norton et al., 1990). In fact, this efficiency is so great that it may be more correct to expand the rhizosphere concept through consideration of effects of CO_2 enrichment on the "mycorrhizosphere" (Rambelli, 1973). Ectomycorrhizal hyphae may extend as much as 2 m beyond the root surface and contain more than 120 lateral branches (Fogel, 1988). The increased surface area represented by this extraradical hyphal system may mean that mycorrhizae exert a much greater effect on the soil microbial biomass than do roots alone. In addition, rhizodeposition in mycorrhizal plants is greater than that produced by ecologically nonmycotrophic plants (Schwab et al., 1984) or by rapidly growing nonmycorrhizal root tips of mycorrhizal plants (Norton et al, 1990).

Zak et al. (1993) have provided the first evidence that elevated CO_2 may increase microbial biomass/activity not only in the rhizosphere, but also in bulk soil. For both rhizosphere and bulk soil of *Populus grandidentata* Michx. grown in elevated or ambient CO_2 in open-top chambers or in chamberless-treatments (ambient CO_2 only), microbial biomass C (measured by chloroform fumigation) was greatest at elevated CO_2. Labile or available C increased in the rhizosphere with CO_2 enrichment, but not in bulk soil. Bulk soil was not assayed for mycorrhizal hyphae; the presence of mycorrhizal hyphae may have increased microbial biomass in areas not directly penetrated by roots. Nitrogen availability was also hypothesized in this study to increase as a result of increased root exudation. In short-term laboratory incubations, net N mineralization was greater in bulk soil from the elevated CO_2 treatments than in either the ambient or chamberless treatments. However, available N was not increased significantly in either bulk or rhizosphere soil by CO_2 enrichment. Although stimulation of nitrifying bacteria in the rhizosphere could increase available N (O'Neill et al., 1987), other mechanisms could be significant in an enriched CO_2 atmosphere. Zak et al. (1993) utilized an hypothesis of Clarholm (1985) to link elevated CO_2-induced increases in rhizosphere microbial populations to increases in N availability. According to Clarholm's hypothesis, as roots grow through the soil, soil microbe activity increases as a result of the pulse of exudate produced. Increased microbial respiration attracts bacteriophagic protozoans. As protozoans feeding on bacteria only assimilate 33% of bacterial N, the rest is excreted as cell wall or NH_4^+ which can then be assimilated by plant roots and mycorrhizae (Clarholm, 1985; Fenchel, 1982). However, Robinson et al. (1989) and Griffiths and Robinson (1992) determined through the use of nitrogen balance models that, even taking bacterial grazing into account, root-induced nitrogen mineralization could not contribute substantially towards plant N acquisition. If these models accurately describe the relationship between root exudation and N mineralization, then an increase in N availability might not occur as a result of CO_2 enrichment unless exudate C-to-N ratios increase. Model predictions are driven strongly by substrate C-to-N ratios; there is some evidence that these ratios may increase under CO_2 enrichment due to partitioning of C into nonstructural carbohydrates (Körner and Arnone, 1992).

Microbial biomass C and N also increased at elevated CO_2 in two herbaceous communities in laboratory microcosms, although there were no data reported on soil N availability (Diaz et al., 1993). In this study, leaf nitrogen concentrations declined in one of the two communities (an acidic grassland) and the authors hypothesized that the decline in leaf N resulted from increased microbial biomass which immobilized available N. It was unclear whether total plant content (as opposed to concentration) declined also. Reduced foliar N concentrations with CO_2 enrichment is commonly observed, although frequently this is an artifact of increased carbohydrate content, with total plant N uptake unchanged (Curtis et al., 1989; Körner and Arnone, 1992; Owensby et al., 1993). Leaf N concentrations may not be a valid indicator of soil N availability (Tschaplinski and Norby, 1993).

Other studies have not shown a CO_2 response in microbial populations. Whipps (1985) found no change in bacterial numbers (dilution plate counts) at elevated CO_2 in wheat rhizospheres. O'Neill et al. (1987), in a pot study with yellow-poplar, found no change in total bacteria in rhizosphere soil over 24 weeks as measured by direct counts. Populations of phosphate-dissolving bacteria (PDB; Azcón et al., 1976) and nitrifying bacteria in the rhizosphere were also quantified. Results were inconclusive; however, there was a strong trend towards higher numbers of PDB in rhizosphere of seedlings grown at elevated CO_2. Phosphorus availability could be enhanced by a preferential increase in PDB in the rhizosphere with CO_2 enrichment. Cotton grown under Free-Air CO_2 Enrichment (FACE) showed a trend towards higher numbers of rhizosphere bacteria (fluoroscein isothiocyanate direct counts) and total microbial respiratory activity (dehydrogenase assay) at elevated CO_2, however no difference was seen in colonization by VAM fungi in this study (Rogers et al., 1992). In a later study, at a different FACE site, similar results were obtained (Runion et al., 1994).

Why are there conflicting results? Rhizosphere responses could vary with plant species, community type, or soils (Zak et al., 1992). Differences in methodology may also lead to opposing conclusions. Chloroform fumigation and the dehydrogenase assay measure total microbial biomass and activity, respectively, whereas direct counting in these studies accounted only for changes in the bacterial component of the microflora. Results may not conflict, for example, if there were a preferential stimulation of the fungal community. Little attention has been paid to community structure in the rhizosphere and how this may change under the influence of rising CO_2. Differences in nutrient availability in the rhizosphere may also constrain responses. In a review of potential effects of elevated CO_2 on C dynamics in soil, van Veen et al. (1991) summarized work of several researchers who showed that more C is incorporated into microbial biomass under conditions of relatively high soil nitrogen status than in low N soils. They concluded that C availability is not the only rate-limiting factor for rhizosphere microbial growth.

Finally, responses to elevated CO_2 may manifest themselves through process rates, rather than effects on population size. Holmes and Zak (1993) investigated seasonal changes in soil microbial biomass and net N mineralization rates and found that biomass remained relatively constant through time although turnover rates increased in response to variable C input. Increased turnover rates in soil microbes as a result of increased root or hyphal exudation under CO_2 enrichment could enhance mineralization and nutrient availability. Increases in microbial population size, on the other hand, could lead to decreased nutrient availability and constrain plant response to elevated CO_2 (Díaz et al., 1993).

Plant pathogens

An increase in severity or frequency of soil-borne plant disease may also be of significant concern as atmospheric CO_2 rises. The infection process is in some measure stochastic, so more extensive root systems would increase the probability for invasion. This could be offset by increased plant vigor and disease resistance (Runion et al., 1994). Very little data are available in this area. The potential for *Rhizoctonia solani* infestation was assessed on soils from a FACE site in Maricopa, Arizona, and although there was a trend towards higher infestation potential at elevated CO_2, a bioassay for damping-off pathogens did not substantiate the trend (Runion et al., 1994).

Soil fauna

Much of the belowground work to date has focused on organisms, that is, responses of rhizosphere microorganisms and symbionts that affect responses of individual plants. Although it is clear that root symbionts and rhizosphere bacteria represent a first-order response to rising CO_2, responses of organisms that are less intimately associated with roots may also be an integral part of ecosystem response. In fact, it is not until the focus shifts away from the root and into soil food webs that ecosystem-level effects can be discerned. Interactions between predator and prey, such as in Clarholm's (1985) hypothesis, influence nutrient cycling rates. Soil- and litter-inhabiting arthropods can regulate whether net immobilization or mineralization occurs during decomposition (Seastedt and Crossley, 1984).

There is evidence to suggest that C supply from roots affects numbers of root-associated fauna. Population numbers and fecundity of a parasitic nematode increased when C partitioning in Valencia orange trees was altered by defruiting (Duncan and Eissenstat, 1993). Populations of saprophagous nematodes increased at elevated CO_2 in a FACE experiment with cotton (Runion et al., 1994). However, Freckman et al. (1991), found no difference in nematode community structure in intact grassland cores after 2 years' exposure to elevated CO_2. Responses to CO_2 enrichment and rates of change in terrestrial soil communities may depend strongly on edaphic factors, particularly C and N content of soil.

Questions remain at the ecosystem level, where soil food web interactions determine the fate of forests, grasslands, deserts, and other ecosystems. For example, will responses by rhizosphere organisms change either rates or patterns of interactions in soil food webs, or will additional C supply to soil leave soil food webs essentially unchanged? Can we predict changes in food web structure based on differing abilities of the organisms involved to respond to additional C supply? Presently, little data are available to determine answers to these questions.

Decomposition

Elevated CO_2 could affect ecosystem nutrient cycling through indirect effects on decomposition processes, primarily those of leaf litter and fine roots. This could occur through (1) changes in chemical composition ('quality') of litter, or (2) changes in plant species composition of ecosystems (Ågren et al., 1991).

Chemical changes observed in green leaf tissue produced under CO_2 enrichment include increases in nonstructural C (Körner and Arnone, 1992) and decreases in leaf N concentrations (Curtis et al., 1989; Körner and Arnone, 1992; Norby et al., 1992; Owensby et al., 1993), as well as changes in concentrations of secondary plant metabolites such as tannins (Johnson et al., 1994; Lindroth and Kinney, 1993). These responses are remarkably similar across species. If differences in green leaf composition are reflected in differences in foliar litter concentrations, decomposition rates could be reduced (Melillo, 1983). Woody plants typically retranslocate nutrients from leaves to storage tissue prior to litterfall, leading to litter nutrient concentrations that are often very different from green leaf concentrations. Potential enhanced efficiency of nutrient use under CO_2 enrichment may increase retranslocation and result in further declines in litter quality (Johnson et al., 1994) Until recently, little data were available on senescent foliage produced under CO_2 enrichment, and still fewer data on leaves grown at variable CO_2 concentrations and allowed to senesce naturally in the field. Tannins and soluble sugars in white oak litter increased, while lignin decreased, in response to CO_2 enrichment in an environmental chamber experiment (Norby et al., 1986b). While lignin:N and lignin:P decreased slightly with CO_2 enrichment, predicted decay rates using these data suggested that no difference in rates of mass loss would occur. In contrast, Melillo (1983) reported that concentrations of soluble phenolics and structural compounds (cellulose and lignin) were higher and N lower in sweetgum (*Liquidambar styraciflua* L.) seedlings grown at 935 μmol mol^{-1} CO_2 after one growing season in pots in open top chambers, suggesting that this litter would decay more slowly. In neither of these two studies was actual mass loss rate measured. In an experiment looking at the interaction of soil food web complexity and CO_2 effects on litter quality, a near doubling of litter C:N resulted when sweet chestnut (*Castanea sativa* Mill.) seedlings were grown in pots at 700 μmol mol^{-1} CO_2 (Couteaux et al., 1991). Decomposition rates (as measured by mass loss and C mineralization) were initially reduced in CO_2-enriched litter, but increased during the latter stages of the experiment. Responses were dependent upon the complexity of the soil faunal community, and cumulative mass loss was only different in the pots with the least complex food web (microflora and protozoans only).

Litter C:N was not affected by CO_2 enrichment in pure or mixed stands of *Scirpus olneyi* and *Spartina patens* in an estuarine marsh (Curtis et al., 1989). On a seasonal basis, CO_2 enrichment was predicted to have no effect on litter decomposition rates in this system. Mass loss rates were also not affected by CO_2 enrichment in yellow-poplar in a field study where leaves were allowed to naturally abscise (O'Neill and Norby, 1991). Litter was collected from field-grown yellow-poplars in the fall of their second year of growth in ambient air supplemented with 0, 150, or 300 μmol mol^{-1} CO_2. Although green-leaf nitrogen concentrations were reduced by CO_2 enrichment, no significant difference in initial lignin:N or C:N of litter was observed and litter N concentration was only slightly reduced. After two years of decomposition in litterbags in the field, no difference was seen in percentage of original dry mass remaining nor in mass loss rates.

Decomposition rates could also be affected by CO_2 enrichment if plant species composition changes within the ecosystem. Plant species composition has a profound effect upon overall decomposition rates (Kelly and Beauchamp, 1987). Shifts in species composition as a result of long-term exposure of the ecosystem to rising CO_2 may alter *ecosystem* litter quality and change decomposition rates. All studies conducted to date on decomposition and CO_2 enrichment have examined dynamics of single species. Although short-term mass loss rates of mixed-species litter may not differ from those of its component species, other parameters, such as N dynamics, C mineralization rates, and long-term mass loss rates may. Blair et al. (1990) found no difference in mass loss when comparing single- and mixed-species litter. However, single-species litter responses proved to be a poor predictor of mixed-litter response for bacterial, fungal and nematode densities. Estimates of N flux based on single-species data resulted in underestimates of N release and overestimates of accumulated N relative to observed responses of mixed litter. Ecosystem-level effects of elevated CO_2 on decomposition should be based upon mixed-litter results, preferably over a time scale long enough to include lignin decomposition dynamics.

To date, no data are available on fine root decomposition at elevated CO_2. Turnover of fine roots represents a large input of C to soils (Vogt et al., 1986). This input could be greatly increased with CO_2 enrichment, as fine root production increases in some species at elevated CO_2 (Norby et al., 1994). N concentrations in roots are reduced (Norby, 1994), and C-to-N ratios may increase (Curtis et al., 1990), suggesting that fine root decomposition rates may decline under CO_2 enrichment. There is a critical need for data on turnover of fine roots that are produced and decomposed under field conditions at elevated CO_2.

Summary

Predictably, the responses of soil biota to CO_2 enrichment and the degree of experimental emphasis on them increase with proximity to, and intimacy with, roots. Symbiotic associations are all stimulated to some degree. Total plant mycorrhization increases with elevated CO_2. VAM fungi increase proportionately with fine root length/mass increase. ECM fungi, however, exhibit greater colonization per unit root length/mass at elevated CO_2 than at current atmospheric levels. Total N-fixation per plant increases in all species examined, although the mechanisms of increase, as well as the eventual benefit to the host relative to N uptake may vary. Microbial responses are unclear. The assumption that changes in root exudation will drive increased mineralization and facilitate nutrient uptake should be examined experimentally, in light of recent models. Microbial results to date suggest that metabolic activity (measured as changes in process rates) is stimulated by root C input, rather than population size (measured by cell or colony counts). Insufficient evidence exists to predict responses of either soil-borne plant pathogens or soil fauna (i.e., food web responses). These are areas requiring attention, the first for its potential to limit ecosystem production through disease and the second because of its importance to nutrient cycling processes. Preliminary data on foliar litter decomposition suggests that neither nutrient ratios nor decomposition rates will be affected by rising CO_2. This is another important area that may be better understood as the number of longer term studies with more realistic CO_2 exposures increase. Evidence continues to mount that C fixation increases with CO_2 enrichment and that the bulk of this C enters the belowground component of ecosystems. The global fate and effects of this additional C may affect all hierarchical levels, from organisms to ecosystems, and will be largely determined by responses of soil biota.

Acknowledgements

I wish to thank Glenn Berntson for technical assistance with mycorrhizal assessment and Nelson T Edwards, Barbara T Walton and two anonymous journal referees for critical reviews of the manuscript. Research sponsored by the Global Change Research Program, Office of Energy Research, U. S. Department of Energy, under Contract No. DE-AC05-840R21400 with Martin Marietta Energy Systems, Inc. Publication No. 4296, Environmental Sciences Division, Oak Ridge National Laboratory.

References

Ågren G I, McMurtie R E, Parton W J, Pastor J and Shugart H H 1991 State-of-the-art of models of production-decomposition linkages in conifer and grassland ecosystems. Ecol. Appl. 1, 118–138.

Azcón R, Barea J M and Hayman D S 1976 Utilization of rock phosphate in alkaline soils by plants inoculated with mycorrhizal fungi and phosphate-solubilizing bacteria. Soil Biol. Biochem. 8, 135–138.

Allen M F 1987 Ecology of vesicular-arbuscular mycorrhizae in an arid ecosystem: use of natural processes promoting dispersal and establishment. *In* Mycorrhizae in the Next Decade: Practical Implications and Research Priorities. Eds. D M Sylvia, L L Hung and J H Graham. pp 133–135. University of Florida, Gainesville, Florida.

Allen M F 1991 The Ecology of Mycorrhizae. Cambridge University Press, New York. 184 p.

Allen M F, Hipps L E and Woolridge G L 1989 Wind dispersal and subsequent establishment of VA mycorrhizal fungi across a successional landscape. Landscape Ecol. 2, 165–167

Arnone J A III and Gordon J C 1990 Effect of nodulation, nitrogen fixation and CO_2 enrichment on physiology, growth and dry mass allocation of seedlings of *Alnus rubra* Bong. New Phytol. 116, 55–66.

Blair J M, Parmalee R W and Beare M H 1990 Decay rates, nitrogen fluxes and decomposer communities of single- and mixed-species foliar litter. Ecology 71, 1976–1985.

Bledsoe, C S 1992 Physiological ecology of ectomycorrhizae: Implications for field application. *In* Mycorrhizal Functioning. Ed. M F Allen. pp 424–438. Chapman and Hall, New York.

Caldwell M M, Eissenstat D M, Richards J H and Allen M F 1985 Competition for phosphorus: differential uptake from dual-isotope-labelled soil interspaces between shrub and grass. Science 229, 384–386.

Clarholm M 1985 Interactions of bacteria, protozoa and plants leading to mineralization of soil nitrogen. Soil Biol. Biochem. 17, 181–187.

Couteaux M–M, Mousseau M, Celerier M–L and Bottner P 1991 Increased atmospheric CO_2 and litter quality: decomposition of sweet chestnut leaf litter with animal food webs of different complexities. Oikos 61, 54–64.

Curtis P S, Drake B G and Whigham D F 1989 Nitrogen and carbon dynamics in C_3 and C_4 estuarine marsh plants grown under elevated CO_2 in situ. Oecologia 78, 297–301.

Curtis P S, Balduman L M, Drake B G and Whigham D F 1990 Elevated atmospheric CO_2 effects on belowground processes in C_3 and C_4 estuarine marsh communities. Ecology 71, 2001–2006.

Díaz S, Grime J P, Harris J and McPherson E 1993 Evidence of a feedback mechanism limiting plant response to elevated carbon dioxide. Nature 364, 616–617.

Duncan L W and Eissenstat D M 1993 Responses of *Tylenchulus semipenetrans* to citrus fruit removal: implications for carbohydrate competition. J. Nematol. 25, 7–14.

Fenchel T 1982 Ecology of heterotrophic microflagellates. II. Bioenergetics and growth. Marine Ecology Progress Series 8, 225–231.

Fenwick L 1973 Studies on the rhizosphere microflora of onion plants in relationship to temperature changes. Soil Biol. Biochem. 5, 315–320.

Finn G A and Brun W A 1982 Effect of atmospheric CO_2 enrichment on growth, nonstructural carbohydrate content and root nodule activity in soybean. Plant Physiol. 69, 327–331.

Fogel R 1988 Interactions among soil biota in coniferous ecosystems. Agric. Ecosys. Environ. 24, 69–85.

Fogel R and Hunt G 1979 Fungal and arboreal biomass in a Western Oregon Douglas–fir ecosystem — distribution patterns and turnover. Can. J. For. Res. 9, 245–256.

Fogel R and Hunt G 1983 Contribution of mycorrhizae and soil fungi to nutrient cycling in a Douglas-fir ecosystem. Can. J. For. Res. 13, 219–232.

Freckman D W, Moore J C, Hunt H W and Elliot E T 1991 The effects of elevated CO_2 and climate change on soil nematode community structure of prairie sod. Bull. Ecol. Soc. Am. 72 (Suppl.), 119.

Griffiths B and Robinson D 1992 Root-induced nitrogen mineralization: A nitrogen balance model. Plant and Soil 139, 253–263.

Hardy R W F and Havelka U D 1975 Photosynthate as a major limiting factor limiting nitrogen fixation by field grown legumes with emphasis on soybeans. *In* Symbiotic Nitrogen Fixation in Plants. Ed. P S Nutman. pp 421–439. International Biology Program Series, Vol. 7. Cambridge University Press, London.

Holmes W E and Zak D R 1993 Soil microbial biomass and net nitrogen mineralization in northern hardwood ecosystems. Soil Sci. Soc. Am. J. (*In press*).

Johnson D W, Ball J T and Walker R F 1994 Effects of elevated CO_2 and nitrogen on nutrient uptake in ponderosa pine seedlings. Plant and Soil (*In press*).

Kelly J M and Beauchamp J J 1987 Mass loss and nutrient dynamics in decomposing oak and mesic mixed-hardwood leaf litter. Soil Sci. Soc. Am. J. 51, 1616–1622.

Kramer P 1981 Carbon dioxide concentration, photosynthesis, and dry matter production. BioScience 31, 29–33.

Körner C and Arnone J A III 1992 Responses to elevated carbon dioxide in artificial tropical ecosystems. Science 257, 1672–1675.

Lamborg M R, Hardy R W F and Paul E A 1983 Microbial effects. *In* CO_2 and Plants. Ed. E R Lemon. pp 131–176. AAAS Selected Symposium no. 84. Westview Press, Boulder, Colorado.

Lemon E R 1983 CO_2 and Plants. AAAS Selected Symposium No. 84. Westview Press, Boulder, Colorado. 280 p.

Lewis J D, Thomas R B and Strain B R 1992 Interactive effects of nutrient supply and elevated CO_2 on response of seedlings from two populations of *Pinus taeda* to mycorrhizal infection. Bull. Ecol. Soc. Am. 73 (Suppl), 249.

Lewis J D, Thomas R B and Strain B R 1994 Effect of elevated CO_2 on mycorrhizal colonization rates of loblolly pine (*Pinus taeda* L.) seedlings. Plant and Soil (This volume).

Lindroth R L and Kinney K K 1993 Effects of atmospheric CO_2 and soil NO_3^- on tree-insect interactions. I. Phytochemical responses. Bull. Ecol. Soc. Am. 74 (Suppl.), 332

Lockwood J L 1981 Exploitation competition. *In* The Fungal Community: Its Organization and Role in the Ecosystem. Eds. D T Wicklow and G C Carroll. pp 319–349. Marcel Dekker, New York.

Luxmoore R J 1981 CO_2 and phytomass. BioScience 31, 626.

Masterson C L and Sherwood M T 1978 Some effects of increased atmospheric carbon dioxide on white clover (*Trifolium repens*) and pea (*Pisum sativum*). Plant and Soil 49, 421–426.

Melillo J M 1983 Will increases in atmospheric concentrations affect litter decay rates? pp 10–11. The Ecosystems Center Annual Report, Marine Biological Laboratory, Woods Hole, Massachusetts, USA.

Monz C A, Hunt H W, Reeves F B and Elliot E T 1994 The response of mycorrhizal colonization to elevated CO_2 and climate change in *Pascopyrum smithii* and *Bouteloua gracilis*. Plant and Soil (This volume).

Newman E I 1985 The rhizosphere: carbon sources and microbial populations. *In* Ecological Interactions in Soil: Plants, Microbes and Animals. Ed. A H Fitter. pp 107–122. London, Blackwell Scientific Publications, London.

Norby R J 1987 Nodulation and nitrogenase activity in nitrogen-fixing woody plants stimulated by CO_2 enrichment of the atmosphere. Physiol. Plant. 71, 77–82.

Norby R J 1994 Issues and perspectives for investigating root responses to elevated atmospheric carbon dioxide. Plant and Soil (This volume).

Norby R J, O'Neill E G and Luxmoore R J 1986a Effects of atmospheric CO_2 enrichment on the growth and mineral nutrition of *Quercus alba* in nutrient-poor soil. Plant Physiol. 82, 83–89.

Norby R J, Pastor J and Melillo J M 1986b Carbon-nitrogen interactions in CO_2-enriched white oak: physiological and long-term perspectives. Tree Physiol. 2, 233–241.

Norby R J, O'Neill E G and Wullschleger S D 1994 Belowground responses to atmospheric carbon dioxide in forests. *In* Carbon Forms and Functions in Forest Soils. Eds. W W McFee and J M Kelly. Soil Sci. Soc. of Am. Madison, WI (*In press*).

Norton J M, Smith J L and Firestone M K 1990 Carbon flow in the rhizosphere of ponderosa pine seedlings. Soil Biol. Biochem. 22, 449–445.

O'Neill E G and Norby R J 1991 First-year decomposition dynamics of yellow-poplar leaves produced under CO_2 enrichment. Bull. Ecol. Soc. Am 72 (Suppl.), 208.

O'Neill E G, Luxmoore R J and R J Norby 1987 Elevated atmospheric CO_2 effects on seedling growth, nutrient uptake, and rhizosphere bacterial populations of *Liriodendron tulipifera* L. Plant and Soil 104, 3–11.

O'Neill E G, Luxmoore R J and Norby R J 1987 Increases in mycorrhizal colonization and seedling growth in *Pinus echinata* and *Quercus alba* in an enriched CO_2 atmosphere. Can. J. For. Res. 17, 878–883.

O'Neill E G, O'Neill R V and Norby R J 1991 Hierarchy theory as a guide to mycorrhizal research on large-scale problems. Env. Poll. 73, 271–284.

Owensby C E, Coyne P I and Auen L M 1993 Nitrogen and phosphorus dynamics of a tallgrass prairie ecosystem exposed to elevated carbon dioxide. Plant Cell Environ. 16, 843–850.

Phillips D A, Newell K D, Hassell S A and Felling C E 1976 The effect of CO_2 enrichment on root nodule development and symbiotic N_2 fixation in *Pisum sativum* L. Am. J. Bot. 63, 356–362.

Rambelli A 1973 The rhizosphere of mycorrhizae. *In* Ectomycorrhizae. Eds. G L Marks and T T Kozlowski. pp 299–343. Academic Press, New York.

Robinson D, Griffiths B, Ritz K and Wheatley R 1989 Root-induced nitrogen mineralization: A theoretical analysis. Plant and Soil 117, 185–193.

Rogers H H, Prior S A and O'Neill E G 1992 Cotton root and rhizosphere responses to free-air CO_2 enrichment. Crit. Rev. Pl. Sci. 11, 251–263.

Rogers H H, Runion G B and Krupa S V 1993 Plant responses to atmospheric enrichment with emphasis on roots and the rhizosphere. Environ. Poll. (*In press*).

Rouatt J W and Katznelson H 1960 Influence of light on bacterial flora of roots. Nature 186, 659–660.

Rouatt J W, Petersen E A Katznelson H and Henderson V E 1963 Microorganisms in the root zone in relation to temperature. Can. J. Microbiol. 9, 227–236.

Runion G B, Curl E A, Rogers H H, Backman P A, Rodriguez-Kabana R and Helms B E 1994 Effects of CO_2 enrichment on microbial populations in the rhizosphere and phyllosphere of cotton. Agric. For. Met. (*In press*).

Schwab S M, Leonard R T and Menge J A 1984 Quantitative and qualitative comparison of root exudates of mycorrhizal and non-mycorrhizal plant species. Can. J. Bot. 62, 1227–1231.

Seastedt T R and Crossley D A Jr 1984 The influence of arthropods on ecosystems. BioScience 34, 157–161.

Strain B R and Bazzaz F A 1983 Terrestrial plant communities. *In* CO_2 and Plants. Ed. E R Lemon. pp 177–222. AAAS Selected Symposium no. 84. Westview Press, Boulder, Colorado.

Thomas R B, Richter D D, Ye H, Heine P R and Strain B R 1991 Nitrogen dynamics and growth of seedlings of an N-fixing tree (*Gliricidia sepium* (Jacq. Walp.)) exposed to elevated atmospheric carbon dioxide. Oecologia 8, 415–421.

Trappe J M 1977 Selection of mycorrhizal fungi for inoculation in nurseries. Ann. Rev. Phytopath. 15, 203–222.

Tschaplinski T J and Norby R J 1993 Physiological indicators of nitrogen response in a short rotation sycamore plantation. II. Nitrogen metabolism. Can. J. Bot. 71, 841–847.

van Veen J A, Liljeroth E, Lekkerkerk L J A and Van de Geun S C 1991 Carbon fluxes in plant-soil systems at elevated atmospheric CO_2 levels. Ecol. Applic. 1, 175–181.

Vogt K A, Grier C C and Vogt D J 1986 Production, turnover, and nutrient dynamics of above- and below ground detritus of world forests. Adv. Ecol. Res. 15, 303–377.

Whipps J M 1985 Effects of CO_2 concentration on growth, carbon distribution and loss of carbon from the roots of maize. J. Exp. Bot. 36, 645–651.

Whitbeck J E 1993 Mycorrhizal response to elevated CO_2 in serpentine grassland communities. Bull. Ecol. Soc. Am. 74 (Suppl.), 484.

Zak D R, Pregitzer K S, Curtis P, Teeri J A, Fogel R and Randlett D L 1993 Elevated atmospheric CO_2 and feedback between carbon and nitrogen cycles in forested ecosystems. Plant and Soil 151, 105–117.

Plant and Soil **165**: 67–74, 1994.

Soil microbial response in tallgrass prairie to elevated CO_2

Charles W. Rice[1], Fernando O. Garcia[2], Colleen O. Hampton[1] and Clenton E. Owensby[1]
[1]*Department of Agronomy, Kansas State University, Manhattan, KS 66501–5501, USA and* [2]*Departamento Agronomia, E.E.A. Inta, Balcarce, Argentina*

Key words: microbial activity, N availability, soil organic matter

Abstract

Terrestrial responses to increasing atmospheric CO_2 are important to the global carbon budget. Increased plant production under elevated CO_2 is expected to increase soil C which may induce N limitations. The objectives of this study were to determine the effects of increased CO_2 on 1) the amount of carbon and nitrogen stored in soil organic matter and microbial biomass and 2) soil microbial activity. A tallgrass prairie ecosystem was exposed to ambient and twice-ambient CO_2 concentrations in open-top chambers in the field from 1989 to 1992 and compared to unchambered ambient CO_2 during the entire growing season. During 1990 and 1991, N fertilizer was included as a treatment. The soil microbial response to CO_2 was measured during 1991 and 1992. Soil organic C and N were not significantly affected by enriched atmospheric CO_2. The response of microbial biomass to CO_2 enrichment was dependent upon soil water conditions. In 1991, a dry year, CO_2 enrichment significantly increased microbial biomass C and N. In 1992, a wet year, microbial biomass C and N were unaffected by the CO_2 treatments. Added N increased microbial C and N under CO_2 enrichment. Microbial activity was consistently greater under CO_2 enrichment because of better soil water conditions. Added N stimulated microbial activity under CO_2 enrichment. Increased microbial N with CO_2 enrichment may indicate plant production could be limited by N availability. The soil system also could compensate for the limited N by increasing the labile pool to support increased plant production with elevated atmospheric CO_2. Longer-term studies are needed to determine how tallgrass prairie will respond to increased C input.

Introduction

Global atmospheric CO_2 concentration is increasing and the consequences of that increase are subject to much speculation (Boden et al., 1990). Organism-level responses to CO_2 enrichment have been documented and usually result in increased plant production (Bazzaz, 1990; Coleman and Bazzaz, 1992; Wray and Strain, 1986). The production response is uncertain at the ecosystem-level. Curtis et al. (1990) reported increased plant production with elevated CO_2 in a C_3 dominated estuarine marsh but no response in a C_4 dominated plant community. Owensby et al. (1993a) reported that in tallgrass prairie, dominated by C_4 species, plant production increased in response to elevated CO_2. Root biomass also increased under elevated CO_2 (Curtis et al., 1990; Owensby et al., 1993a), thus increasing C allocation belowground.

In temperate grasslands, primary and secondary productivity of organisms can be substantially limited by both nutrients and water (Owensby et al., 1969). Limits to ecosystem productivity in natural systems often are from scarce nutrients, particularly nitrogen. Increased plant productivity observed under elevated CO_2 generally results in reduced nutrient concentrations relative to carbon (Bazzaz, 1990; Curtis et al., 1990). In tallgrass prairie, elevated CO_2 decreased plant N concentrations of both the above- and belowground components (Owensby et al., 1993b).

The ramifications of increased plant production with lower C:N ratios under elevated CO_2 could affect carbon storage, nutrient cycling, herbivory, and productivity in temperate terrestrial ecosystems. How the ecosystem responds depends on microbial activity and N cycling. The C:N ratio of plant litter and roots regulate the decomposition rate. Reduced litter quality slows decomposition (Melillo et al., 1982) and increases N immobilization. If the soil serves as a sink for increased CO_2, microbial growth will increase thus increasing the proportion of soil organic matter as microbial biomass. Without an increase in N supply or a reduction in plant N requirement, plant productivity could decrease due to N limitations. Diaz et al. (1993) suggested a feedback where increased C input into the soil from plant growth would sequester limited nutrients by microorganisms. The nutrient limitation would limit further plant response to elevated CO_2. Zak et al. (1993) suggested increased C input would stimulate microbial activity and enhance N cycling and availability.

Microbial biomass could be a more sensitive indicator than total soil organic matter to changes in the rate of detritus input or decomposition (Follett and Schimel, 1989; Ladd et al., 1981; Powlson et al., 1987; van Veen et al., 1989); climatic regimes (Insam et al., 1989; Insam, 1990; Wardle and Parkinson, 1990); and soil factors (Burke et al., 1989; Gregorich et al., 1991; Parton et al., 1987). The rapid response of active soil organic matter, including microbial biomass, to differences in detritus inputs suggests that these fractions could be used to indicate long-term trends in total soil organic matter levels. Inferences also can be made about soil organic matter quality from the microbial biomass C:N ratio and the ratio of microbial biomass to total organic C. The soil microbial biomass serves as a source and sink for energy and nutrients. Several models of soil organic matter dynamics of terrestrial ecosystems include microbial biomass for ecosystem processes (Parnas, 1975; Parton et al., 1988).

Thus, the soil microbial response to increased atmospheric CO_2 will impact N availability and plant production. The objectives of this study were to determine the effect of increased atmospheric CO_2 on 1) the amount of C stored in the soil organic matter and the microbial biomass and 2) soil, microbial activity and N availability.

Materials and methods

Study site

The experimental site is located in pristine tallgrass prairie north of Manhattan, KS (39.12°N, 96.35°W, 324 m above M.S.L.). Vegetation on the site is a mixture of C_3 and C_4 species, dominated by big bluestem (*Andropogon gerardii* Vitman) and indiangrass (*Sorghastrum nutans* (L.) Nash). Subdominants include Kentucky bluegrass (*Poa pratensis* L.), sideoats grama (*Bouteloua curtipendula* (Michx.) Torr.), and tall dropseed (*Sporobolus asper* var. *asper* (Michx.) Kunth). Members of the sedge family make up 5–10% of the composition. Principal forbs include ironweed (*Vernonia baldwinii* var. *interior* (Small) Schub.), western ragweed (*Ambrosia psilostachya* DC.), Louisiana sagewort (*Artemesia ludoviciana* Nutt.), and manyflower scurfpea (*Psoralea tenuiflora* var. *floribunda* (Nutt.) Rydb.). Average peak biomass of 425 g m^{-2} occurs in early August, of which 35 g m^{-2} is from forbs (Owensby and Anderson, 1967). Soils in the area are transitional from Ustolls to Udolls (Tully series: fine, mixed, mesic, montmorillonitic, Pachic Argiustolls). The 30-year average annual precipitation is 840 mm, with 520 mm occurring during the growing season.

Treatments

Circular plots (4.5 m dia.) were established in early May 1989. Treatments were ambient CO_2-no chamber, ambient CO_2-with chamber, and 2 × ambient CO_2-with chamber. Each treatment was replicated three times. The aluminum structural framework of the open-top chambers was a scaled-up version of a design described by Heagle et al. (1979) as modified by the USDA-ARS Air Quality Field Laboratory at Raleigh, NC. The chambers were 4.5 m in diameter by 3.25 m in height, with a cone-top baffle that reduced the top opening to 3 m. The baffle added 0.75 m to the height of the open-top chamber for a total height of 4 m. The structural framework was covered by 6 mil, UV-resistant, polyethylene film. The cone-top baffle placed atop each chamber reduced the opening by 54%, thus restricting the precipitation that entered the chamber. Within 24 hours following each rainfall event, water equal to 54% of the rainfall amount for an unchambered plot was added using a rotating sprinkler adjusted to cover the diameter of the chamber. Aluminum edging was placed around the upslope bottom edge of the chamber to prevent run-off from entering the chamber. No edging was placed on the lower half of the chamber.

Enrichment of CO_2 was supplied constantly until late October 1989. In 1990 through 1992, the same plots were subjected to the same treatments, but the enrichment period was from 1 April to 1 November. Duplicate chambers received no fertilizer N or 45 kg N ha^{-1} as NH_4NO_3 during 1990 and 1991. The fertilizer N treatment was discontinued in 1992. The soil was sampled at 0 to 5 and 5 to 15 cm depths during the fumigation periods in 1991 and 1992. All samples were sieved through a 6.3-mm mesh and stored at 4°C until analysis.

Measurements

Microbial biomass C and N were determined by the fumigation-incubation method (Jenkinson and Powlson, 1976). Soil (25 g) was added to duplicate 125-mL erlenmeyer flasks. When the gravimetric soil water content was less than 0.28 g g^{-1}, water was added to this level, and the samples were preincubated at 25°C for 5 days. At the end of the preincubation period, one of the samples was fumigated with chloroform. Samples were placed in a vacuum desiccator containing a wet paper towel and a beaker containing approximately 50 mL of ethanol-free chloroform and nonvolatile granules for distillation. Vacuum was applied three times for approximately 30 s each to boil the chloroform. Immediately after the third evacuation, the desiccator was closed tightly to allow diffusion of the chloroform into the soil. After 20–24 h, the beaker of chloroform and towel were removed, and the desiccator was evacuated eight times for 3 minutes each. Fumigated and unfumigated samples were placed into 940-mL mason jars containing enough water to maintain a highly humidified environment. Jars were closed tightly and incubated for 10 days at 25°C. At the end of the incubation period, the headspace CO_2-C concentration was measured using a Shimadzu GC–8A gas chromatograph (Shimadzu Scientific Instruments Inc., Columbia, MD) equipped with a 2m Porapak Q column operated at 70°C and He carrier gas at 14 mL min^{-1}. After measuring CO_2-C, 100 mL of 1*M* KCl was added to each flask, and the flasks were shaken for 1 h on an orbital shaker at 300 rpm. The suspension was transferred to a 250-mL centrifuge bottle and centrifuged at 16,000 g for 10 min. The supernatant was filtered through a nylon mesh (10μm) and stored in the freezer until analyzed for NH_4^+-N and NO_3^--N colorimetrically on an Alpkem Autoanalyzer (Alpkem Corp., Clackamas, OR). Ammonium-N was determined by the salicylate-hypochlorite method (Crooke and Simpson, 1971) and NO_3^--N+NO_2^--N by the Groess-Ilosvay technique (Keeney and Nelson, 1982). We expressed microbial biomass C and N as suggested by Voroney and Paul (1984) where:

$$\text{Microbial biomass C as mg C g}^{-1} = \frac{C_f - C_u}{0.41}$$

$$\text{Microbial biomass N as mg N g}^{-1} = \frac{N_p - N_u}{k_N}$$

where

C_f = CO_2-C evolved from fumigated sample
C_u = CO_2-C evolved from unfumigated sample
N_f = NH_4^+ – N and NO_3^--N mineralized from fumigated sample
N_u = NH_4^+ – N and NO_3^--N mineralized from unfumigated sample
k_N = $(-0.014 \times (C_f/N_f)) + 0.39$

Inorganic N (NH_4^+ – N and NO_3^- – N) was determined by extracting 20 g moist soil with 100 mL 1*M* KCl as indicated above for microbial biomass N. Concentration is expressed on the basis of dry soil weight. Soil water content was determined by oven drying at 105°C for 48 h.

Potential mineralizable C and N were calculated from the unfumigated samples for microbial biomass and are expressed as the amount of CO_2-C and inorganic N released during the 10-d incubation period. Total organic C and N was measured by direct combustion on a Carlo-Erba C and N analyzer (Carlo Erba Strumentazione, Rodana, MI, Italy).

Soil respiration as an index of microbial activity was measured from soil samples taken from 0–5 cm. After collection in the field, the samples were immediately sieved to pass a 6.3 mm mesh to discard large roots. Field moist soil (20 g) was added to 160 mL serum bottles, which were sealed with rubber stoppers and aluminum seals and incubated at 25°C. The headspace CO_2-C concentration was measured four times during a 48-h incubation as described previously. The linear rate of increase was calculated as mg CO_2-C g^{-1} h^{-1}.

Data analysis

Data for each year and depth were analyzed separately using either Proc ANOVA or Proc GLM (SAS Institute, 1988) as a randomized complete block design. The model included replication, treatment (CO_2 and N fertilization), date, and the date by treatment interaction. Differences between means were tested using Duncan's Multiple Range Test ($p < 0.10$). For the ratios of microbial and total C and N we were not able to calculate an appropriate variance, so these data are presented without statistics.

Table 1. Soil organic C and N after three seasons of CO_2 enrichment and N addition

Treatment	Organic C 0–5	5–15	Organic N 0–5	5–15 cm
	g C kg^{-1}		g N kg^{-1}	
Ambient				
Unfertilized	51.5 bc*	27.4 bc	2.04 bc	2.31 bc
+N	51.8 bc	26.1 c	1.96 bc	2.19 c
Chamber Ambient				
Unfertilized	64.2 ab	30.8 ab	2.57 ab	2.52 ab
+N	44.4 c	27.3 bc	1.58 c	2.23 c
Chamber Enriched				
Unfertilized	61.5 ab	32.9 a	2.58 ab	2.71 a
+N	70.4 a	31.9 a	3.19 a	2.60 a

* Different letters within a column represent significant difference at $\alpha = 0.1$ by Duncan's Multiple Range Test.

Results

After three seasons, total organic C and N in the surface 5 cm of soil generally increased as a result of the chambers, but there was no additional effect from CO_2 enrichment (Table 1). In the 5–15 cm layer organic C and N was higher under CO_2 enrichment. When N was added, CO_2 enrichment significantly increased organic C and N levels, compared to ambient conditions. The higher levels of organic C in the surface 5 cm compared to the 5 to 15 cm layer probably reflected partially decomposed leaf litter and fine roots, even though the surface litter was removed from the surface before sampling. Organic N was not stratified in the soil; similar concentrations occurred between the 0 to 5 and 5 to 15 cm depths.

Microbial biomass

Approximately 46% of microbial C and 50% of microbial N in the 0 to 15 cm soil layer were concentrated in the surface 5 cm. The time of sampling significantly affected microbial biomass C and N levels throughout the sampling period. In 1991, microbial biomass C increased significantly ($p = 0.001$) in the chambers regardless of CO_2 enrichment (Table 2). However, the chambers did not affect microbial biomass C in 1992

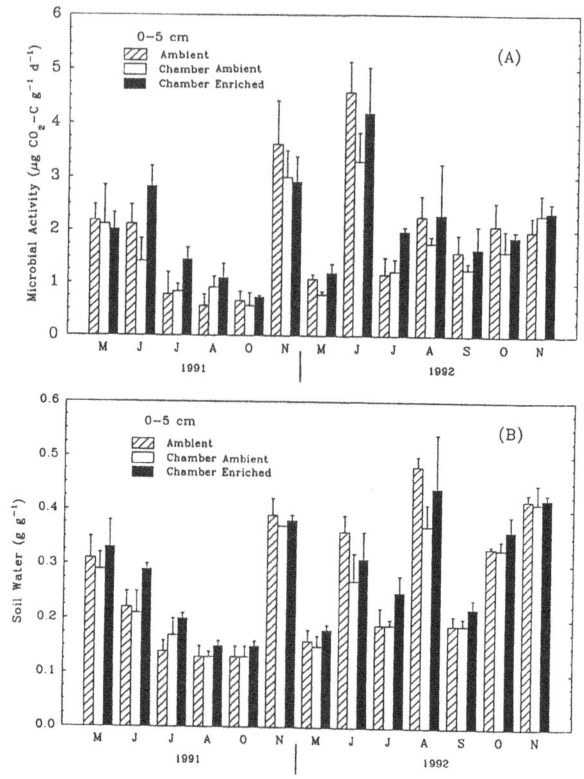

Fig. 1. Temporal dynamics of microbial activity (**A**) as measured by soil respiration and soil water contents (**B**) at the 0–5 cm depth in the different CO_2 treatments.

(Table 3). The difference between years may be due to precipitation and soil moisture (Fig. 1B). Precipitation in 1991, a dry year, was 669 mm while in 1992, a wet year, 1028 mm was recorded. Microbial biomass N was significantly greater (0–5 cm, $p = 0.0001$; 5–15 cm, $p = 0.0012$) in the chambers at both soil depths in 1991 (Table 2). At 0–5 cm, CO_2 enrichment significantly increased microbial biomass N. However, no significant treatment effect on microbial biomass N occurred in 1992 (Table 3). Soil moisture affected the level of microbial biomass N, with greater levels occurring in a dry year.

In 1991, we were able to determine the effect of added N on microbial biomass C and N levels (Table 2). Added N increased microbial biomass under the CO_2 enrichment in the surface 5 cm of soil which was only significant for microbial biomass N. For the 5 to 15 cm depth, microbial biomass C and N were higher in the chamber regardless of CO_2 levels.

Table 2. Average microbial biomass C and N and respiration for 1991 in the 0 to 5 and 5–15 cm soil layers

Treatment	Microbial Biomass C 0–5	Microbial Biomass C 5–15	Microbial Biomass N 0–5	Microbial Biomass N 5–15	Microbial Activity 0–5	Microbial Activity 5–15
	μg C g^{-1}		μg N g^{-1}		μg CO_2-C g^{-1} h^{-1}	
Ambient						
Unfertilized	1418 bc*	867 b	397 c	185 b	1.65 bc	0.56 bc
+N	1478 bc	862 b	383 c	175 b	1.40 cd	0.55 c
Chamber Ambient						
Unfertilized	1487 bc	958 a	408 c	209 a	1.65 bc	0.66 bc
+N	1354 c	982 a	378 c	216 a	1.40 cd	0.53 c
Chamber Enriched						
Unfertilized	1555 ab	993 a	445 b	224 a	1.83 b	0.74 ab
+N	1674 a	1026 a	500 a	226 a	2.23 a	0.86 a

* Different letters within a column represent significant difference at $\alpha = 0.1$ by Duncan's Multiple Range Test.

Table 3. Average microbial biomass C and N and respiration for 1992 in the 0 to 5 and 5–15 cm soil layers

Treatment	Microbial Biomass C 0–5	Microbial Biomass C 5–15	Microbial Biomass N 0–5	Microbial Biomass N 5–15	Microbial Activity 0–5	Microbial Activity 5–15
	μg C g^{-1}		μg N g^{-1}		μg CO_2-C g^{-1} h^{-1}	
Ambient	1269 a*	781 a	267 a	125 a	2.10 a	0.83 ab
Chamber Ambient	1424 a	764 a	246 a	122 a	1.75 b	0.73 b
Chamber Enriched	1402 a	799 a	254 a	141 a	2.22 a	0.93 a

* Different letters within a column represent significant difference at $\alpha = 0.1$ by Duncan's Multiple Range Test.

Microbial activity

Microbial activity tended to be greater under CO_2 enrichment than under ambient conditions at the 0 to 5 cm depth but the differences were only significant when N fertilizer was added (Fig. 1A, Tables 2, 3). In the 5 to 15 cm depth, differences in microbial activity were not as consistent, but it tended to be greater under CO_2 enrichment. The surface 5 cm accounted for approximately 53% of the activity in the entire 15 cm of soil. The date of sampling significantly affected microbial activity, which was partially regulated by soil moisture. Microbial activity was correlated highly with soil water content in the surface 5 cm with a correlation coefficient of 0.68. For the 5–15 cm depth, the correlation was only 0.28, indicating that other factors were more important in controlling microbial activity. During the two seasons when microbial measurements were made, soil water contents were lower in 1991 because of below average precipitation, but were greater in 1992 because of more normal precipitation (Fig. 1B). In the surface 5 cm, yearly estimates of microbial activity were 1.66 and 2.03 mg CO_2-C g^{-1} h^{-1} for 1991 and 1992, respectively. Soil water contents were generally greater under CO_2 enrichment. The lower precipitation in 1991 depleted soil moisture to a level that may have limited plant production and microbial activity.

Soil organic pools

Ratios of microbial biomass to total organic C and N have been suggested as indicators of soil organic matter quality and nutrient cycling (Insam et al., 1989; Insam 1990). In our study, microbial biomass C as a proportion of the total organic pool was not affected by CO_2 enrichment (Table 4). The proportion of organic N as microbial biomass decreased with CO_2 enrichment, but increased in the chambers when N was added. Potentially mineralizable organic pools also can indicate nutrient cycling and soil organic matter quality. Potentially mineralizable C and N as measured in the unfumigated samples for microbial biomass determinations were not significantly affected by CO_2 enrichment (data not shown). Potentially mineralizable C averaged 21.7 mg C kg^{-1} d^{-1} and potentially mineralizable N averaged 0.45 mg N kg^{-1} d^{-1} across all treatments and dates sampled in 1992. It is important to note that these measurements are from short-term (10 d) incubations and may not adequately reflect the entire mineralizable organic pools (Cabrera and Kissel, 1988). Nevertheless, during the 4-y exposure to elevated CO_2, the mineralizable pools were not significantly affected by the treatments. The C:N ratio of soil organic matter and microbial biomass did not change with CO_2 enrichment during the short exposure period of this study. No large shift occurred in the quality of the soil organic fractions.

Table 4. Proportion of organic C and N as microbial biomass and C : N ratio of the soil organic matter (SOM) and microbial fractions in the 0 to 5 cm soil layer

Treatment	MBC/TC	MBN/TN	C : N SOM	C : N Microbial biomass
	%	%		
Ambient				
Unfertilized	2.75	19.46	25.2	3.66
+N	2.85	19.54	26.4	3.87
Chamber Ambient				
Unfertilized	2.32	15.88	25.0	3.74
+N	3.05	23.92	27.8	3.63
Chamber Enriched				
Unfertilized	2.52	13.52	23.84	3.63
+N	2.38	15.67	22.07	3.40

Discussion

The increase in total organic C and N in the chambers was probably due to enhanced plant production. Owensby et al. (1993a) reported increased plant biomass production of above- and belowground components. Aggradation of soil organic C and N occurs slowly so differences because of CO_2 will be difficult to detect. There was a tendency for organic C and N to increase. Microbial biomass also increased under CO_2 enrichment; however, the response was dependent upon the year and addition of N. Differences in microbial biomass were accentuated in a dry year, such as 1991. Plant production responded similarly (Owensby et al., 1993b). Owensby et al. (1993b) attributed the differential response to soil water to greater water use efficiency under CO_2 enrichment. Thus, in a year with water stress, such as 1991, plant production would be greater under CO_2 enrichment. This would feed back into the soil system by providing greater C inputs and less water stress under elevated atmospheric CO_2 than under ambient conditions. This suggests a strong relationship between plant production and microbial biomass. Other studies have noted that microbial biomass is more sensitive to changes in litter inputs and climatic conditions and may reflect long-term trends in soil organic matter dynamics (Follett and Schimel, 1989; Insam et al., 1989; Insam, 1990).

Increased levels of microbial N under CO_2 enrichment were probably due to greater input of plant biomass into the soil. Two processes could explain the response of microbial N to CO_2 enrichment. Increased total N returned as litter could increase the amount of substrate available to the microorganisms. Second, the higher C:N ratio of the plant litter could slow the decomposition rate and require more soil N to be assimilated and retained by the microbes to complete decomposition. Owensby et al. (1993b) reported increased C:N ratios of above- and belowground plant components under CO_2 enrichment. Nitrogen assimilation and retention by the soil microbes could exacerbate the N limitation commonly observed in tallgrass prairie (Owensby et al., 1969). In this situation the plant could increase N use efficiency to compensate for lower soil N availability. In native tallgrass prairie, fire can cause an effect similar to that effect observed in this study. Long-term annual burning increases plant production under apparent lower N availability (Ojima et al., 1990). The active pool of organic N also is increased under long-term annual burning suggesting faster turnover of N and potentially increased N availability (Hunt et al., 1991; Rice and Garcia, 1993). The long-term effect of elevated CO_2 on ecosystem production is not known and requires further study.

The microbial response to added N under CO_2 enrichment suggests that N could be limiting microbial processes and eventually could affect plant N availability. Increased microbial N could result in decreased N availability; however, a greater amount of the N in the microbial pool also could maintain N availability because of more rapid turnover of this fraction compared to other organic pools (Paul and Juma, 1981; Zak et al., 1993). Hocking and Meyer (1985) suggested that plant N requirement is reduced by CO_2 enrichment. Microbial activity was greater under CO_2 enrichment, indicating more rapid decomposition. The enhanced activity might have been due to the higher soil water contents measured under CO_2 enrichment. Garcia (1992) found a similar relationship between microbial activity and soil water. The response to N further supports the N limitations with CO_2 enrichment.

At this time it is unclear what the long-term response to increased atmospheric CO_2 will be. A tendency for increased organic C under CO_2 enrichment may indicate future storage of atmospheric C in the tallgrass prairie ecosystem. However, the impact on N availability is uncertain. Microbial N increased and added N resulted in increased microbial biomass. Microbial activity also was stimulated by added N under CO_2 enrichment. Further studies are required to address the N competition and availability under increased atmospheric CO_2 to determine the long-term response of the ecosystem.

Acknowledgements

This research was supported by the U S Department of Energy, Carbon Dioxide Research Division. Contribution No. 94–98-J from the Kansas Agricultural Experiment Station.

References

Bazzaz F A 1990 The response of natural ecosystems to the rising global CO_2 levels. Ann. Rev. Ecol. Syst. 21, 167–196.

Boden T A, Kanciruk P and Farrell M P 1990 Trends '90: A compendium of data on global climate change. Carbon Dioxide Information Analysis Center, Oak Ridge National Laboratory, Oak Ridge TN. 257 p.

Burke I C, Yonkers C, Parton W J, Cole C V, Flach K and Schimel D S 1989 Texture, climate, and cultivation effects on organic matter in U.S. Grassland Soils. Soil Sci. Soc. Am. J. 53, 800–805.

Cabrera M L and Kissel D E 1988 Length of incubation time affects the parameter values of the double exponential model of nitrogen mineralization. Soil Sci. Soc. Am. J. 52, 1186–1187.

Coleman J S and Bazzaz F A 1992 Effects of CO_2 and temperature on growth and resource use of co-occurring C_3 and C_4 annuals. Ecology 73, 1244–1259.

Crooke W M and Simpson W E 1971 Determination of ammonium in Kjeldahl digests of crops by an automated procedure. J. Sci. Food Agric. 22, 9–10.

Curtis P S, Balduman L M, Drake B G and Whigham D F 1990 Elevated atmospheric CO_2 effects on belowground processes in C_3 and C_4 estuarine marsh communities. Ecology 71, 2001–2006.

Diaz S, Grime J P, Harris J and McPherson J 1993 Evidence of a feedback mechanism limiting plant response to elevated CO_2. Nature 364, 616–617.

Follett R F and Schimel D S 1989 Effect of tillage practices on microbial biomass dynamics. Soil Sci. Soc. Am. J. 53, 1091–1096.

Garcia F O 1992 Carbon and nitrogen dynamics and microbial ecology in tallgrass prairie. Ph.D. Dissertation. Kansas State Univ., Manhattan. 194 p.

Gregorich E G, Voroney R P and Kachanoski R G 1991 Turnover of carbon through the microbial biomass in soils with different textures. Soil Biol. Biochem. 23, 799–805.

Heagle A S, Philbeck R B and Letchworth M B 1979 Injury and yield responses of spinach cultivars to chronic doses of ozone in open-top field chambers. J. Environ. Qual. 8, 368–373.

Hocking P J and Meyer C P 1985 Responses of noogoora burr (*Xanthium occidentale* Bertol.) to nitrogen supply and carbon dioxide enrichment. Ann. Bot. 55, 835–844.

Hunt H W, Trlica M J, Redente E F, Moore J C, Detling J K, Kittell T G F, Walter D E, Fowler M C, Klein D A and Elliott E T 1991 Simulation model for the effects of climate change on temperate grassland ecosystems. Ecol. Modeling 53, 205–246.

Insam H, Parkinson D and Domsch K H 1989 Influence of macroclimate on soil microbial biomass. Soil Biol. Biochem. 21, 211–221.

Insam H 1990 Are the soil microbial biomass and basal respiration governed by the climatic regime? Soil Biol. Biochem. 22, 525–532.

Jenkinson D S and Powlson D S 1976 The effects of biocidal treatments on metabolism in soil. V. A method for measuring soil biomass. Soil Biol. Biochem. 8, 209–213.

Keeney D R and Nelson D W 1982 Nitrogen-inorganic forms. *In* Methods of Soil Analysis. Part 2. 2nd ed. Ed. A L Page. Agronomy 9, 643–689.

Ladd J N, Oades J M and Amato M 1981 Microbial biomass formed from ^{14}C, ^{15}N-labelled plant material decomposing in soils in the field. Soil Biol. Biochem. 13, 119–126.

Melillo J M 1983 Will increases in atmospheric CO_2 affect decay processes? Ecosys. Center Ann. Rpt., Marine Biological Laboratory, Woods Hole, MA, USA. pp 10–11.

Melillo J M, Aber J D and Muratore J F 1982 Nitrogen and lignin control of hardwood leaf litter decomposition dynamics. Ecology 62, 621–633.

Ojima D S, Parton W J, Schimel D S and Owensby C E 1990 Simulated impacts of annual burning on prairie ecosystems. *In* Fire in North American Tallgrass Prairies. Eds. S L Collins and L L Wallace. pp 118–132. Univ. of Oklahoma Press, Norman, OK.

Owensby C E and Anderson K L 1967 Yield responses to time of burning in the Kansas Flint Hills. J. Range Manage. 20, 12–16.

Owensby C E, Hyde R M and Anderson K L 1969 Effect of clipping and added moisture and nitrogen on loamy upland bluestem range. J. Range Manage. 23, 341–347.

Owensby C E, Coyne P I, Ham J M, Auen L and Knapp A K 1993a Biomass production in a tallgrass prairie ecosystem exposed to ambient elevated CO_2. Ecol. Appl. 3, 644–653.

Owensby C E, Coyne P I and Auen L 1993b Nitrogen and phosphorus dynamics of a tallgrass prairie ecosystem exposed to elevated carbon dioxide. Plant Cell Environ. 16, 843–850.

Parnas H 1975 Model for decomposition of organic material by microorganisms. Soil Biol. Biochem. 7, 161–169.

Parton W J, Schimel D S, Cole C V and Ojima D S 1987 Analysis of factors controlling soil organic matter levels in Great Plains grasslands. Soil Sci. Soc. Am. J. 51, 1173–1179.

Parton W J, Stewart J W B and Cole C V 1988 Dynamics of C, N, P, and S in grassland soils: A model. Biogeochemistry 5, 109–131.

Paul E A and Juma N G 1981 Mineralization and immobilization of soil nitrogen by microoganisms. *In* Terrestrial Nitrogen Cycles. Eds. F E Clark and T Rosswall. Ecol. Bull. (Stockholm) 33, 179–195.

Powlson D S, Brookes P C and Christensen B T 1987 Measurement of soil microbial biomass provides an early indication of changes in total soil organic matter due to straw incorporation. Soil Biol. Biochem. 19, 159–164.

Rice C W and Garcia F O 1994 Biologically active pools of soil C and N in tallgrass prairie soil. *In* Defining Soil Quality for a Sustainable Environment. Eds. Doran et al. pp 201–208. Soil Sci. Soc. Am., Madison, WI. Spec. Publ. 35.

SAS Institute Inc 1988 SAS procedures guide, Release 6.03 Edition. Cary, N.C.

van Veen J A, Merckx R and Van De Geijn S C 1989 Plant- and soil-related controls of the flow of carbon from roots through the soil microbial biomass. *In* Ecology of Arable Land. Eds. M Clarholm and L Bergstrom, pp 43–52. Kluwer Academic Publishers, Dordrecht.

Voroney R P and Paul E A 1984 Determination of k_C and k_N in situ for calibration of the chloroform fumigation-incubation method. Soil Biol. Biochem. 16, 4–14.

Wardle D A and Parkinson D 1990 Interactions between microclimatic variables and the soil microbial biomass. Biol. Fert. Soils 9, 273–280.

Wray S M and Strain B R 1986 Response of two old field perennials to interactions of CO_2 enrichment and drought stress. Am. J. Bot. 73, 1486–1491.

Zak D R, Pregitzer K R, Curtis P S, Teeri J A, Fogel R and Randlett D L 1993 Elevated atmospheric CO_2 and feedback between carbon and nitrogen cycles. Plant and Soil 151, 105–107.

Plant and Soil **165**: 75–80, 1994.

The response of mycorrhizal colonization to elevated CO_2 and climate change in *Pascopyrum smithii* and *Bouteloua gracilis*

C. A. Monz[1,4], H. W. Hunt[1,2], F. B. Reeves[3] and E. T. Elliott[1]
[1]*Natural Resource Ecology Laboratory,* [2]*Department of Science and* [3]*Department of Biology, Colorado State University, Fort Collins, CO 80523, USA.* [4]*Present address: National Outdoor Leadership School, Research Department, 288 Main Street, Lander, WY 82520, USA*

Key words: *Bouteloua gracilis*, climate change, elevated CO_2, *Pascopyrum smithii*, VA mycorrhiza

Abstract

Large intact soil cores of nearly pure stands of *Pascopyrum smithii* (western wheatgrass, C_3) and *Bouteloua gracilis* (blue grama, C_4) were extracted from the Central Plains Experimental Range in northeastern Colorado, USA and transferred to controlled environment chambers. Cores were exposed to a variety of water, temperature and CO_2 regimes for a total of four annual growth cycles. Root subsamples were harvested after the completion of the second and fourth growth cycles at a time corresponding to late winter, and were examined microscopically for the presence of mycorrhizae. After two growth cycles in the growth chambers, 54% of the root length was colonized in *P. smithii*, compared to 35% in blue grama. Field control plants had significantly lower colonization. Elevation of CO_2 increased mycorrhizal colonization in *B. gracilis* by 46% but had no effect in *P. smithii*. Temperatures 4°C higher than normal decreased colonization in *P. smithii* by 15%. Increased annual precipitation decreased colonization in both species. Simulated climate change conditions of elevated CO_2, elevated temperature and lowered precipitation decreased colonization in *P. smithii* but had less effect on *B. gracilis*. After four growth cycles in *P. smithii*, trends of treatments remained similar, but overall colonization rate decreased.

Introduction

Extensive research using crop species has shown a number of plant responses to elevated CO_2, with increases in photosynthetic rate and growth being common (Allen, 1990; Bazzaz, 1990). These responses are species specific, are dependent on environmental factors, and have been studied primarily in systems in which plant resources are not limiting (Bazzaz, 1990). Currently, there is great interest in examining the effects of elevated CO_2 and predicted climate change on native systems in which water and nutrients are more limiting. Mechanisms may be present that would enable native species to respond to elevated CO_2 in spite of these limiting conditions (Luxmore, 1981). Strain and Bazzaz (1983) suggest that the high availability of photosynthate resulting from elevated CO_2 increases root growth and root exudation. In turn, plant nutrition is affected by increasing root density and mycorrhizal colonization. In recent experimental work, total biomass, root biomass, and ectomycorrhizal colonization of *Pinus echinata* and *Quercus alba* seedlings increased under elevated CO_2 in unfertilized forest soil (O'Neill et al., 1987).

The objective of this study was to investigate the response of vesicular-arbuscular mycorrhiza (VAM) formation in a C_3 and a C_4 native perennial grass subjected to elevated CO_2 and simulated climate change. We know of no reports of the VAM response to atmospheric CO_2 in intact native grasses. In addition, examination of the interacting and possibly offsetting effects of CO_2, temperature and precipitation is important in predicting the response to potential climate change.

Materials and methods

Plant material

Intact soil cores, 24 cm (w) × 45 cm (h), were collected from the Central Plains Experimental Range (CPER), approximately 12 km north of Nunn, Colorado, USA (40°42' north, 104°42' west). Two soil-plant associations were sampled in late May 1989; one a sandy clay loam dominated by *Pascopyrum smithii* (western wheatgrass, C_3) and the other a sandy loam dominated by *Bouteloua gracilis* (blue grama, C_4). The two associations were located on the same site approximately 150 m apart within an ungrazed enclosure. Soil cores were obtained by driving steel irrigation pipe (25.4 cm × 45.7 cm) into the sod with a backhoe and then removing the entire intact soil core in the pipe. Sods were retained within the irrigation pipe throughout the experiment.

Experimental design

The field design consisted of 8 field blocks with 24 cores each for each soil-plant association. A total of 456 cores were removed and transported by refrigerated truck to the Duke University Phytotron (Durham, NC, USA). All cores were maintained at 2–5°C during transit and arrived at the Phytotron with minimal disturbance. Cores were randomly assigned to treatment combinations within the 8 replicate blocks, in a complete factorial design consisting of the two plant species, two CO_2 levels (350 and 700 μL L^{-1}), two temperature regimes (field average and 4°C elevated), three precipitation regimes (15 cm, 25 cm and 36 cm per year), and two destructive harvest dates (after 2 and 4 growth cycles) totaling forty-eight treatment combinations. In the field, one control cylinder was driven and left in place and one non-cylinder control plot was designated in each block for each of the two destructive harvest dates.

This design occupied 4 large environmental chambers, each regulated for CO_2 and temperature regime with an average light intensity of 550 μE m^{-2} s^{-1} at core-top level. Precipitation regimes approximated seasonal patterns, with applications of distilled water in 2.5 cm events. The 25 cm precipitation treatment was selected to approximate field production levels. Total seasonal precipitation was controlled by the number of events. Growing season temperature regime and photoperiod closely followed field seasonal and diurnal patterns and were adjusted weekly to correspond to weekly field averages. Temperature maximum of each season was reached on simulated week 29 with highs of 31°C and lows of 13°C (field average treatment) while the photoperiod maximum of 15 h 8 min. was reached on simulated week 25. Late fall and winter seasons were shortened by thirteen weeks, compared with the field, with a temperature minimum of 3°C (no diurnal fluctuation in winter) and a photoperiod minimum of 9 h 13 min. The positions of all cores within chambers were rerandomized three times per season.

Sampling for mycorrhizae

After two and four growth cycles, three small soil cores (2.0 × 10.0 cm) were taken along a random transect within each of the larger cores. Field samples for both cylinder and non-cylinder control plots were taken after two seasons only. Soil was then passed through a 2 mm sieve, with the majority of the plant material (mainly roots with minimal litter) being retained on the sieve. Subsamples of fine roots ($\leq$ 0.5 mm diameter) were handpicked from the material retained on the sieve for mycorrhizal counts. Any obvious weed and senescent roots were eliminated. Roots were placed in a formalin acetic acid preservative solution (1 part glacial acetic acid, 6 parts 40% formalin, 20 parts 95% Et OH, and 40 parts DI H_2O, by volume) until staining. Root cytoplasm was cleared using a 10% KOH solution at 90°C for 90 minutes, and mycorrhizae were stained with 0.01% trypan blue. Mycorrhizal colonization was assayed with the gridline intercept method (Giovanetti and Mosse, 1980) under a dissecting microscope at 40–63×. Occasional verification of particular intercepts was made with a Zeiss compound scope at 200×. Intercept points were categorized as to the relative degree of colonization (heavy or light) and senescent roots (without cortexes) were counted to correct percent colonization to a live root basis. Vesicles and arbuscules were noted where present.

Statistical analysis was performed with Systat software (Wilkenson, 1989). Bartlet's test for homogeneity of group variances was employed throughout and arcsine transformations and Kruskal-Wallis nonparametric tests were utilized to supplement original ANOVA results where applicable.

Results

After two growth cycles under controlled conditions, both species showed high VAM colonization: 54%

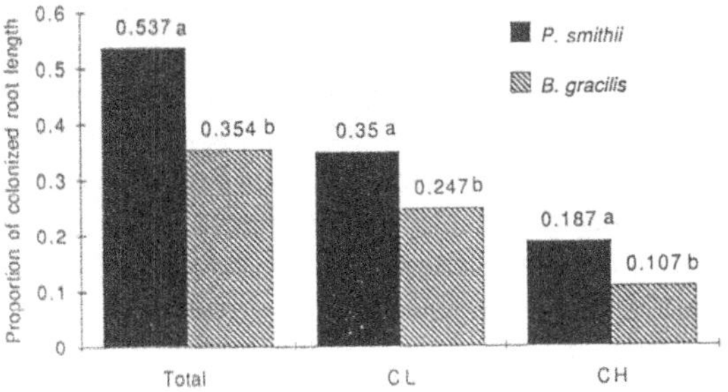

Fig. 1. Comparison of the proportion of the root length colonized by mycorrhizal fungi after two growth cycles in *Pascopyrum smithii* and *Bouteloua gracilis*. Categories are total colonization, lightly colonized (CL), and heavily colonized (CH). Values are means across all treatments. Within infection categories, means followed by the same letters are not significantly different at $p < 0.05$.

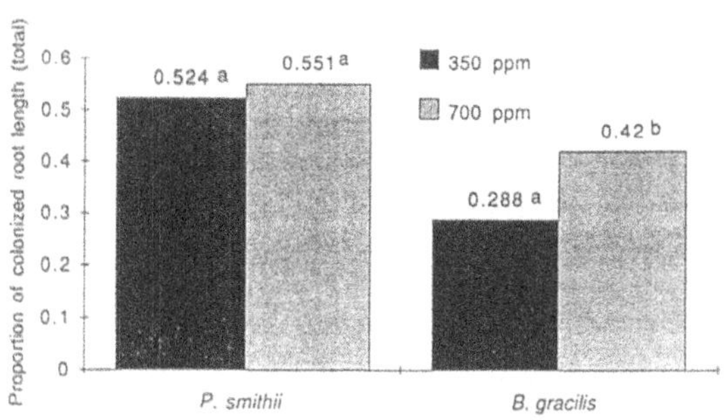

Fig. 2. Effect of atmospheric CO_2 on the proportion of the root length colonized by mycorrhizal fungi in *P. smithii* and *B. gracilis* after two growth cycles. Values are means. Within species, means followed by the same letters are not significantly different at $p < 0.05$.

Table 1. The effects of CO_2, precipitation, temperature, and climate change on mycorrhizal colonization in *Pascopyrum smithii* after two and four growth cycles. Values are means of the total proportion of the root length colonized. Within harvest date and treatment factor, means followed by the same letter are not significantly different at $p < 0.05$. For the climate change comparison, current ambient = 350 $\mu L\ L^{-1}$ CO_2, field average temperature, and 25 cm yr^{-1} precipitation; climate change = 700 $\mu L\ L^{-1}$ CO_2, 4°C elevated temperature, and 15 cm yr^{-1} precipitation

	Harvest date	
Treatment	2 years	4 years
CO_2		
350 $\mu L\ L^{-1}$	0.524a	0.259a
700 $\mu L\ L^{-1}$	0.551a	0.230a
Temperature		
Field average	0.698a	0.273a
4°C elevated	0.517b	0.216b
Precipitation (cm)		
15	0.558a	0.197b
25	0.623a	0.294a
36	0.392b	0.244a
Climate change comparison		
Current ambient	0.734a	0.330a
Climate change	0.512b	0.150b

of the subsampled root length was colonized in *P. smithii* and 35% in *B. gracilis* (Fig. 1). Species differences remained similar regardless of the degree of colonization (i.e., heavy or light). Field samples were less colonized than the 'current ambient' conditions simulated in the chambers, with colonization at 12% in *B. gracilis* and 9% in *P. smithii* in the field, and 39% in *B. gracilis* and 74% in *P. smithii* in the chambers. Hyphae were the predominant infective structure observed, with vesicles being rare and mature arbuscules being absent regardless of treatment and harvest date. ANOVA indicated a significant stimulation of colonization by elevated CO_2 in *B. gracilis* after two growth cycles ($p < 0.05$, species × CO_2 interaction $p = 0.077$, Fig. 2) but no effect on *P. smithii* after two or four years (Table 1, Fig. 2).

After two growth cycles, plants grown under the high precipitation treatment had significantly lower colonization than the plants grown under normal and low precipitation treatments (Fig. 3a). Here, the species × precipitation interaction was not significant ($p = 0.55$) and therefore means were averaged across species. Lower than normal precipitation had little effect on colonization of either species. After four growth cycles in *P. smithii*, however, this trend was not consistent, with the low precipitation treatment having significantly lower colonization (Table 1).

Higher temperatures had no effect on VAM colonization in *B. gracilis* but decreased colonization in *P. smithii* after both two and four growth cycles (Fig. 3b, Table 1). The climate change scenario of elevated CO_2, elevated temperature, and decreased precipitation resulted in a significant decrease in colonization in *P. smithii* after two and four growth cycles (Fig. 4, Table 1). In contrast, climate change led to increased colonization in *B. gracilis* (Fig. 4). The effects of climate change are compatible with the main effects, since colonization in *B. gracilis* increased with elevated CO_2 and did not respond to temperature or reduction

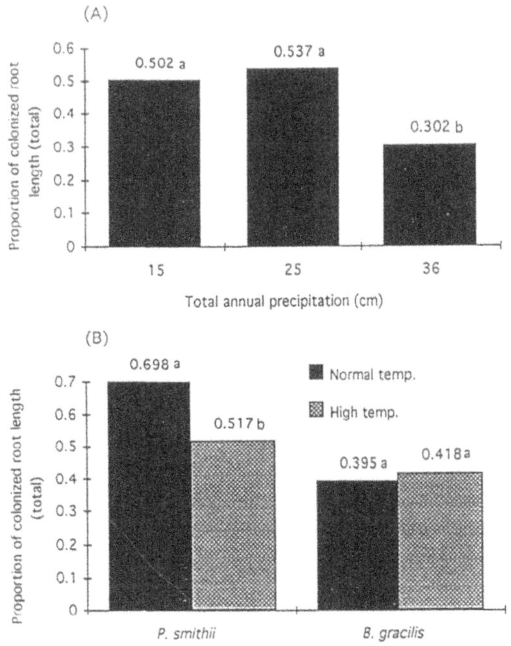

Fig. 3. Effects of annual precipitation and seasonal temperature regime on the proportion of root length colonized by mycorrhizal fungi after two growth cycles. Annual precipitation (**A**) is averaged across both species while seasonal temperature regime (**B**) is by species. Values are means. Within species and between precipitation levels, means followed by the same letter are not significantly different at $p < 0.05$.

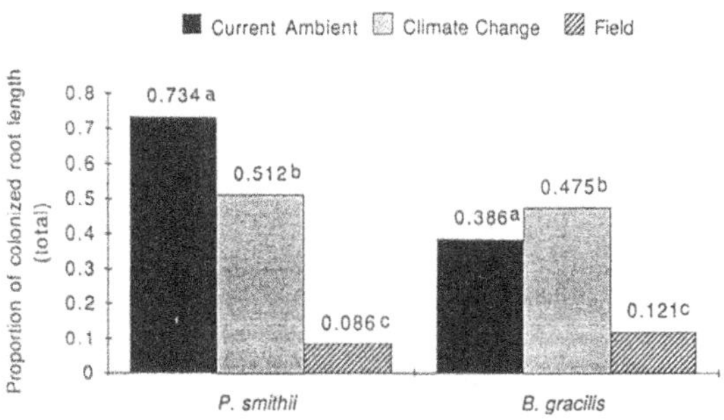

Fig. 4. Comparison of possible climate change effects on mycorrhizal colonization in *P. smithii* and *B. gracilis* after two growth cycles. 'Current ambient' = growth chamber conditions designed to mimic field levels of CO_2, temperature and precipitation; 'Climate Change' = Elevated CO_2, elevated temperature, and reduced precipitation; 'Field' = samples obtained from field control plots. Within species, means followed by the same letters are not significantly different at $p < 0.05$.

of precipitation from 25 to 15 cm, whereas colonization in *P. smithii* either decreased or did not respond to these same factors.

Discussion

C_4 grasses are usually considered more likely to be mycorrhizal dependent than C_3 grasses (Hetrick et al., 1990). However, greater colonization in *P. smithii* (C_3) in chamber-grown plants may indicate greater mycorrhizal dependence than in *B. gracilis* (C_4) (Graham et al., 1991). Although *B. gracilis* had a fairly high colonization rate, there is some evidence that mycorrhizae are parasitic rather than mutualistic in this C_4 species (Hayes et al., 1982). In field samples, *B. gracilis* exhibited slightly higher colonization than *P. smithii*, and overall colonization was significantly lower than in the chamber grown plants. These field levels appear lower than have been previously reported for these species (Allen et al., 1984; Davidson and Christensen, 1977) but direct comparisons are difficult due to differences in time of sampling and enumeration methodology.

Despite C_4 photosynthetic physiology, *B. gracilis* has been shown to respond to elevated CO_2 with enhanced photosynthesis (Morgan et al., 1991) and growth (Hunt et al., 1990; Reichers and Strain, 1988). In this species, increased carbon availability from the host plant may play a significant role in stimulating mycorrhizal colonization. Primary production of both species apparently responded to elevated CO_2, with *B. gracilis* yielding 19% and *P. smithii* 14% greater plant size after two seasons under elevated CO_2 (Hunt et al., 1992). *P. smithii* is also photosynthetically responsive to elevated CO_2 (Morgan et al., 1991) but mycorrhizal colonization was not significantly affected.

Mycorrhizae have been shown to play a role in the nitrogen uptake of some plants (Ames et al., 1983; Jakobsen and Jensen, 1992; Raven et al., 1978; Van Kessel et al., 1985) and N nutrition may be a more important factor in this system which appears to be limited by nitrogen and not phosphorus (Clark et al., 1980). After two growth cycles, colonization was negatively correlated with nitrogen concentration of the shoot ($r = -0.494$, $p < 0.001$) across all treatments, which may indicate the formation of additional mycorrhizae under nitrogen limited conditions.

Hunt et al. (1992) reported a significant increase in root nitrogen with elevated temperature in *P. smithii* but not in *B. gracilis*, a difference that may be related to photosynthetic temperature optima in *P. smithii*, a cool season grass and *B. gracilis*, a warm season grass (Monson et al., 1986). In *B. gracilis*, both plant growth and N mineralization increased with temperature, with little net change in degree of N limitation (Hunt et al., 1992). In *P. smithii*, growth showed no response to temperature, perhaps because of a lower optimum temperature for photosynthesis, while mineralization still increased. Therefore, in *P. smithii*, the adverse direct effect of elevated temperature may have been offset by reduced N limitation, and mycorrhizal colonization could be responding to this lack of a nitrogen stress.

Elevated soil water has been shown to decrease mycorrhizal colonization in some plants such as sorghum (Sieverding, 1981) perhaps due to the limitation of oxygen diffusion to the rhizosphere. In addition, simulation modeling of temperate grassland systems (Hunt et al., 1991) has suggested decreases in colonization with increased precipitation. Reid (1984) suggested this is possibly due to a decrease in the advantage of mycorrhizae over roots in nutrient uptake in wetter soil. On the other hand, N limitation would be expected to be greater with greater precipitation, as evidenced by lower plant N concentration in the high water treatment (Hunt et al., in prep), which should lead to greater VAM colonization. Our observations are inconsistent in this regard with elevated precipitation yielding lower colonization after two years in both species and decreased precipitation yielding lower colonization after 4 years in *P. smithii*.

Colonization rates declined after four growth cycles in *P. smithii* in all treatments. This may have resulted from a pulse of decomposition and subsequent release of nitrogen stress in the system, an hypothesis that we can test with data forthcoming.

Our chamber grown plants exhibited higher colonization than found in the field. Compared to the field, the chambers had lower light intensity, less extreme temperatures with no freezing, and higher relative humidity. Total plant growth in the field was comparable with the reduced precipitation treatment in the chambers, indicating a decrease in water stress in the chambers. These conditions somehow favored VAM development, perhaps by increasing nutrient stress or carbon availability from the host plant.

Currently, there is little information on the ecophysiology of the mycorrhizal symbiosis in these species. There is some evidence that mycorrhizae can be parasitic in *B. gracilis* (Hayes, 1982), but there are no data in this regard for *P. smithii*. Consequently, it is difficult to infer specific response mechanisms from these data. In addition, we know of no published data on the response of a parasitic symbiosis to elevated CO_2.

In conclusion, our results indicate several important mycorrhizal responses to elevated CO_2 and climate change in these grasses: colonization in intact native sods of *B. gracilis* was stimulated by elevated CO_2, alterations of annual precipitation yielded inconclusive results, and elevation of temperature yielded lower colonization in *P. smithii*. We plan to undertake a thorough analysis of the nutrient and production dynamics of this system with data forthcoming in an attempt to further interpret all of these responses. In addition, further work on the nature of the symbiosis in these grasses (i.e., parasitism or mutualism?) may assist in interpreting these results mechanistically.

Acknowledgements

The authors wish to thank D E Reuss for many helpful suggestions, L J Schedivy for technical assistance, and W K Knight and S D Frey for careful reviews of this manuscript. This research was supported by National Science Foundation grants BSR-8818269, BSR-8706429 and DEB-9112571.

References

Allen L H 1990 Plant responses to rising carbon dioxide and potential interactions with air pollutants. J. Environ. Qual. 19, 15–34.

Allen M F, Allen E B and Stahl P D 1984 Differential niche response of *Bouteloua gracilis* and *Pascopyrum smithii* to VA mycorrhizae. Bull. Torrey Bot. Club 111, 361–365.

Ames R N, Reid C P P, Porter L K and Cambardella C 1983 Hyphal uptake and transport of nitrogen from 2 ^{15}N-labeled sources by *Glomus mosseae*, a vesicular-arbuscular mycorrhizal fungus. New Phytol. 95, 381–396.

Davidson D E and Christensen M 1977 Root-microfungal and mycorrhizal associations in a shortgrass prairie. *In* The Belowground Ecosystem: A Synthesis of Plant-associated Processes. Ed. J K Marshall. pp 279–287. Range Science Department Series No. 26. Colorado State University, Fort Collins.

Bazzaz F A 1990 The response of natural systems to the rising of global CO_2. Annu. Rev. Ecol. Syst. 21, 167–196.

Clark F E, Cole C V and Bowman R A 1980 Nutrient Cycling. *In* Grasslands, Systems Analysis and Man. Eds. A J Breymeyer and G M Van Dyne. pp 659–712. International Biological Programme, Cambridge University Press, Cambridge.

Giovanetti M and Mosse B 1980 An evaluation of the techniques for measuring vesicular arbuscular mycorrhizal infection in roots. New Phytol. 84, 489–500.

Graham J H, Eissenstat D M and Drouillard D L 1991 On the relationship between plant mycorrhizal dependency and rate of vesicular-arbuscular mycorrhizal colonization. Functional Ecol. 5, 773–779.

Hayes R, Reid C P P, St. John T V and Coleman D C 1982 Effects of nitrogen and phosphorus on blue grama growth and mycorrhizal infection. Oecologia 54, 260–265.

Hetrick B A D, Wilson G W T and Todd T C 1990 Differential responses of C_3 and C_4 grasses to mycorrhizal symbiosis, phosphorus fertilization, and soil microorganisms. Can. J. Bot. 68, 461–464.

Hunt H W, Detling J K, Elliott E T, Monz C A and Strain B R 1990 The effects of elevated CO_2 and climate change on grasslands. I. Response of aboveground primary production in intact sods of native shortgrass prairie. Bull. Ecol. Soc. Am. 71,196.

Hunt H W, Trlica M J, Redente E F, Moore J C, Detling J K, Kittel T G F, Walter D E, Fowler M C, Klein D A and Elliott E T 1991 Simulation model for the effects of climate change on temperate grassland ecosystems. Ecol. Model. 53, 205–246.

Hunt H W, Elliott E T, Detling J K, Monz C A and Reuss D E 1992 Plant size and shoot to root ratio in intact sods of native shortgrass prairie exposed to elevated CO_2 for two growing seasons. Bull. Ecol. Soc. Am. 73, 254.

Jakobsen J I and Jensen E S 1992 Hyphal transport of ^{15}N-labeled nitrogen by a vesicular-arbuscular mycorrhizal fungus and its effect on depletion of inorganic soil N. New Phytol. 122, 281–288.

Luxmore R J 1981 CO_2 and phytomass. Bioscience 31, 626.

Monson R K, Sackschewsky M R and Williams III G J 1986 Field measurements of photosynthesis, water-use efficiency and growth in *Agropyron smithii* (C_3) and *Bouteloua gracilis* (C_4) in the Colorado shortgrass steppe. Oecologia 68, 400–409.

Morgan J, Hunt H W, Monz C A, LeCain D and Detling J K 1991 The effects of elevated CO_2 and climate change on the photosynthetic response of *Agropyron smithii* and *Bouteloua gracilis*. Bull. Ecol. Soc. Am. 72, 200.

O'Neill E G, Luxmore R J and Norby R J 1987 Increases in mycorrhizal colonization and seedling growth in *Pinus echinata* and *Quercus alba* in an enriched CO_2 atmosphere. Can. J. Forest Res. 17, 878–883.

Raven J A, Smith S E and Smith F A 1978 Ammonium assimilation and the role of mycorrhizas in climax communities in Scotland. Trans. Bot. Soc. Edinburgh 43, 27–35.

Reichers G H and Strain B R 1988 Growth of blue grama (*Bouteloua gracilis*) in response to atmospheric CO_2 enrichment. Can. J. Bot. 66, 1570–1573.

Reid C P P 1984 Mycorrhizae: A root-soil interface in plant nutrition. *In* Microbial-Plant Interactions. pp 29–50. Soil Science Society of America, Madison.

Sieverding E 1981 Influence of soil water regimes on VA-mycorrhiza. I. Effect on plant growth, water utilization, and development of mycorrhiza. J. Agron. Crop. Sci. 150, 400–411.

Strain B R and Bazzaz F A 1983 Terrestrial plant communities. *In* CO_2 and Plants: The Response of Plants to Rising Levels of Carbon Dioxide. Ed. E R Lemon. pp 177–222. Westview, Boulder.

Van Kessel C, Singleton P W and Hoben H J 1985 Enhanced N-transfer from a soybean to maize by vesicular arbuscular mycorrhizal (VAM) fungi. Plant Physiol. 79, 562–563.

Wilkenson L 1989 SYSTAT: The system for statistics. SYSTAT, Inc., Evanston. 638p.

Plant and Soil **165**: 81–88, 1994.

Effect of elevated CO_2 on mycorrhizal colonization of loblolly pine (*Pinus taeda* L.) seedlings

J. D. Lewis, R. B. Thomas and B. R. Strain
Department of Botany, Duke University, Durham, NC 27708-0339, USA

Key words: carbohydrates, ectomycorrhizae, elevated CO_2, phosphorus, *Pinus taeda* L., *Pisolithus tinctorius* (Pers.) Coker & Couch

Abstract

Interactive effects of elevated atmospheric CO_2 and phosphorus supply on mycorrhizal colonization rates were investigated using loblolly pine (*Pinus taeda* L.) seedlings from Florida and coastal North Carolina. Seedlings from both populations were grown in greenhouses maintained at either 35.5 Pa or 71.0 Pa CO_2. In both CO_2 treatments, seedlings were grown in a full factorial experiment with or without mycorrhizal inoculum and with an adequate or a limiting supply of phosphorus. Seedlings were harvested 60, 90 and 120 days after emergence and at each harvest root subsamples were examined to determine the percent of fine roots that were mycorrhizal. Additionally, root carbohydrate and nutrient levels were measured at each harvest. Root starch, sugar and total non-structural carbohydrate (TNC) concentrations were increased by growth in elevated CO_2 and decreased by mycorrhizal colonization. Phosphorus stress decreased root starch concentrations, increased root sugar concentrations and did not significantly affect TNC concentrations. However, despite significant effects on root carbohydrate levels, there were generally no significant treatment effects on mycorrhizal colonization. Additionally, at all harvests, root starch and sugar concentrations were not correlated with percent of fine roots that were mycorrhizal. These results suggest that although elevated CO_2 may significantly increase root carbohydrate levels, the increases may not affect the percent of fine roots that are mycorrhizal.

Abbreviations: M - mycorrhizal, NM - non-mycorrhizal, TNC - total non-structural carbohydrates

Introduction

Increased photosynthetic rates are frequently found in C_3 plants exposed to elevated partial pressures of CO_2, and can result in increased photosynthate allocation to roots, particularly fine roots (Strain and Cure, 1985). Similarly, increased allocation to roots due to many other abiotic factors has been shown to benefit both the host and symbiont in mycorrhizal associations, suggesting that factors affecting photosynthesis may affect mycorrhizae (O'Neill et al., 1987). Increasing photosynthate allocation to mycorrhizae may be particularly beneficial during seedling establishment, when ectomycorrhizae may be a substantial photosynthate sink for the host plant (Hacskaylo, 1973). Elevated CO_2 may result in larger, more vigorous plants in part by reducing the relative carbohydrate cost of the symbiosis and by increasing the functional area of the mycorrhizae (O'Neill et al., 1987). However, although mycorrhizal colonization and proliferation have been shown to increase on trees exposed to elevated CO_2 in some studies (Conroy et al., 1990; Norby et al., 1986, 1987; O'Neill et al., 1987), exposure of the host plant to elevated CO_2 does not always result in increased colonization (Conroy et al., 1990). A possible reason for these conflicting results is that root nutrient status regulates the effect of elevated CO_2 on mycorrhizal colonization rates.

Elevated CO_2 may not uniformly affect mycorrhizal colonization rates across a range of nutrient availabilities (Conroy et al., 1990). Data from recent experiments suggest that mycorrhizal colonization is increased with CO_2 enrichment under conditions where soil nutrients are limiting (Norby et al., 1986, 1987; O'Neill et al., 1987; but see Conroy et al., 1990). Under ambient CO_2, colonization is frequently higher on roots grown with lower phosphorus fertilization than on roots from similar plants with greater phosphorus fertilization (Hatch, 1937; Jones et al., 1990). This may be due to effects such as increased photosynthate allocation to roots (Eamus and Jarvis, 1989) or increased carbohydrate concentrations in root cells (Douds and Schenck, 1990; Hacskaylo, 1973; Marx et al., 1977; Same et al., 1983; Thomson et al., 1986); conditions often associated with low root nutrient contents. Because elevated CO_2 and root nutrient concentrations may independently affect root carbohydrate levels, elevated CO_2 and soil nutrient availability may affect mycorrhizal colonization rates in an additive fashion. If, however, mycorrhizal colonization rates are regulated directly by root nutrient concentrations, then increased CO_2 may not affect colonization rates.

In this study, the interactive effects of elevated CO_2 and nutrient supply on mycorrhizal colonization rates were examined on seedlings from two populations of loblolly pine (*Pinus taeda* L.). We hypothesized that: 1), elevated CO_2 would result in increased mycorrhizal colonization rates; 2), nutrient stress would result in increased mycorrhizal colonization rates; 3), the highest mycorrhizal colonization rates would occur on nutrient-limited seedlings growing in elevated partial pressures of CO_2.

Materials and methods

Seeds of mixed maternity from a Florida population and from a North Carolina coastal population of loblolly pine (*Pinus taeda* L.) were sown in fine vermiculite in seed flats. Seeds were germinated in growth chambers maintained either at 35.0 Pa or 67.5 Pa CO_2. Day/night temperatures were 28°/17°C in both chambers. Seed flats were watered every other day with deionized water, and were fertilized weekly with modified one-half strength Hoagland's nutrient solution. Approximately 21 days after emergence, uniformly-sized seedlings were transplanted individually into 2.75 liter pots filled with a 1:1 (v:v) mixture of gravel and vermiculite. Seedlings in the mycorrhizal treatment (M) were inoculated with vermiculite containing hyphae of *Pisolithus tinctorius* (Pers.) Coker and Couch by placing inoculum directly on the roots during transplanting. Non-mycorrhizal seedlings (NM) were not inoculated. After transplanting, seedlings were placed into one of four greenhouses maintained at either 35.5 Pa or 71.0 Pa CO_2. The day/night temperatures were initially 28°/17°C, and the night temperature was increased to 22°C after five weeks.

Two phosphorus treatments were initiated three days after transplanting. In the low phosphorus (Low P) treatment, seedlings were fertilized with a modified Hoagland's solution containing 0.083 m*M* KH_2PO_4. This concentration has been shown to be limiting to loblolly pine growth (Lewis, Thomas and Strain, unpublished data). Seedlings in the high phosphorus (High P) treatment were fertilized with a modified Hoagland's solution containing 0.5 m*M* KH_2PO_4. Both nutrient solutions contained 7.0 m*M* NH_4NO_3, 3.0 m*M* K, 3.0 m*M* Ca, 2.0 m*M* S, 1.5 m*M* Mg, 0.14 m*M* Fe, 0.05 m*M* B, 0.01 m*M* Mn, 0.001 m*M* Zn, 0.001 m*M* Cu and 0.05 μM Mo. Seedlings were watered every third day with the appropriate nutrient solution and were watered with deionized water on days when the nutrient solution was not applied. Seedlings were harvested and examined for levels of mycorrhizal colonization 60, 90 and 120 days after emergence. The experimental design was a full factorial with six replicates per treatment for a total of 288 seedlings (2 CO_2 levels $\times$ 2 population sources $\times$ 2 mycorrhizal treatments $\times$ 2 phosphorus levels $\times$ 6 replicates $\times$ 3 harvests).

At each harvest, seedlings were separated into roots and shoots. The extent of mycorrhizal colonization was estimated for each seedling on a subsample of root. The total number of root tips and the number of root tips that were mycorrhizal were counted for each subsample, and then used to calculate the percent of root tips that were mycorrhizal for the subsample. Roots were then digested and ground in a Wiley mill. Kjeldahl N and total P were measured by digesting root tissue using a sulfuric acid-hydrogen peroxide flux (Lowther, 1980). The digest was analyzed using a TRACCS 800 Continuous Flow Analytical System. To determine starch and soluble sugar contents, root tissue was extracted three times with a methanol:chloroform:water solution (Dickson, 1979). The pellet was digested for 1 hr. in perchloric acid to hydrolyze the starch (Sutton et al., 1981). Soluble sugar and starch concentrations were then measured colorimetrically using a phenol-sulfuric acid technique (Dubois et al., 1956). Total non-structural carbohydrates (TNC) were calculated as the sum of starch and sugar, expressed as a percent of root dry mass.

Root and total plant dry mass from each harvest were analyzed separately using seed source, CO_2 supply, phosphorus supply and mycorrhizal status as main effects. Mycorrhizal colonization was analyzed for the 60 and 90 day harvest using all four main effects. Because no significant mycorrhizal formation was found on uninoculated seedlings at any harvest, mycorrhizal seedlings were then analyzed independently using seed source, CO_2 supply and phosphorus supply as main effects. All mycorrhizal seedlings at the 120 day harvest were over 90 percent colonized, so effects on mycorrhizal colonization were not analyzed for these seedlings. Root nutrient and carbohydrate levels were analyzed for seedlings at all harvests using seed source, CO_2 supply, phosphorus supply and mycorrhizal status as main effects. Least square means of treatments were compared using ANOVA and Scheffe's test for pairwise comparisons. The relationships between mycorrhizal colonization and root carbohydrates were analyzed using the regression analysis procedure in SYSTAT (SYSTAT, Evanston, IL).

Results

Growth in elevated CO_2 resulted in significantly greater total plant dry mass at all three harvests (Table 1, $p < 0.0001$ in all cases). Total seedling dry mass was increased 18, 50 and 40% at the 60, 90 and 120 day harvests, respectively. Additionally, seedlings grown in elevated CO_2 had significantly greater root dry mass (Table 1) than seedlings grown in ambient CO_2 at all three harvests ($p < 0.0001$ in all cases). Seedlings from the FL seed source had significantly greater root dry mass compared to seedlings from NC at all three harvests ($p < 0.0001$ in all cases). The only significant interactions between main effects on root dry mass were observed between mycorrhizal status and nutrient supply at the 90 and 120 day harvests ($p < 0.0001$ in both cases; Fig. 1). Seedlings in the Low P, non-mycorrhizal treatment had significantly less root dry mass than seedlings in the other three treatments. Additionally, at the 120 day harvest, High P, mycorrhizal seedlings had significantly greater root dry mass than seedlings in the other three treatments. There were no significant differences among seedlings in the High P treatments at the 90 day harvest. The interaction between mycorrhizal status and nutrient supply on root dry mass at the 60 day harvest was not significant ($p = 0.074$).

Inoculation with *P. tinctorius* resulted in significant mycorrhizal colonization at all three harvests ($p < 0.0001$); no seedlings in the non-mycorrhizal treatment developed mycorrhizae. At the 60 day harvest, there were no significant treatment effects on the mycorrhizal colonization rate (Fig. 2). At the 90 day harvest, there was a significant interaction between seed source and CO_2 treatment on percent of fine roots colonized. Mycorrhizal seedlings from NC grown in elevated CO_2 had a significantly greater percent of fine roots colonized than seedlings in the other three treatments. There were no significant differences among FL seedlings in either CO_2 treatment and NC seedlings in the ambient CO_2 treatment. At the 120 day harvest, all

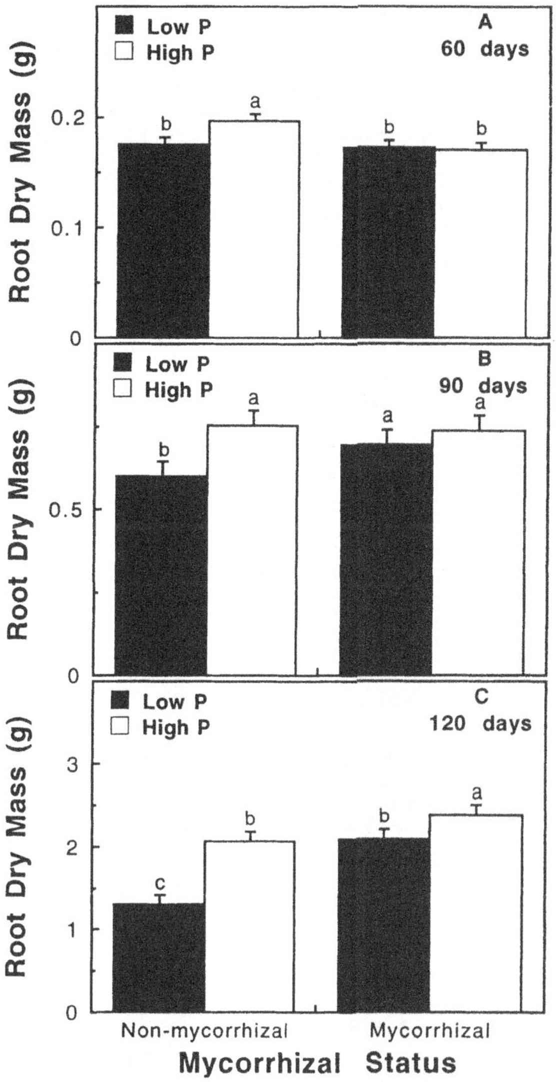

Fig. 1. Interactive effects of phosphorus supply and mycorrhizal status on root dry mass 60 (**A**), 90 (**B**) and 120 (**C**) days after emergence. Note different scales on the ordinate axis. Seedlings from the different seed sources and from the different CO_2 treatments are grouped together. Different letters indicate means that are significantly different at $p \leq 0.05$ using Scheffe's test for pairwise comparisons. n = 23 or 24 for all treatments.

Table 1. Least square means (± SE) for effects of CO_2 supply and seed source on root and total plant dry mass at the 60, 90 and 120 day harvests. Different letters within main effects indicate means that are significantly different at $p \leq 0.05$ using Scheffe's test for multiple comparisons (n = 47 or 48)

Main effects	Root dry mass (g)			Total plant dry mass (g)		
	60 days	90 days	120 days	60 days	90 days	120 days
CO_2 partial pressure						
35.5 Pa	0.16 (0.005)b	0.52 (0.04)b	2.04 (0.21)b	0.40 (0.01)b	2.01 (0.15)b	7.51 (0.57)b
71.0 Pa	0.20 (0.005)a	0.87 (0.04)a	3.02 (0.21)a	0.47 (0.01)a	3.02 (0.15)a	10.53 (0.57)a
Seed source						
Florida	0.20 (0.005)a	0.75 (0.04)a	2.89 (0.21)a	0.48 (0.01)a	2.75 (0.15)a	9.91 (0.57)a
North Carolina	0.16 (0.005)b	0.65 (0.04)b	2.16 (0.21)b	0.39 (0.01)b	2.29 (0.15)b	8.12 (0.57)b

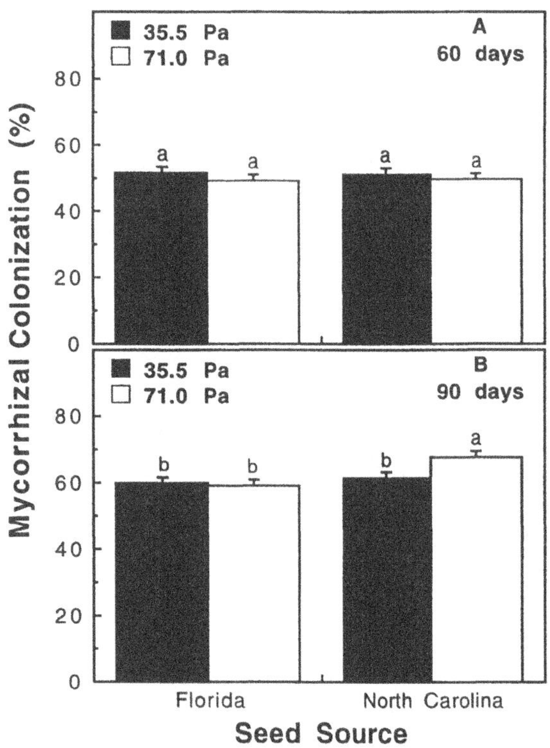

Fig. 2. Percent of fine roots that were mycorrhizal at 60 (**A**) and 90 (**B**) days on inoculated seedlings. All uninoculated seedlings had no mycorrhizal fine roots at either harvest. Seedlings from the different nutrient treatments are grouped together. Different letters indicate means that are significantly different at $p \leq 0.05$ using Scheffe's test for pairwise comparisons. n = 11 or 12 for all treatments.

inoculated seedlings had over 90 percent of fine roots colonized.

Whole-root concentrations of starch, soluble sugars and total non-structural carbohydrates (TNC) decreased approximately 35% between the 60 and the 120 day harvests, and were not significantly affected by seed source at any harvest. Additionally, there were no significant interactions between the main effects on root carbohydrate levels. Results from the 90 day harvest are representative of all three harvests and are shown in Table 2. Growth in elevated CO_2 resulted in significantly increased root starch, sugar and TNC concentrations, while all three parameters were significantly decreased by mycorrhizal colonization ($p < 0.001$ in all cases). Low P fertilization significantly decreased root starch concentrations and increased sugar concentrations but did not affect TNC concentrations. Root carbohydrate concentrations were not correlated with the percent of fine roots that were mycorrhizal at any harvest; only data from the 90 day harvest is shown (Fig. 3). Additionally, while root carbohydrate concentrations declined between the 60 and 120 day harvests, mycorrhizal colonization increased.

Data in Table 3 represents combined results for root nutrient concentration and content for seedlings from NC and FL at the 120 day harvest. Root N content was significantly greater in Florida seedlings than in North Carolina seedlings ($p < 0.001$). However, seed source did not significantly affect root P levels or root N concentration, and there were no significant interactions between the effect of seed source and other main effects on root nutrient levels. There were significant three-way interactions between CO_2 supply, phosphorus supply and mycorrhizal status on root P content and concentration. Mycorrhizal colonization significantly increased root P concentration in both phosphorus treatments at ambient CO_2; however, at elevated CO_2, non-mycorrhizal, High P seedlings had significantly higher root P concentrations than mycorrhizal, High

Table 2. Least square means (± SE) for effects of CO_2, phosphorus supply and mycorrhizal inoculation on root starch, soluble sugar and total non-structural carbohydrate (TNC) levels at the 90 day harvest. Means are expressed as a percent of root dry weight. Seedlings from different seed sources are combined within treatments. FL seedlings were not significantly different than NC seedlings for any measure ($p > 0.18$ in all cases), and there were no significant interactions between the effect of seed source and other main effects. Different letters within columns indicate means that are significantly different at $p < 0.05$ using Scheffe's test for multiple comparisons (n = 34 or 35). Due to rounding errors, TNC levels may differ slightly from the combined values for starch and soluble sugars

Main effects	Starch	Soluble sugars	TNC
	(% of DW)		
Mycorrhizal status			
Mycorrhizal	6.37 (0.07)b	2.99 (0.08)b	9.38 (0.13)b
Non-mycorrhizal	6.88 (0.07)a	4.34 (0.08)a	11.24 (0.13)a
Nutrient treatment			
Low P	6.43 (0.07)b	3.91 (0.08)a	10.36 (0.13)a
High P	6.82 (0.07)a	3.42 (0.08)b	10.25 (0.13)a
CO_2 partial pressure			
35.5 Pa	6.35 (0.07)b	3.34 (0.08)b	9.70 (0.13)b
71.0 Pa	6.88 (0.11)a	4.00 (0.08)a	10.91 (0.13)a

Table 3. Least square means (± SE) for effects of CO_2, phosphorus supply and mycorrhizal inoculation on root nitrogen and phosphorus levels at the 120 day harvest. Means are expressed on a dry weight basis. Seedlings from different seed sources are combined within treatments. Different letters within columns indicate means that are significantly different at $p \leq 0.05$ using Scheffe's test for multiple comparisons (n = 11 or 12)

Main effects	P conc. (mg/g)	P content (mg)	N conc. (mg/g)	N content (mg)
35.5 Pa CO_2				
Low P, NM	1.14 (0.13)cd	1.15 (0.30)d	17.96 (0.70)b	18.98 (2.18)d
High P, NM	1.54 (0.11)c	2.56 (0.25)cd	24.31 (0.70)a	39.36 (2.18)b
Low P, M	2.12 (0.11)a	3.42 (0.25)bc	15.76 (0.70)c	25.73 (2.18)cd
High P, M	1.95 (0.12)b	3.74 (0.26)bc	22.67 (0.70)a	43.36 (2.18)b
71.0 Pa CO_2				
Low P, NM	0.94 (0.12)d	1.50 (0.28)d	16.40 (0.70)b	25.23 (2.18)cd
High P, NM	2.17 (0.12)a	5.70 (0.26)a	23.61 (0.70)a	61.90 (2.18)a
Low P, M	2.06 (0.21)ab	5.35 (0.48)ab	15.25 (1.33)bc	41.84 (4.14)b
High P, M	1.71 (0.11)b	4.73 (0.25)ab	22.59 (0.70)a	63.43 (2.18)a

P seedlings. Additionally, elevated CO_2 significantly increased root P concentrations and contents only in High P, non-mycorrhizal seedlings. Mycorrhizal status significantly increased root N concentration and content, while increased P supply resulted in significantly increased root N content but significantly decreased

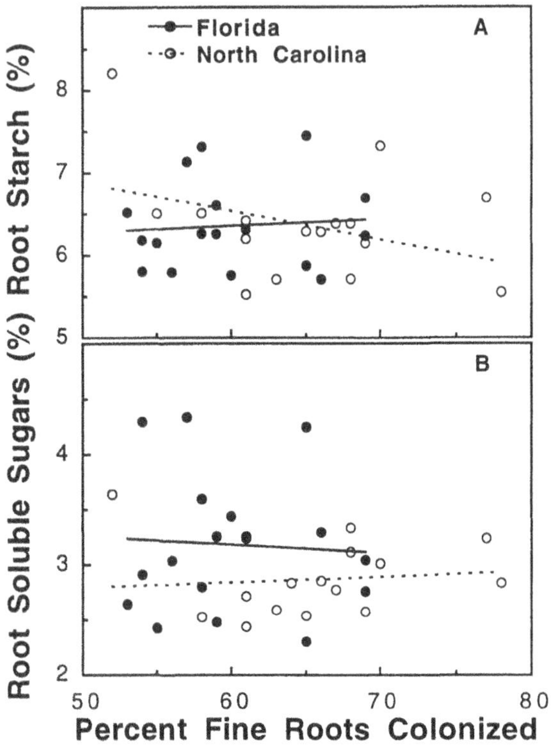

Fig. 3. Relationships between percent of fine roots that were mycorrhizal and root starch (**A**) and root soluble sugar (**B**) concentrations at the 90 day harvest for mycorrhizal seedlings from Florida and North Carolina. Similar patterns were observed at 60 days. $r^2 < 0.15$ in all cases. n = 23 for both seed sources.

root N concentration. Treatment effects were similar when root nutrient concentrations were expressed on a starch-free basis.

Discussion

Exposure of loblolly pine seedlings to elevated CO_2 generally did not result in significant increases in mycorrhizal colonization rates. This was unexpected, given that the carbohydrate concentrations and the total carbohydrate content of the root system were greater in seedlings grown in elevated CO_2 regardless of seed source. Root carbohydrate levels, particularly sucrose, have frequently been tied to mycorrhizal colonization rates (Douds and Schenck, 1990; Hacskaylo, 1973; Marx et al., 1977), and are one mechanism through which mycorrhizae are expected to increase in response to elevated atmospheric CO_2. It is possible that, in the present study, mycorrhizal colonization was not a significant sink for photosynthate, and thus did not benefit from the increased allocation of carbon to the roots. Mycorrhizal roots, however, had significantly lower carbohydrate concentrations than non-mycorrhizal roots, indicating that the fungus functioned as an active photosynthate sink (Borges and Chaney, 1989) during the course of the study. In addition, colonization rates were much faster in this study than those reported for other studies where a stimulatory effect of elevated CO_2 on mycorrhizal formation was observed (Conroy et al., 1990; Norby et al., 1986), further indicating that mycorrhizae were an active photosynthate sink.

Root and soil phosphorus and nitrogen levels have frequently been found to be inversely correlated with the rate and extent of mycorrhizal formation, with mycorrhizal formation often ceasing at high concentrations of phosphorus (Koide, 1991). Despite a threefold variation in root phosphorus levels at the 120 day harvest, and the highest concentrations occurring in seedlings inoculated with mycorrhizae, all inoculated seedlings had over 90 percent of fine roots colonized by mycorrhizae. Nitrogen levels also varied significantly between phosphorus treatments, yet were not associated with variation in colonization levels. Thus, phosphorus fertilization significantly affected root nutrient concentrations but did not affect colonization.

Another potential reason for the lack of effect of elevated CO_2 on mycorrhizal colonization is that colonization was measured as the percent of fine roots that were mycorrhizal. This technique may not have adequately assessed the spread of mycorrhizae on the root system (Marx, 1990). Because the quantifying technique used in this study only compared the number of mycorrhizal fine root tips to the number of non-mycorrhizal fine root tips, an increase in the total number of mycorrhizal tips could have been missed if there was a concomitant increase in non-mycorrhizal tips. If a small part of the increase in root dry mass observed in response to elevated CO_2 was the result of increased fine root production, this would indicate increased mycorrhizal dry mass, since the total number of fine roots would increase while the proportion of fine roots that were mycorrhizal remained the same. It is, therefore, possible that increased mycorrhizal colonization under elevated CO_2 occurred in this study but was not detected using our methodology. In addition, increases in activity, proliferation and turnover of the mycorrhizal fungus in the rooting medium may have occurred in response to increased photosynthate supply, but were not quantified by the technique used in this study.

Increased root carbohydrate concentrations and contents observed in response to elevated CO_2 suggest that the root system has an important role in storage of photosynthate as photosynthate production continues to increase in response to rising atmospheric CO_2. Increased carbohydrate pools are likely to affect both the host plant and the rhizosphere community. Pathogenic fungi and diseases may become more persistent and facultative symbionts may become parasitic as more fixed carbon is released into the rhizosphere. Although increases in root carbohydrates were not associated with significant decreases in root N concentrations in this study, carbohydrate accumulation may lead to increases in root C:N ratios. These changes may alter behavior or the composition of the saprophytic community and the rate of litter degradation and accumulation much as changes in the C:N ratio affect herbivory on leaf tissue (Lincoln et al., 1986).

These results suggest that mycorrhizal colonization, measured as percent of fine roots colonized, is relatively unaffected by root carbohydrate levels. This is surprising given that other studies have shown that increased root carbohydrate levels or increased atmospheric CO_2 levels are associated with increased mycorrhizal colonization rates (see above). However, the relative insensitivity of mycorrhizal colonization to root carbohydrate and nutrient concentrations suggests that the fungus plays an active role in regulating mycorrhizal formation. Both inter- and intraspecific variation in colonization rates have repeatedly been found for mycorrhizal fungi when grown under ambient CO_2 conditions, but little work has been done to compare genotypes and species under elevated CO_2 conditions. Also, little work has been done comparing different combinations of host and symbiont genotypes and species to determine how communities may respond to elevated CO_2. Greater understanding of the interaction between host and symbiont, and the importance of intraspecific variation within each, in regulating mycorrhizal formation is necessary to determine how forest species will respond to increasing atmospheric CO_2. Longer-term experiments, under field conditions, and seasonal monitoring of mycorrhizal communities will be important in providing information on the relative impact of the association to the host physiology after the seedling stage, and may determine stages where the fungal symbiont plays a greater or lesser role in regulating the host response to elevated CO_2.

Acknowledgements

We thank Elizabeth Thomas, David Tissue and the Duke Phytotron staff for technical assistance. Kevin Griffin and two anonymous reviewers provided helpful comments on earlier versions of the manuscript. Seeds were generously provided by Drs. Paul Belonger of Georgia-Pacific and David Todd of Weyerhauser. This study was supported by the U.S. Department of Energy, CO_2 Research Division, grant no. DE-FG05-87ER60575 and the Electric Power Research Institute Forest Response to CO_2 program. The Phytotron is supported by National Science Foundation grant no. BSR 87-06429.

References

Borges R G and Chaney W R 1989 Root temperature affects mycorrhizal efficacy in *Fraxinus pennsylvanica* Marsh. New Phytol. 112, 411–417.

Conroy J P, Milham P J, Reed M L and Barlow E W 1990 Increases in phosphorus requirements for CO_2-enriched pine species. Plant Physiol. 92, 977–982.

Dickson R E 1979 Analytical procedures for the sequential extraction of ^{14}C labelled constituents from leaves, bark and wood of cottonwood plants. Physiol. Plant. 45, 480–488.

Douds D D Jr. and Schenck N C 1990 Relationship of colonization and sporulation by VA mycorrhizal fungi to plant nutrient and carbohydrate contents. New Phytol. 116, 621–627.

Dubois M, Gilles K A, Hamilton J K, Rebers P A and Smith F 1956 Colorimetric method for determination of sugars and related substances. Anal. Chem. 28, 350–356.

Hacskaylo E 1973 Carbohydrate physiology of ectomycorrhizae. *In Ectomycorrhizae*. Eds. G C Marks and T T Kozlowski. pp 207–231. Academic Press, New York.

Hatch A B 1937 The physical basis of mycotrophy in *Pinus*. Black Rock For. Bull. 6, 1–168.

Jones M D, Durall D M and Tinker P B 1990 Phosphorus relationships and production of extramatrical hyphae by two types of willow ectomycorrhizas at different soil phosphorus levels. New Phytol. 115, 259–267.

Koide R T 1991 Nutrient supply, nutrient demand and plant response to mycorrhizal infection. New Phytol. 117, 365–386.

Lincoln D E, Couvet D and Sionit N 1986 Response of an insect herbivore to host plants grown in carbon dioxide enriched atmospheres. Oecologia 69, 556–560.

Marx D H 1990 Soil pH and nitrogen influence *Pisolithus* ectomycorrhizal development and growth of loblolly pine seedlings. For. Sci. 36, 224–245.

Marx D H, Hatch A B and Mendicino J F 1977 High soil fertility decreases sucrose content and susceptibility of loblolly pine roots to ectomycorrhizal infection by *Pisolithus tinctorius*. Can. J. Bot. 55, 1569–1574.

Norby R J, O'Neill E G and Luxmoore R J 1986 Effects of atmospheric CO_2 enrichment on the growth and mineral nutrition of *Quercus alba* seedlings in nutrient-poor soil. Plant Physiol. 82, 83–89.

Norby R J, O'Neill E G, Hood W G and Luxmoore R J 1987 Carbon allocation, root exudation and mycorrhizal colonization of *Pinus echinata* seedlings grown under CO_2 enrichment. Tree Physiol. 3, 203–210.

O'Neill E G, Luxmoore R J and Norby R J 1987 Increases in mycorrhizal colonization and seedling growth in *Pinus echinata* and *Quercus alba* in an enriched CO_2 atmosphere. Can. J. For. Res. 17, 878–883.

Same B I, Robson A D and Abbott L K 1983 Phosphorus, soluble carbohydrates and endomycorrhizal infection. Soil Biol. Biochem. 15, 593–597.

Strain B R and Cure J 1985 Direct effects of increasing carbon dioxide on vegetation. Publication ER-0238, United States Department of Energy, Washington, DC.

Sutton B G, Ting I P and Sutton R 1981 Carbohydrate metabolism of cactus in a desert environment. Plant Physiol. 68, 784–787.

Thomson J P, Robson A D and Abbott L K 1986 Effects of phosphorus on the formation of mycorrhizas by *Gigaspora calospora* and *Glomus fasciculatum* in relation to root carbohydrates. New Phytol. 103, 751–765.

Plant and Soil **165:** 89–101, 1994.

Evaluation of the use of a model rhizodeposition technique to separate root and microbial respiration in soil*

J. Swinnen
DLO-Research Institute for Agrobiology and Soil Fertility (AB-DLO), P.O. Box 14, NL-6700 AA Wageningen, The Netherlands

Key words: barley, glucose, microbial respiration, pulse-labelling, rhizodeposition, root respiration, wheat

Abstract

A model rhizodeposition technique to estimate the root and microbial components of ^{14}C soil/root respiration in pulse-labelling experiments is described. The method involves the injection of model rhizodeposits, consisting of ^{14}C-labelled glucose, root extract or root cell wall material, into the rooted soil of an unlabelled plant, simultaneously with the pulse-labelling of a separate but similar plant with $^{14}CO_2$. In a growth chamber experiment with 30 day old wheat and barley the contribution of direct root respiration to ^{14}C soil/root respiration over a 26 day period after labelling was estimated 89–95%. Estimates of direct root respiration in field-grown wheat and barley at different development stages in most cases accounted for at least 75% of ^{14}C soil/root respiration over a 21 day period after labelling. The mineralization rate of injected ^{14}C-glucose was positively correlated with the concentration of glucose-C established in soil. The use of the method in rhizosphere carbon budget estimations is evaluated.

Introduction

Since the release of organic compounds from roots during plant growth strongly influences microbial biomass in soil (Whipps and Lynch, 1985) and since the soil microbial biomass is the driving force in nutrient cycling and soil organic matter decomposition (Van Veen et al., 1989), the study of root-derived organic matter input to soil has received considerable attention (e.g. Barber and Martin, 1976; Liljeroth et al., 1990; Merckx et al., 1987). All the losses of organic substances arising from the plant root are described by the term 'rhizodeposition' (Whipps and Lynch, 1985; Whipps, 1990). It covers a whole gamut of compounds which can be classified according to their mode of arrival: (1) exudates are water-soluble compounds leaking from the root without involvement of metabolic energy, (2) secretions depend upon metabolic processes, and (3) lysates are released when cells autolyse and include cell walls, sloughed cells and decaying whole roots.

Labelling of plants with ^{14}C has been widely used to quantify rhizodeposition (e.g. Liljeroth et al., 1990; Merckx et al., 1985, 1987), as it allows to distinguish root-derived organic C from native soil organic C. Compared to continuous labelling, pulse-labelling of plants with ^{14}C has the advantage of being more easily applicable under field conditions (Lynch and Whipps, 1990) and giving more information on photosynthate distribution at specific development stages of the plants (Meharg and Killham, 1990). Recently, we have used ^{14}C pulse-labelling to study C fluxes in the rhizosphere of wheat and barley under field conditions (Swinnen, 1994; Swinnen et al., 1994a, b). An allocation period of 3 weeks between labelling and determination of the ^{14}C distribution over shoots, roots, soil/root respiration and soil residue was found satisfactory.

After labelling ^{14}C soil/root respiration originates from 2 processes that cannot be separated directly: direct root respiration and microbial respiration of rhizodeposition substrates. Thus, to quantify gross ^{14}C rhizodeposition during the 3 week period after pulse-labelling an estimate is required of the components of ^{14}C soil/root respiration. To provide such estimates, a model rhizodeposition method was introduced (Swinnen and Van Veen, 1993). Summarized results of different experiments were already used in the pulse-labelling studies mentioned above. This paper presents

* Communication No. 73 of the Dutch Programme on Soil Ecology of Arable Farming Systems.

a more detailed description of both the experimental conditions and results and, furthermore, evaluates the use of the method in rhizosphere carbon budget estimations.

Materials and methods

The model rhizodeposition technique consists of injecting ^{14}C-labelled organic compounds (model rhizodeposits) into the rooted soil of an unlabelled plant, simultaneously with the pulse-labelling of a physically separate but similar plant with $^{14}CO_2$ (Fig. 1). The assumption is made that the model rhizodeposits create a model rhizosphere and that the microbial transformation of the introduced compounds is representative for the transformation of ^{14}C-labelled rhizodeposits in the system pulse-labelled with $^{14}CO_2$. Prior to labelling, the soil in both systems is separated from the atmosphere by a silicone rubber seal. Soil/root-derived CO_2 is trapped during a 3 week period after which the distribution of ^{14}C over shoots, root, soil/root respiration and soil organic C in the pulse-labelled system is determined. By rewriting the equation given in Fig. 1, the amount of microbial ^{14}C respiration in the pulse-labelled system ($^{14}C_{micr.resp.}$) is calculated:

$$^{14}C_{micr.resp.} = {}^{14}C_{soil} \times \frac{MRR}{100 - MRR} \quad (1)$$

where $^{14}C_{soil}$ is the amount of ^{14}C recovered as soil organic C in the pulse-labelled system, and MRR is the percentage of ^{14}C model rhizodeposit respired. The ^{14}C soil residue in the pulse-labelled system is first corrected for losses of ^{14}C from roots during isolation from soil. ^{14}C root respiration ($^{14}C_{rootresp.}$ is calculated by difference from ^{14}C soil/root respiration ($^{14}C_{soil/rootresp.}$):

$$^{14}C_{rootresp.} = {}^{14}C_{soil/rootresp.} - {}^{14}C_{micr.resp.} \quad (2)$$

Since the chemical composition of actual ^{14}C-labelled rhizodeposits cannot be determined *in situ*, and may moreover fluctuate due to environmental and plant-related conditions, a range of uniformly ^{14}C-labelled compounds differing in degradability was used as model rhizodeposit: (1) glucose, (2) an ice-water extract of oven-dried and ground wheat roots, and (3) the residual cell wall material of this extraction.

This method was applied in 3 different experiments, performed either in a growth chamber (experiment 1 and 3) or in the field (experiment 2). Unless explicitly mentioned, oven-drying of samples was done at 100°C for 72 h, and grinding was carried out using a 0.5 mm mesh size. Total C and ^{14}C content of solid samples was determined by wet digestion (modified from Amato, 1983). NaOH samples were analyzed for total CO_2 by titration with excess $BaCl_2$ and for ^{14}C by liquid scintillation counting. ANOVA was performed using SAS regression procedures (SAS Institute Inc., Cary, NC) with a 95% confidence level in the LSD-test for differences between means. Decimal codes to indicate plant development stages are described by Zadoks et al. (1974).

Experiment 1

Uniformly ^{14}C-labelled wheat roots (specific activity 3.35 MBq g^{-1} C), obtained after plant growth in a continuously ^{14}C-labelled atmosphere (Lekkerkerk et al., 1990), were oven-dried and ground (< 0.3 mm). A 4.37 g aliquot was shaken in ice-water (110 mL water and 100 g ice) for 1 h on a table shaker (150 rpm). The suspension was filtered (8 μm), washed with 200 mL cold water, and both the extract and the residue were freeze-dried. The dry residue was analyzed for total C and ^{14}C, and the ^{14}C concentration of the freeze-dried extract determined by liquid scintillation counting after dilution in water. A glucose solution with the same specific activity as the wheat roots was prepared from D-[U-^{14}C]-glucose (Amersham) and unlabelled glucose. 5 mL aliquots of the glucose, root extract and root cell wall substrates containing 5 kBq ^{14}C and 1.5 mg C were prepared in weighed vials and immediately frozen.

Seeds of spring wheat (variety Vedette) and spring barley (variety Prisma) were pregerminated on wet paper for 3 days at 20°C. Calcareous silt loam soil (Calcaric Fluvisol (FAO-Unesco, 1981), pH-KCl 7.5, $CaCO_3$ 10%, organic matter content 2.1%) was taken from the A_p horizon of a field at the Lovinkhoeve experimental farm (Kooistra et al., 1989), sieved (ϕ 4 mm) and mixed with commercial fertilizer Sporumix (described by Merckx et al., 1987) at a rate of 2.5 g kg^{-1} dry soil. The soil moisture content was 23% on dry wt basis, as determined by oven drying at 105°C for 24 h. Columns with a closed lower end (25 cm height, 9 cm i.d.) were filled with a 22 cm layer of soil (1220 g dry wt) and closed with a lid that was provided with openings for shoot growth, aeration and water addition. The columns were planted with a 3 day old seedling before they were placed in a growth cham-

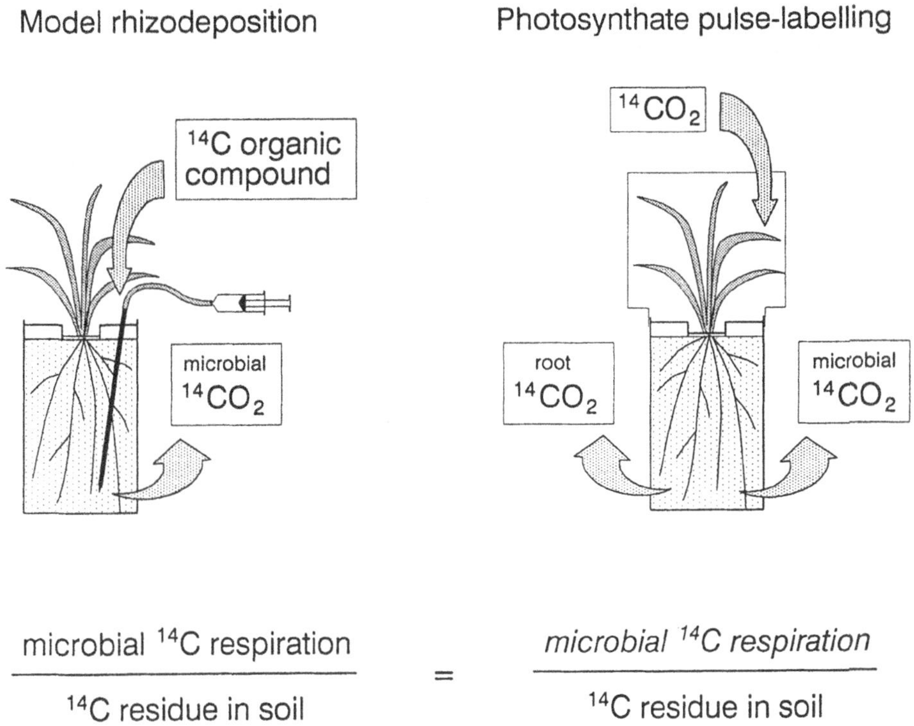

Fig. 1. Outline of the model rhizodeposition method to separate root and microbial ^{14}C respiration after pulse-labelling of plants with $^{14}CO_2$. The equation reflects the assumption of equal microbial transformation of model rhizodeposits (left) and actual ^{14}C rhizodeposits (right).

ber with 20/15°C day/night temperature and 16 h day length. Approximately every 3 days the columns were weighed and the soil moisture content was adjusted in such a way that its fluctuation in time was around an average of 23% (dry wt basis), which corresponds to a pF-value of 2.8.

When the plants were 28 days old the soil was separated from the atmosphere by pouring a silicone rubber seal (Dow Corning, Q3-3481, Base + Catalyst) on the lid and around the shoot base. When the plants were 30 days old (late tillering stage, decimal code 29) 4 replicate columns with wheat and 4 columns with barley were pulse-labelled with 500 kBq $^{14}CO_2$ using the method described by Swinnen et al. (1994a). The pulse-labelling started 70–90 min after the beginning of the day period and lasted for 1.5 h. After labelling a vial with an appropriate volume of 5*M* NaOH was placed in the column headspace and the aeration opening was closed. This soil/root respiration trap was renewed every day during the first 2 weeks after labelling and at larger time intervals thereafter. The headspace of the columns was flushed with oxygen gas every time that water was added, starting 8 days after labelling. The ^{14}C was left to be allocated in the plant-soil system for 26 days.

During the labelling period separate columns were treated with model rhizodeposits, which were thawed and kept cool (5°C) prior to the application. The respiration trap was removed and a 22 cm deep hole made in the soil column using a stainless steel pricker (3.0 mm diameter). The solution or suspension was sucked into a 5 mL syringe equipped with a special needle (40 cm length, 1.0 mm i.d., opening at the side of the tip). The tip was placed at the bottom of the hole and slowly withdrawn up to the surface while injecting the substrate. The syringe and needle were rinsed with 5 mL of cold water, which was put back in the original vial and immediately frozen. Before the next injection the syringe was further rinsed several times. 2 replicate columns with wheat and barley were treated with each of the 3 model rhizodeposits. Soil respiration was monitored for 57 days.

26 days after labelling (booting stage, decimal code 47) the shoots from the pulse-labelled columns were clipped. After hand-picking large roots a subsample of the soil was taken. The remaining soil was washed

with tap water on a 0.25 mm sieve to isolate small roots. Shoots, handpicked roots, the soil subsample, and the roots recovered on the sieve were dried at 80°C for 48 h, ground and analyzed for total C and ^{14}C. NaOH samples from the respiration traps were analyzed for total CO_2 and ^{14}C. The water used to rinse the syringe was analyzed for remaining ^{14}C. Where root cell wall material was used this was done by freeze-drying and wet digestion.

Experiment 2

Some details of the field were described by Swinnen (1994). Briefly, winter wheat and spring barley were grown in 1990 on calcareous silt loam soil described for experiment 1. There were 4 plots per crop which differed in agricultural management and soil organic matter content. In spring, PVC columns (100 cm length; 8.5 cm i.d.) with a cutting rim at the open lower end (Swinnen et al., 1994a) were slowly pushed into soil around individual seedlings. At different development stages 1 column per plot was pulse-labelled with 5 MBq of $^{14}CO_2$ for 2 h, starting around midday. 2 days earlier it was sealed at the soil surface with silicone rubber. The methods of sealing and labelling were similar to those described by Swinnen et al. (1994a). The ^{14}C was allowed to be allocated in the plant-soil system for 21 days. Soil/root-derived CO_2 was trapped at the soil surface in 10 g of sodalime (8–14 mesh) contained in a tube (2.1 cm i.d.) connected to the column headspace. To prevent absorption of atmospheric CO_2, separated by a cotton plug a second layer of sodalime was placed on top of the first one. This respiration trap allowed diffusion of oxygen into the soil column (Anderson, 1982). Precipitation was added automatically through a system that consisted of a water-lock and a funnel with the same diameter as the column.

Labelling was carried out at tillering, ear emergence, and dough ripening; decimal codes 22, 58, and 81 respectively. Simultaneously additional columns were treated with model rhizodeposits. The amounts of ^{14}C and total C in the 5 mL volume injected were 3 times as high as in experiment 1 (15 kBq ^{14}C and 4.5 mg C). Instead of applying all possible combinations of crop, plot, development stage and model rhizodeposit compound, only 32 columns were used (4 on each plot). The treatments were distributed in such a way that main effects and interactions between compound, crop and development stage could be analyzed statistically (incomplete block design). The respiration trap was normally left on for 21 day, but for wheat columns treated with model rhizodeposits at tillering, soil/root-derived CO_2 was trapped for 47 days, with renewals of the trap after 21 and 43 days. The samples of sodalime were put in septum cap bottles and dissolved with excess injected 2.5*M* HCl. Released $^{14}CO_2$ was trapped overnight in 5*M* NaOH contained in test tubes inside the bottles, and analyzed for CO_2 and ^{14}C. Controls without CO_2 were treated similarly.

21 days after labelling columns were transferred to the laboratory. On the next day the respiration trap was removed, shoots were clipped, and the soil core divided into a small part (1/8 sector of the cylindrical core), from which large roots were handpicked, and a big part (7/8 sector). Shoots, handpicked roots, and soil core parts were oven-dried and shoots were ground. The big part of the soil core was pretreated with sodium pyrophosphate (NaPP) as described by Swinnen et al. (1994a) to disrupt the soil structure and washed on a double sieve (0.6 and 0.3 mm) using an apparatus that consisted of a tank for sedimentation of the soil and a pump to recycle the NaPP solution. Roots and organic material recovered on the 0.6 mm sieve were separated from shells and resistant soil aggregates, then oven-dried and ground. Subsamples of the material on the 0.3 mm sieve and of the soil suspension were freeze-dried. Total C and ^{14}C content of all samples was determined. Root ^{14}C losses during the washing procedure were estimated using dry handpicked roots, which were put into 100 g of unlabelled dry soil, pretreated with NaPP and washed on a 0.3 mm sieve by shaking in NaPP solution. Recovered roots were oven-dried and ground, the soil suspension was subsampled and freeze-dried, and roots and soil were analyzed for ^{14}C. Finally, the soil from which roots were handpicked was pretreated with NaPP and washed (0.3 mm sieve) to collect remaining roots, which were oven-dried, ground and analyzed for total C and ^{14}C.

Experiment 3

A calcareous silt loam soil (described in experiment 1; 20.6% moisture on dry wt basis) and a loamy sand soil (Humic Podzol (FAO-Unesco, 1981), pH-KCl 6.4, organic matter content 2.8%, 11.7% moisture) taken from the A_p horizon of a field near Ede, the Netherlands was sieved (ϕ 4 mm) and stored at 20°C. Seeds of spring wheat (variety Vedette) were pregerminated as in experiment 1. Columns as in experiment 1, but with a perforated lower end and 40 cm height, were successively filled with a 13.5 cm layer of expanded clay pellets (Asef, Didam, The Netherlands; 10–16 mm diameter), a nylon gauze (0.25 mm mesh size) and a 23.5 cm layer of moist soil (1658 g calcareous silt loam soil or 1779 g loamy sand soil). For each soil 8 columns were planted with a 2 day old seedling and 8 left unplanted, and transferred to a growth chamber with 20/15°C day/night temperature and 16 h day length. To supply water to the plants by capillary action through the expanded clay pellets, each column was placed in a dish with a water level adjusted automatically to a fixed distance from the lower end of the soil layer. When the plants were 24 days old 30 mL of fertilizer solution was added on the soil surface, containing 280 mg NO_3-N, and with loamy sand also 140 mg P and 300 mg K.

When the plants were 29 days old the columns were sealed as in experiment 1. When the plants were 31 days old (late tillering stage, decimal code 29) 50 kBq ^{14}C-glucose was injected using the method of experiment 1. The solutions differed in amount of C (1.5 or 15 mg C), volume (5 or 1 mL), and C/N-ratio (no N or $(NH)_4SO_2$ added to a C/N-ratio of 20). These treatments were applied to both soil types with or without a wheat plant. Thus the experiment was a 5-fold balanced factorial without replicates. Soil/root-derived CO_2 was trapped in 18 g of sodalime as in experiment 2. 21 days after model rhizodeposition (plants at the booting stage, decimal code 44) the soil was divided into 3 layers of equal thickness and after hand-picking large roots, soil moisture content was determined by oven-drying. Respiration traps were analyzed for total CO_2 and ^{14}C as in experiment 2.

Results

Experiment 1

Wheat plants developed normally and produced 8–15 tillers and 5–10 ears. Soil moisture content fluctuated in time around 23% (dry wt basis) but due to plant growth the range of fluctuation broadened from 20–26% around labelling to 12–40% 30 days after labelling and narrowed again to 17–25% 57 days after labelling. Barley plants produced 15–23 tillers. Initially, the fluctuation in time in soil moisture content was similar to that in wheat columns. Ca. 20 days after labelling, however, when the soil moisture content had to be adjusted to 40% to provide sufficient water until the next addition, the water use decreased. Ca, 1 week later the shoots gradually became yellow and eventually died at the late booting stage (decimal code 49). At the end of the experiment the soil moisture content was 31%.

96 ± 1% of the model rhizodeposit was injected in the case of glucose, 92 ± 0% with root extract and 57 ± 14% with root cell wall material. There was significant difference between wheat and barley in dynamics of ^{14}C respiration from model rhizodeposits (Fig. 2a). Accumulated respiration of glucose and root extract was not significantly different. The respiration rate of these 2 compounds was the highest immediately after injection. By contrast, root cell wall material respiration showed lag-phase-like dynamics, with the highest rate ca. 8–11 days after injection. Initially the accumulated respiration was lower compared to glucose and root extract but the difference was no longer significant by day 57. 26 days after injection, when the pulse-labelled columns were harvested, the respired proportion of all model rhizodeposits ranged between 21.3 and 39.2% (Table 1a). Since there was no significant effect of the crop, this minimum and maximum value was used to calculate the components of ^{14}C soil/root respiration in both crops. The distribution of ^{14}C in the pulse-labelled columns was generally similar for the 2 crops (Table 1b), with slightly higher proportions allocated to the shoots and slightly lower proportions recovered in roots, soil/root respiration and soil residue in wheat than in barley. The contribution of direct root respiration to ^{14}C soil/root respiration at this particular instant of plant growth was estimated to be far more important (average 92%) than the contribution of microbial respiration rhizodeposits (average 8%).

Experiment 2

The plants developed normally within the field columns and produced on average 3 ears per plant. As determined visually the soil moisture content inside and outside the columns was similar. Of the model rhizodeposits 94 ± 2% was injected in the case of glucose, 89 ± 3% with root extract and 62 ± 10% with root cell wall material. The differences in ^{14}C respiration dynamics between model rhizodeposits were comparable to experiment 1, with the respiration of root cell wall material at first lagging behind but afterwards being compensated (Fig. 2b). 21 days after injection, the accumulated respiration was ca 63% of that in experiment 1. Due to the less pronounced decrease in respiration rate this figure was ca. 97% 47 days after injection. The proportion of model rhizodeposits respired after 21 days was not significantly influenced

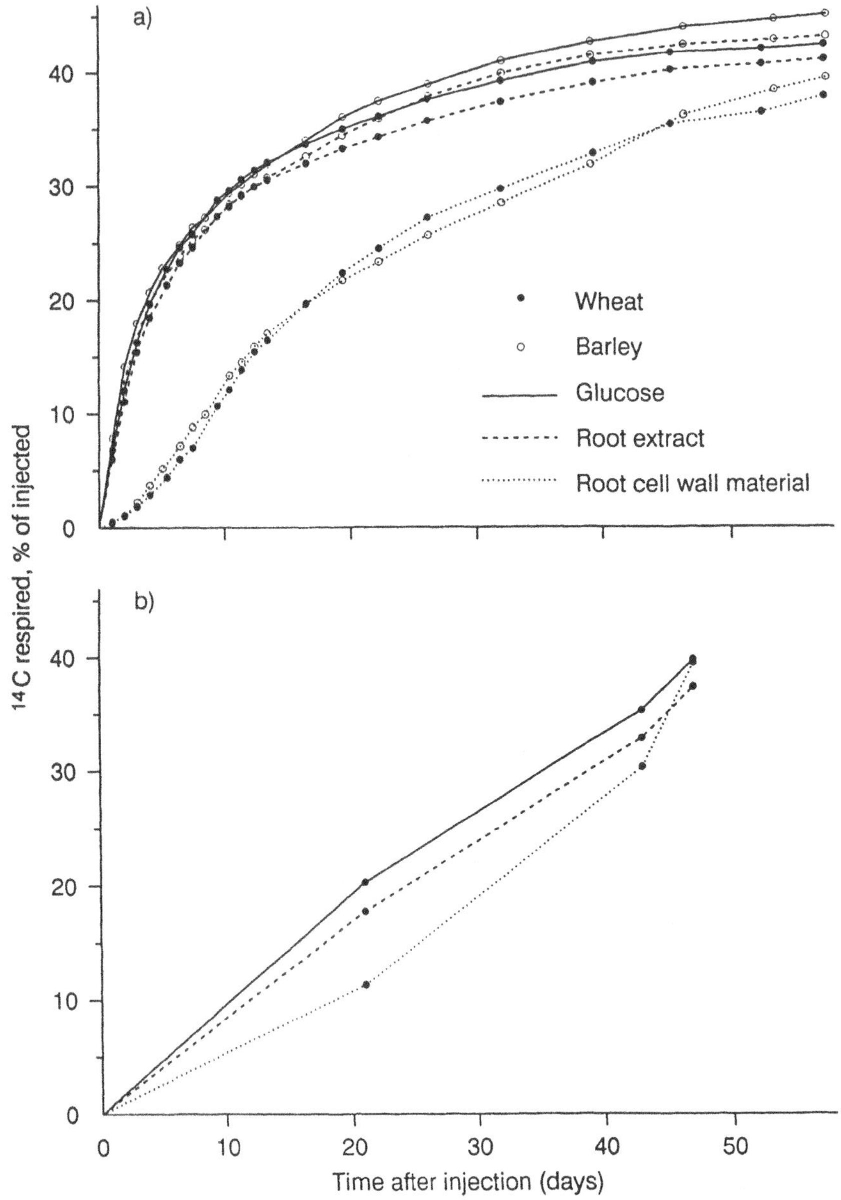

Fig. 2. Accumulated ^{14}C respiration from different model rhizodeposits injected in the rooted soil of **a**) spring wheat and spring barley grown in a growth chamber (experiment l; 1.5 mg C injected with 5 mL solution), and **b**) field-grown winter wheat at the tillering stage (experiment 2; 4.5 mg C injected with 5 mL solution). Data points are means of 2 replicates. Data from ANOVA are given in Table 1 and 2.

by the crop species but changed with the development stage of the crop, with lower values at ear emergence than at tillering and dough ripening (Table 2a). Therefore, the minimum and maximum value of each development stage were used to calculate the components of ^{14}C soil/root respiration.

The ^{14}C distribution in the pulse-labelled columns depended on the development stage of the crop at labelling (Table 2b). ^{14}C allocated to the shoots increased from ca. 41% of net ^{14}C assimilation at tillering to ca. 94% at dough ripening, whereas the proportions in each belowground compartment generally decreased dramatically. The ^{14}C contents of roots

Table 1. a) Proportion of ^{14}C respired from model rhizodeposits (1.5 mg C in 5 mL) after 26 days of incubation in the rooted soil of spring wheat and spring barley (2 columns per crop) in a growth chamber (experiment 1). b) ^{14}C distribution 26 days after pulse-labelling of 30 day old plants (4 columns per crop).

a) Model rhizodeposit respiration*			
	Crop		
Compound	Wheat	Barley	mean
	Percentage ^{14}C Respired (MRR)		
Glucose	37.6	38.9	38.3^{a}
Root extract	35.7	37.9	36.8^{a}
Root cell wall material	27.2	25.7	26.4^{b}
mean	34.1^{NS}	33.5	33.8
minimum			21.3
maximum			39.2
b) ^{14}C distribution*			
Compartment			
	Percentage of Net ^{14}C Assimilation		
Shoots	64.0^{NS}	57.9	61.0
Roots (> 0.25 mm)	11.6^{NS}	12.8	12.2
Soil/root respiration	21.4^{NS}	24.4	22.9
Soil residue (< 0.25 mm)	3.1^{b}	4.8^{a}	4.0
	Percentage of Soil/Root ^{14}C Respiration		
Root respiration (range)	91–96	87–95	89–95
Microbial respiration (range)	4– 9	5–13	5–11

*Means followed by different letters are significantly different (LSD-test, $p < 0.05$); NS = no significant differences. The interaction between crop and compound was not significant with respect to the proportion ^{14}C respired.

and soil residue were corrected for losses of ^{14}C from roots during the washing procedure, as estimated using handpicked roots. They increased with the development stage at labelling, from on average 10 at tillering to 41 at dough ripening. To take into account that exudates and mucilages adhering to handpicked roots may cause an overestimation of root losses, the values were divided by a factor 2. The estimated contribution of direct root respiration to ^{14}C soil/root respiration was at least 75, except in the case of winter wheat at tillering, where the range of estimates was rather large (42–90%) due to the high value of the ^{14}C soil residue relative to the ^{14}C soil/root respiration.

Experiment 3

Plants developed normally and produced 7–12 tillers. Soil moisture content decreased with the distance from the water table. In unplanted silt loam soil it was on average 23.1% at the lower end and 18.8% in the top layer. The corresponding values in unplanted loamy sand soil were 17.2% and 11.0%. The capillary action through the soil could only meet the water demand of plants until ca. 35 days after germination (4 days after injection of model rhizodeposits). Thereafter the soil moisture content gradually decreased in time and, at the end of the experiment (21 days after model rhizodeposition), in silt loam soil it was 20.8% in the bottom layer and 13.1% at the top, and in loamy sand soil 11.8% at the bottom and 5.7% at the top. This did not affect plant growth since roots were abundantly present in the bottom soil layer.

97 ± 2% of the glucose was injected when the volume of the solution was 5 mL and 91 ± 1% when 1 mL volumes were used. The accumulated respiration over a 21 day period ranged from 29 to 49% (Table 3).

Table 2. a) Proportion of ^{14}C respired from model rhizodeposits (4.5 mg C in 5 mL) after 21 days of incubation in the rooted soil of field-grown winter wheat and spring barley in 1990 (experiment 2). Values are means of from a factorial split-plot design with a total of 32 observations. b) ^{14}C distribution 21 days after pulse-labelling. Values are means from a factorial split-plot design with a total of 24 observations*

a) Model rhizodeposit respiration	**Averaged over wheat and barley**			
	Development stage			
Compound	Tillering	Ear emergence	Dough ripening	mean
		Percentage ^{14}C Respired (MRR)		
Glucose	24.9	21.0	25.5	23.4[a]
Root extract	24.7	21.4	28.4	24.8[a]
Root cell wall material	13.4	8.5	12.9	11.5[b]
Mean	21.0[a]	16.6[b]	21.9[a]	19.7
Minimum	7.8	4.4	7.0	
Maximum	33.6	24.9	32.4	

b) ^{14}C distribution	**Crop**					
	Wheat			**Barley**		
	Development stage			Development stage		
Compartment	Tillering	Ear emergence	Dough ripening	Tillering	Ear emergence	Dough ripening
			Percentage of Net ^{14}C Assimilation			
Shoots	45.0[b]	91.5[a]	91.6[a]	36.4[b]	87.3[a]	95.7[a]
Roots	19.9[a]	3.7[b]	2.4[b]	21.7[a]	4.5[b]	1.1[b]
Soil/root respiration	16.6[a]	3.3[b]	3.6[b]	35.7[a]	5.8[b]	2.3[b]
Soil residue	18.5[a]	1.5[b]	2.0[b]	6.3[a]	2.3[b]	1.0[b]
			Percentage of ^{14}C Soil/Root Respiration			
Root respiration (range)	42–90	84–98	78–96	91–99	88–98	75–96
Microbial respiration (range)	10–58	2–16	4–22	1– 9	2–12	4–25

*Within the same experimental factor means followed by different letters are significantly different (LSD-test, $p < 0.05$).

It was lower in the presence of plant roots compared to unplanted soil and higher in the loamy sand soil than in the silt loam soil. The largest increase in glucose respiration was induced by increasing the amount of glucose C injected. Further, decreasing the volume of water in which the glucose was dissolved increased the proportion of ^{14}C respired, which indicates that the concentration of C established in soil after injection determined the mineralization of the substrate. Finally, the C/N-ratio of the model rhizodeposit had no significant effect, nor had the interactions between parameters.

Discussion

Experimental conditions

Watering the plants every 3 days in experiment 1 was clearly insufficient to keep the soil moisture content constant in time. The use of expanded clay pellets in experiment 3 to provide water automatically, was only successful as long as plants were small, and resulted in a vertical gradient in soil moisture content. The death of the barley plants in experiment 1 was correlated with temporary high soil moisture contents. However, this seemed to have had little effect on the mineralization of model rhizodeposits, possibly because the water stress took place more than 20 days after the start of the incubation, when the added substrate was already fixed in more resistant microbial metabolites. The effect of water stress on the ^{14}C distribution was probably negligible, since the allocation of ^{14}C after pulse-labelling

Table 3. Effect of different parameters on the proportion of ^{14}C respired from glucose model rhizodeposits after 21 days of incubation in soil in a growth chamber (experiment 3). Values are means from a balanced factorial design with a total of 32 pots

Parameter	Treatment	Mean	Volume added	
			5 mL	1 mL
		Percentage ^{14}C Respired*		
Wheat plant present	yes	36.2**		
	no	39.8		
Soil type	silt loam	36.0***		
	loamy sand	39.9		
C added	1.5 mg	33.7**	32.0	35.4
	15 mg	42.3	40.7	43.8
Volume added	5 mL	36.3**		
	1 mL	39.6		
N added	0	38.2NS		
	C/N-ratio - 20	37.7		
mean		38.0		
minimum		29.1		
maximum		49.2		

*The difference between means of the 2 treatments of the same parameter is significant at: *, $p < 0.05$; **, $p < 0.01$; ***, $p < 0.001$; NS, not significant. Interactions between parameters were not significant.

is known to be completed within 3 weeks (Swinnen et al., 1994a) and most of the incorporation of ^{14}C into structural plant tissue occurs during the first few days (Gordon et al., 1977). This is supported by the ^{14}C soil/root respiration curve of pulse-labelled barley columns which showed no effect of the water stress (data not shown).

Since the development stage at labelling in experiment 1 was in between the first and the second labelling of experiment 2, the ^{14}C allocation to shoots and roots in both experiments is in good agreement. The dependency on the development stage in experiment 2 is also comparable to the one of spring wheat in a similar field experiment (Swinnen et al., 1994a). Winter wheat in experiment 2 had lower proportions of ^{14}C recovered in soil/root respiration and higher proportions of ^{14}C in soil residue at the tillering stage compared to barley and compared to experiment 1. A possible explanation is that a considerable amount of roots was already formed during the winter, which requires less investment in root growth in spring and can lead to greater root leakage. For the calculation of microbial ^{14}C respiration the amount of ^{14}C recovered in soil organic C after pulse-labelling is used (equation 1). Although the soil residue ^{14}C used in experiment 1 and 2, consists of soil organic and inorganic ^{14}C, the latter was shown to represent less than 3% of the soil residue ^{14}C (Swinnen et al., 1994a); thus this distinction is negligible. In experiment 1 root ^{14}C contents were not corrected for losses of ^{14}C with a size smaller than 0.25 mm. As estimated losses at the corresponding development stage in experiment 2 were ca. 5% (10%/2), with a more aggressive method to isolate roots and with a 0.3 mm sieve, the error in experiment 1 is considered small.

Model rhizodeposit representativity

The principle of the model rhizodeposition technique is based on the assumption that respiration of model rhizodeposits injected into soil is representative for respiration of actual ^{14}C rhizodeposits after pulse-labelling. Consequently, it is necessary that the injected model substrates resemble actual ^{14}C rhizodeposits in a number of ways. First, the degradability should be similar. The degradability of real rhizodeposits depends on their chemical composition, which is known to fluctuate with the age of the root (Rovira, 1973). Root tips mainly release soluble compounds and mucilages while older roots lose root hairs, epidermis and cortex cells, even within 3 weeks after root formation

(Deacon and Lewis, 1982; Holden, 1975). Moreover, the relative contribution of root tips and older roots to rhizodeposition probably varies with the development stage of the plants. To circumvent the uncertainty about the chemical composition, 3 different substrates covering a wide range of degradability were used as model rhizodeposits. In view of the origin of rhizodeposits it is likely that glucose and root cell wall material are appropriate extremes of a degradability range. The inevitable disadvantage of this solution is that the calculated components of ^{14}C soil/root respiration are expressed as ranges rather than as single values. As observed in experiments 1 and 2 the differences in respiration between glucose and root cell wall material decreased with the time after injection (Fig. 2). This was due to the lag-phase in root cell wall material respiration, which suggests that microbes needed an adaptation period to decompose this substrate. Microbes present in the vicinity of the substrate may first have to produce enzymes, e.g. cellulase, to decompose specific cell wall constituents, or specific groups of microorganisms may have to proliferate before a maximum decomposition rate is reached. Also the particle form of the substrate, which was not immediately accessible to microbes, may have resulted in a lag-phase in ^{14}C respiration.

A second condition to be met by model rhizodeposits is that the time of arrival in the soil is similar to that of actual ^{14}C rhizodeposits. It is known that most of the ^{14}C fixed by the shoots during pulse-labelling of different plant species is allocated in the plant-soil system within a few days (e.g. Swinnen et al., 1994a; Warembourg et al., 1982). Therefore, the release of C as exudates, secretions, mucilages and sloughed root cap cells is likely to occur within a few days after assimilation by the shoots (Swinnen et al., 1994b). Consequently, with respect to the timing of these types of rhizodeposition, the simultaneous application of model rhizodeposition and pulse-labelling is a reasonable approximation. However, ^{14}C rhizodeposition through decay of root hairs, epidermis and cortex cells, and whole roots, may occur during the entire allocation period of 3 weeks, which would plead for a delayed application of the root cell wall material model rhizodeposit. This was not done because the proportion respired over shorter incubation period-can be derived from the respiration curve (Fig. 2) and because simultaneous application has the advantage of a better comparison of the 3 different substrates. Using the proportion of root cell wall material respired over shorter incubation periods would result in a broader range of the estimated ^{14}C soil/root respiration components, but it is uncertain to which extent this may be done.

A third condition required for a good representativity of model rhizodeposits is that the place of arrival in the soil has rhizosphere-like conditions. It is very difficult practically to apply the substrates on the actual root surface without disturbing the system, especially under field conditions. The method of injecting the substrates into the soil while withdrawing a needle up to the soil surface is a very simple one and can be applied in the field. As the soil is rooted by a similar plant as in the pulse-labelled system, its water and nutrient status is identical. It was assumed that under these conditions the injected substrate itself would create a temporary model rhizosphere similarly to the pulse-event of a root growing through the soil (Clarholm, 1985).

This assumption implies that the established concentration of injected C in soil, is representative for the concentration of rhizodeposits in a real rhizosphere. The latter can be roughly estimated from reported rhizodeposition rates expressed per unit root weight. In a review Newman (1985) reported values for various plant species of 10–100 mg soluble and 20–50 mg insoluble dry matter in rhizodeposition g^{-1} root dry wt produced. Trofymow et al. (1987) measured 436 mg C in rhizodeposition g^{-1} root C produced by oat. Since these values are related to young sterile roots they do not include lysates. After continuous ^{14}C labelling of wheat and barley grown in non-sterile soil (Barber and Martin, 1976; Johnen and Sauerbeck, 1977; Merckx et al., 1985; Whipps, 1984; Whipps and Lynch, 1983) the amount of rhizodeposits recovered in soil was 30–910 mg C g^{-1} root C produced as calculated from data reviewed by Whipps (1990). It may be overestimated due to incomplete removal of roots from soil or underestimated because part of the rhizodeposits was already respired by microorganisms. As the age of the plants varied from 21 to 143 days, root lysates are included in the estimates. Assuming a similar C content per unit dry wt in rhizodeposits and roots, the reported range of rhizodeposition rates is 30–910 mg C g^{-1} root C produced. To convert this into actual concentrations (μg C g^{-1} soil) the following simplifications were made: rhizodeposits are released by the daily produced root segment behind the root tip (Trofymow et al., 1987); root dry wt contains 35% C; the specific root length for wheat is 200 m g^{-1} dry wt and the average root diameter is 0.2 mm (Van Noordwijk and Van De Geijn, 1994); a 2 mm layer of soil around the root is affected (Darrah, 1991; Martens,

1982) and the bulk density of the soil is 1.2 g cm^{-3}. The resulting range of C concentrations is 3–96 μg C g^{-1} dry soil.

The established concentration of added C in soil in the model rhizodeposition experiments was estimated based on the amount of soil that can be wetted to field capacity by the volume of solution injected. It is independent of the efficiency of injection and in the case of root cell wall material particles it is obviously meaningless. In experiment 1 the estimated concentration of added C, was ca 90 μg C g^{-1} dry soil. In experiment 2 the amount of C added was 3 times as high, but due to a higher soil density the resulting estimated concentration was only 90–160 μg C g^{-1} dry soil, varying with the development stage of the crop. These values are in agreement with the estimated concentration of real rhizodeposits. In experiment 3 the injection of different amounts of C with different volumes of solution resulted in a wide range of glucose-C concentrations: 84–4200 μg C g^{-1} dry silt loam soil and 24–1200 μg C g^{-1} dry loamy sand soil.

The positive correlation between glucose respiration and the added substrate concentration in experiment 3 was also found by Sparling et al. (1981) and Bremer and Van Kessel (1990). Using the same 2 soils as in experiment 3, Bremer and Kuikman (1994) observed that the mineralization of glucose-^{14}C after 3 days was ca 42% with concentrations of added glucose-C higher than 144 μg C g^{-1} dry soil. With lower concentrations the proportion respired declined steeply to ca 30% with 36 μg C g^{-1} dry soil. In experiment 3 the dependency of ^{14}C respiration on the glucose-C concentration seemed to follow a similar pattern (data not shown), except that the respiration was independent of the concentration above ca 800 μg C g^{-1} dry soil. This would indicate that the amount of soil affected by the solution injected, was underestimated by a factor 5. After division by this factor the concentrations applied in experiment 1 and 2 still fall within the range of real rhizodeposition rates found in literature. Therefore, it may be concluded that the model rhizodeposition method approaches rhizosphere conditions with respect to the degradability, the time of arrival and the concentration of substrates.

Substrate concentration effect

The proportions of glucose-^{14}C mineralized in experiment 1 and 2 are low compared to literature data, which can be explained by the dependency of ^{14}C respiration on substrate concentration. With 1000 μg glucose-C g^{-1} dry soil added, Martens (1985) found that 52–67% was respired over a 14 day period. Sparling and West (1988) reported values of 23–29% respiration in 2 days after addition of 4000 μg glucose-C g^{-1} soil. Bremer and Kuikman (1994) hypothesized that low amounts of substrate would not result in proliferation of C-starved cells and that assimilated glucose-^{14}C would be incompletely metabolized and stored among cytoplasmic constituents with little release of CO_2. In this case both the concentrations derived from literature and the ones applied in experiments 1 and 2 could not explain the abundance of microorganisms in the rhizosphere, which suggests that rhizodeposition may be much more localized than assumed above. Thus more work is necessary to determine substrate concentrations in the rhizosphere and its use by microorganisms.

The sensitivity of the separation of root and microbial respiration for the concentration of added C can be tested by using the proportion ^{14}C respired with the highest glucose concentration in planted silt loam soil in experiment 3 (39.6%) in the calculations of experiments 1 and 2. In experiment 1 obviously nothing changed, since the calculations were already done with a value of 39.2%. In experiment 2 the contribution of root respiration to ^{14}C soil/root respiration changed from at least 75% to at least 65%, except for winter wheat at tillering where it became very uncertain (25–90%) due to the high ratio between ^{14}C recovered in soil organic C and ^{14}C in soil/root respiration. This shows that the sensitivity of the calculations for the added C concentration is rather strongly correlated with this ratio and that the method is less useful with a ratio larger than unity.

The lower glucose respiration rate in experiment 2 compared to experiment 1, in spite of the higher substrate concentrations, was probably due to lower temperatures in the field than in the growth chamber. The negative effect of planting on ^{14}C respiration in experiment 3 was not related to C concentrations since the soil moisture content in planted columns at the time of injection was equal or even slightly lower than in unplanted columns, but it may be due to the decreasing soil moisture content in planted columns during the incubation. Despite the lower C concentrations in loamy sand compared to silt loam, ^{14}C respiration was higher. This corresponds to earlier observations of the effect of soil texture on substrate decomposition (Merckx et al., 1985). Soils with finer texture tend to be more 'preservative' than soils with coarser texture.

Comparison with other methods

Using the model rhizodeposition technique the estimated proportion of root respiration in ^{14}C soil/root respiration was mostly greater than 75%. This is clearly in contrast to results obtained from other methods, which showed a higher proportion of microbial respiration. However, unlike attempts by using nutrient solution culture with antibiotics (Helal and Sauerbeck, 1991; Minchin and McNaughton, 1984) or by comparing sterile versus non-sterile systems (Barber and Martin, 1976; Whipps and Lynch, 1983), the model rhizodeposition method is performed in normal soil conditions, and thus takes into account that root respiration may be influenced by physical resistance and by microorganisms. Attempts to discern 2 subsequent peaks in $^{14}CO_2$ production rate after pulse-labelling, one from root respiration and one from microbial respiration (Warembourg, 1975a, b; Warembourg and Billés, 1979) are difficult to apply in the field since they require a constant air flow through the soil column. The method proposed by Johansson (1992) calculates gross rhizodeposition in continuously ^{14}C-labelled plants from the amount of ^{14}C residue after a long-term incubation. This is divided by a stabilization factor, which is approximated through comparison with the humification of known organic materials. The obtained 32% microbial respiration is a cumulative value for a 7 week growth period, and therefore cannot be compared to our data from pulse-labelling. However, the technique could be adapted to pulse-labelling experiments and then a comparison with the model rhizodeposition technique would be interesting. Cheng et al. (1993) eliminated ^{14}C microbial respiration after pulse-labelling by saturating the soil solution with ^{12}C-glucose immediately before labelling. This method of isotopic trapping allowed to estimate root respiration (41%) and rhizomicrobial respiration (59%), as well as soluble C concentrations in the rhizosphere of 3 weeks old wheat plants. Its disadvantage, however, is that the allocation period is necessarily kept short (4 h), which means that the ^{14}C root respiration is incomplete. In conclusion, although further study should be done to determine which rate of C addition can establish conditions to create an artificial rhizosphere in a rooted soil column, for the present the model rhizodeposition method is a valuable approach to estimate the fraction of rhizodeposits respired by microorganisms. It causes minimal disturbance of rhizosphere interactions and due to its simplicity, it can be applied under field conditions.

Acknowledgements

The technical assistance of I M Van Veelen and H Kuipers is gratefully acknowledged. I also wish to thank J Koning, A Van Der Lijn, E Hummelink and J Winters for technical assistance on the Lovinkhoeve experimental farm, A W H C F M Wolfs, C Van De Peppel, E Van Der Kruk and J W Dalenberg for supervision on radioactive work, S L G E Burgers for helping with statistical analysis, and E Bremer, P J Kuikman, R Merckx and J A Van Veen for useful discussions.

References

Amato M 1983 Determination of carbon ^{12}C and ^{14}C in plant and soil. Soil Biol. Biochem. 15, 611–612.

Anderson J P E 1982 Soil Respiration. *In* Methods of Soil Analysis, Part 2, Chemical and Microbiological Properties, Second Edition. Eds. A L Page, R H Miller and D R Keeney. pp 852–871. Am. Soc. Agron. Inc., Soil Sci. Soc. Am. Inc., Madison, Wisconsin, USA.

Barber D A and Martin J K 1976 The release of organic substances by cereal roots into soil. New Phytol. 76, 69–80.

Bremer E and Kuikman P J 1994 Microbial utilization of ^{14}C[U]glucose in soil over a wide range of glucose addition rates. Soil Biol. Biochem. 26, 511–517.

Bremer E and Van Kessel C 1990 Extractability of microbial ^{14}C and ^{15}N following addition of variable rates of labelled glucose and $(NH_4)_2SO_4$ to soil. Soil Biol. Biochem. 22, 707–713.

Cheng W, Coleman D C, Carroll C R and Hoffman C A 1993 *In situ* measurement of root respiration and soluble C concentrations in the rhizosphere. Soil Biol. Biochem. 25, 1189–1196.

Clarholm M 1985 Possible roles for roots, bacteria, protozoa and fungi in supplying nitrogen to plants. *In* Ecological Interactions in Soil. Plants, Microbes an Animals. Ed. A H Fitter. pp 355–365. Blackwell Scientific Publications, Oxford, UK.

Darrah P R 1991 Models of the rhizosphere I. Microbial population dynamics around a root releasing soluble and insoluble carbon. Plant and Soil 133, 187–199.

Deacon J W and Lewis S J 1982 Natural senescence of the root cortex of spring wheat in relation to susceptibility to common root rot (*Cochliobolus sativus*) and growth of a free-living nitrogen-fixing bacterium. Plant and Soil 66, 13–20.

Fao-Unesco 1981 Soil Map of the World, Volume V, Europe. Unesco, Paris, France. 199 p.

Gordon A J, Ryle G J A and Powell C E 1977 The strategy of carbon utilization in uniculm barley I. The chemical fate of photosynthetically assimilated ^{14}C. J. Exp. Bot. 28, 1258–1269.

Helal H M and Sauerbeck D 1991 Short-term determination of the actual respiration rate of intact plant roots. *In* Plant Roots and Their Environment. Eds. B L McMichael and H Persson. pp 88–92. Elsevier, Amsterdam, The Netherlands.

Holden J 1975 Use of nuclear staining to assess rates of cell death in cortices of cereal roots. Soil Biol. Biochem. 7, 333–334.

Johansson G 1992 Release of organic C from growing roots of meadow fescue (*Festuca pratensis* L.). Soil Biol. Biochem. 24, 427–433.

Johnen B G and Sauerbeck D R 1977 A tracer technique for measuring growth, mass and microbial breakdown of plant roots during vegetation. *In* Soil Organisms as Components of Ecosystems. Eds. V Lohm and T Persson. pp 366–373. Ecological Bulletins 25, Stockholm, Sweden.

Kooistra M J, Lebbink G and Brussaard L 1989 The Dutch programme on soil ecology of arable farming systems. 2. Geogenesis, agricultural history, field site characteristics and present farming systems at the Lovinkhoeve experimental farm. Agric. Ecosystems Environ. 27, 361–387.

Lekkerkerk L J A, Van De Geijn S C and Van Veen J A 1990 Effects of elevated atmospheric CO_2-levels on the carbon economy of a soil planted with wheat. *In* Soils and the Greenhouse Effect. Ed. A F Bouwman. pp 423–429. Wiley and Sons, Chichester, UK.

Liljeroth E, Van Veen J A and Miller H J 1990 Assimilate translocation to the rhizosphere of two wheat cultivars and subsequent utilization by microorganisms at two soil nitrogen levels. Soil Biol. Biochem. 22, 1015–1021.

Lynch J M and Whipps J M 1990 Substrate flow in the rhizosphere. Plant and Soil 129, 1–10.

Martens R 1982 Apparatus to study the quantitative relationships between root exudates and microbial populations in the rhizosphere. Soil Biol. Biochem. 14, 315–317.

Martens R 1985 Limitations in the application of the fumigation technique for biomass estimations in amended soils. Soil Biol. Biochem. 17, 57–63.

Meharg A A and Killham K 1990 Carbon distribution within the plant and rhizosphere in laboratory and field-grown *Lolium perenne* at different stages of development. Soil Biol. Biochem. 22, 471–477.

Merckx R, Den Hartog A and Van Veen J A 1985 Turnover of root-derived material and related microbial biomass formation in soils of different texture. Soil Biol. Biochem., 17, 565–569.

Merckx R, Dijkstra A, Den Hartog A and Van Veen J A 1987 Production of root-derived material and associated microbial growth in soil at different nutrient levels. Biol. Fertil. Soils 5, 126–132.

Minchin P E H and McNaughton G S 1984 Exudation of recently fixed carbon by non-sterile roots. J. Exp. Bot. 35, 74–82.

Newman E I 1985 The rhizosphere: carbon sources and microbial populations. *In* Ecological Interactions in Soil. Plants, Microbes and Animals. Ed. A H Fitter. pp 107–121. Blackwell Scientific Publications, Oxford, UK.

Rovira A D 1973 Zones of exudation along plant roots and spatial distribution of micro-organisms in the rhizosphere. Pestic. Sci. 4, 361–366.

Sparling G P and West A W 1988 A direct extraction method to estimate soil microbial C: calibration *in situ* using microbial respiration and ^{14}C labelled cells. Soil Biol. Biochem. 20, 337–343.

Sparling G P, Ord B G and Vaughan D 1981 Microbial biomass and activity in soils amended with glucose. Soil Biol. Biochem. 13, 99–104.

Swinnen J 1994 Rhizodeposition and turnover of root-derived organic material in barley and wheat under conventional and integrated management. Agric. Ecosystems Environ. (*In press*).

Swinnen J and Van Veen J A 1993 Unterscheidung von Wurzel- und microbieller Atmung im Boden durch Exudatmarkierung. *In* Bodennutzung und Bodenfruchtbarkeit, Vol. 4. Humushaushalt. Berichte über Landwirtschaft 206. Sonderh. pp 114–116. Parey, Hamburg, Berlin.

Swinnen J, Van Veen J A and Merckx R 1994a ^{14}C pulse-labelling of field-grown spring wheat: an evaluation of its use in rhizosphere carbon budget estimations. Soil Biol. Biochem. 26, 161–170.

Swinnen J, Van Veen J A and Merckx R 1994b Rhizosphere carbon fluxes in field-grown spring wheat: model calculations based on ^{14}C partitioning after pulse-labelling. Soil Biol. Biochem. 26, 171–182.

Trofymow J A, Coleman D C and Cambardella C 1987 Rates of rhizodeposition and ammonium depletion in the rhizosphere of axenic oat roots. Plant and Soil 97, 333–344.

Van Noordwijk M and Van De Geijn S C 1994 Root, shoot and soil parameters required for process-oriented models of crop growth limited by water or nutrients. *In* New Methods of Root Research. Ed. A Smucker. A.S.A. Special Publication. (*In press*).

Van Veen J A, Merckx R and Van De Geijn S C 1989 Plant- and soil related controls of the flow of carbon from roots through the soil microbial biomass. Plant and Soil 116, 167–175.

Warembourg F R 1975a Application de techniques radioisotopiques à l'étude de l'activité biologique dans la rhizosphère des plantes. Rev. Écol. Biol. Sol 12, 261–272.

Warembourg F R 1975b Le dégagement de CO_2 dans la rhizosphère des plantes. Soc. Bot. Fr., Coll. Rhizosphère, 77–87.

Warembourg F R and Billes G 1979 Estimating carbon transfers in the plant rhizosphere. *In* The Soil-Root Interface. Eds. J L Harley and R S Russel. pp 183–196. Academic Press, London, UK.

Warembourg F R, Montagne D and Bardin R 1982 The simultaneous use of $^{14}CO_2$ and $^{15}N_2$ labelling techniques to study the carbon and nitrogen economy of legumes grown under natural conditions. Physiol. Plant. 56, 46–55.

Whipps J M 1984 Environmental factors affecting the loss of carbon from the roots of wheat and barley seedlings. J. Exp. Bot 35, 767–773.

Whipps J M 1990 Carbon economy. *In* The Rhizosphere. Ed. J M Lynch. pp 59–97. Wiley and Sons, Chichester, UK.

Whipps J M and Lynch J M 1983 Substrate flow and utilization in the rhizosphere of cereals. New Phytol. 95, 605–623.

Whipps J M and Lynch J M 1985 Energy losses by the plant in rhizodeposition. Ann. Proc. Phytochem. Soc. Eur. 26, 59–71.

Zadoks J C, Chang T T and Konzak C F 1974 A decimal code for the growth stages of cereals. Weed Res. 14, 415–421.

ECOSYSTEM PROCESSES

Plant and Soil **165**: 105–113, 1994.

Biomass production in a nitrogen-fertilized, tallgrass prairie ecosystem exposed to ambient and elevated levels of CO_2

Clenton E. Owensby[1], Lisa M. Auen[1] and Patrick I. Coyne[2]
[1]*Department of Agronomy, Kansas State University, Manhattan, KS 66506-5501, USA and* [2]*Ft Hays Branch Experiment Station, Kansas State University, Hays, KS 67601, USA*

Key words: aboveground biomass, carbon dioxide, elevated CO_2, N limitation, nitrogen fertilization, root biomass, tallgrass prairie, water-use efficiency

Abstract

Increased biomass production in terrestrial ecosystems with elevated atmospheric CO_2 may be constrained by nutrient limitations as a result of increased requirement or reduced availability caused by reduced turnover rates of nutrients. To determine the short-term impact of nitrogen (N) fertilization on plant biomass production under elevated CO_2, we compared the response of N-fertilized tallgrass prairie at ambient and twice-ambient CO_2 levels over a 2-year period. Native tallgrass prairie plots (4.5 m diameter) were exposed continuously (24 h) to ambient and twice-ambient CO_2 from 1 April to 26 October. We compared our results to an unfertilized companion experiment on the same research site. Above- and belowground biomass production and leaf area of fertilized plots were greater with elevated than ambient CO_2 in both years. The increase in biomass at high CO_2 occurred mainly aboveground in 1991, a dry year, and belowground in 1990, a wet year. Nitrogen concentration was lower in plants exposed to elevated CO_2, but total standing crop N was greater at high CO_2. Increased root biomass under elevated CO_2 apparently increased N uptake. The biomass production response to elevated CO_2 was much greater on N-fertilized than unfertilized prairie, particularly in the dry year. We conclude that biomass production response to elevated CO_2 was suppressed by N limitation in years with below-normal precipitation. Reduced N concentration in above- and belowground biomass could slow microbial degradation of soil organic matter and surface litter, thereby exacerbating N limitation in the long term.

Introduction

Atmospheric carbon dioxide concentration is increasing (Boden et al., 1990) and is expected to double by the middle of the next century. Ecosystem-level experiments in natural terrestrial systems are limited, and Mooney et al. (1991) emphasized the need for additional research on terrestrial ecosystem response to elevated CO_2. Responses of individual plants and assemblages of plants to elevated CO_2 in controlled-environment studies have been summarized by various authors (Bazzaz, 1990; Newton, 1991; Strain and Cure, 1985). Productivity responses to CO_2 enrichment at the single plant level in controlled environments usually have been dependent on photosynthetic pathway. Carbon fixation rates in plants with the C_3 pathway generally show a greater response to increasing CO_2 levels than rates in C_4 plants (Kimball, 1983; Nijs et al., 1988; Reichers and Strain, 1988; Wray and Strain, 1986).

The response of natural ecosystems to CO_2 enrichment has been researched for estuarine saltmarsh communities (Curtis et al., 1989a), an Arctic tundra tussock sedge ecosystem (Oechel and Strain, 1985), and a tallgrass prairie ecosystem (Owensby et al., 1993a,b). Curtis et al. (1989a) concluded that communities dominated by *Scirpus olneyi* (C_3) had greater productivity, and that senescence was delayed at high compared to ambient CO_2. Production in *Spartina patens* (C_4) communities was not increased with CO_2 enrichment. Oechel and Strain (1985) reported that an Arctic tundra tussock sedge ecosystem initially responded to CO_2 enrichment with increased productivity, but the increase disappeared within the first year. They reasoned that an acclimation response had occurred in the photosynthetic mechanism. Owensby et al. (1993a) concluded that tallgrass prairie productivity was enhanced with twice-ambient CO_2 primarily through increased water-use efficiency of the C_4 perennial grass dominants.

Productivity in most temperate terrestrial ecosystems usually is limited by moisture and/or nutrient availability. Increased water-use efficiency under elevated CO_2 has been documented, but effects of supplemental N in natural systems have not been studied. Because essentially all nutrients are cycled within the ecosystem and nutrient supplies in natural systems are relatively constant, increased productivity indicates an increased nutrient-use efficiency. Curtis et al. (1989a) and Owensby et al. (1993a) reported increased productivity for natural plant communities under elevated CO_2 with the same nutrient resources as communities with ambient CO_2 after 4 and 3 years, respectively. In those natural ecosystems, CO_2 enrichment reduced N concentration for both C_3 and C_4 species, regardless of biomass production response (Curtis et al., 1989b; Owensby et al., 1993b). However, nutrient limitations may negate any long-term increase in productivity in elevated-CO_2 environments.

We assessed effects of ambient and elevated (double ambient) atmospheric CO_2 concentrations on above- and belowground biomass production, leaf area, and N concentration of above- and belowground biomass in a N-fertilized, tallgrass prairie. Effects of CO_2 enrichment on fertilized and unfertilized prairie were compared to test the prediction that increased biomass production was limited by N on CO_2-enriched tallgrass prairie.

Materials and methods

Study site

The experimental site was located in pristine tallgrass prairie north of Manhattan, KS, USA (39.12°N, 96.35°W, 324 m above mean sea level). Vegetation on the site was a mixture of C_3 and C_4 species, dominated by the C_4 grasses, *Andropogon gerardii* Vitman and *Sorghastrum nutans* (L.) Nash. Subdominants included *Poa pratensis* L. (C_3), *Bouteloua curtipendula* (Michx.) Torr. (C_4), and *Sporobolus asper* var. *asper* (Michx.) Kunth (C_4). Members of the sedge family (C_3) made up 5–10% of the composition. Principal forbs (all C_3) included *Vernonia baldwinii* var. *interior* (Small) Schub., *Ambrosia psilostachya* DC., *Artemesia ludoviciana* Nutt., and *Psoralea tenuiflora* var. *floribunda* (Nutt.) Rydb.. Average peak aboveground biomass (dry wt.) of 425 g m^{-2} occurs in early August, of which 35 g m^{-2} is from forbs (Owensby and Anderson, 1967). Soils in the area are transitional from Ustolls to Udolls (Tully series: fine, mixed, mesic, montmorillonitic, Pachic Argiustolls). Slope on the area is 5%. Fire has occurred 2–3 times in 10 years. Past history has included primarily winter grazing by cow-calf pairs. The 30-year average annual precipitation is 840 mm, with 520 mm occurring during the growing season.

Fumigation chambers were placed over the natural vegetation in late March, 1990 and retained on the same area for 2 years. Twice-replicated treatments consisted of ambient CO_2 plus N (**AN**), chamber plus ambient CO_2 plus N (**CAN**), and chamber plus CO_2-enriched plus N (**CEN**). CO_2-enriched treatments received approximately twice the ambient concentration. N was applied as ammonium nitrate at 56 kg ha^{-1} in late March of both years. Data from an unfertilized companion experiment (Owensby et al., 1993a) with the same CO_2 treatments (**A, CA, CE**) were used to compare the interaction of N with plant response to CO_2.

Fumigation chambers

Each open-top chamber (4.5 m in diameter by 3.25 m in height) had a cone-top baffle that reduced the top opening to 3 m. The baffle added 0.75 m to the height Of the chamber for a total height of 4 m. An aluminum structural framework was covered by 0.15 mm thick, UV-resistant, polyethylene film. The cone-top baffle reduced the opening by 54%, thereby restricting the precipitation that entered the chamber. Within 24 hours following each rainfall event, water equal to 54% of the rainfall amount for an unchambered plot was added using a rotating sprinkler adjusted to cover the diameter of the chamber. Aluminum edging was placed around the upslope bottom edge of the chamber to prevent runoff from entering the chamber. No edging was placed on the lower half of the chamber. One half of each plot was used to estimate biomass production and nutrient concentrations, and the remaining half was grazed by esophageally fistulated sheep to determine forage quality differences among treatments. Data on forage quality are not reported here.

CO_2 treatment

Mass flow controllers, interfaced to a computer, were used to regulate CO_2 flow rate into the enriched chambers based on real-time measurement of CO_2 in the chambers. CO_2-enrichment began on 1 April in both 1990 and 1991. Carbon dioxide enrichment and environmental data acquisition were continuous until late October of each year. The polyethylene film covering the chambers was removed in late October and replaced in late March. During periods of high photosynthetic activity, CO_2 concentrations in the nonenriched treatments (**A, CA, AN, CAN**) measured at 1 m above the soil surface were 330–340 $\mu L\ L^{-1}$, but during nighttime hours, CO_2 levels reached 400+ $\mu L\ L^{-1}$. At the beginning of the sampling period for each chamber or unchambered plot, a delay of 20 seconds allowed for the IRGA and sample lines to be purged of previous gasses. Ten readings were then taken for all measured parameters and discarded. A paired-t comparison was made on the next 10 readings of CO_2 concentration and the mean of those readings was accepted if the data fell within a 5% confidence interval. Otherwise, readings in sets of 10 were tested until they did not differ. CO_2 concentration was determined using an infra-red CO_2 analyser (LCA-2, Analytic Development Co., Hoddesdon, UK.) which was calibrated initially and following three samplings of the 6 plots using a high resolution CO_2 zero gas (300 $\mu L\ L^{-1}$) and span gas (800 $\mu L\ L^{-1}$) (Scott Specialty Gases, Inc., Troy, MI, USA - $\pm$ 1% accuracy). Following each 6-plot sampling period, a baseline gas was sampled and used to correct for instrument drift during the previous sampling periods. The coefficient of variation for the CO_2 measurement was always $\leqq 0.2\%$. Each plot was sampled once per hour.

Chamber environment

Air (T_a) and soil (T_s) temperatures, air dew point temperature (T_{dp}), and photosynthetic photon flux density (PPFD) were measured an average of once per hour during each day of the growing season. Soil water content was measured weekly using a CPN - 503 DR Hydroprobe (Campbell Pacific Nuclear Corp., Martinez, CA 94553). PPFD was reduced by approximately 11% inside the chambers as determined by quantum sensors (LI-COR, Lincoln, NE, USA; Model L1-190SB) mounted 1 m above the soil surface within and outside a chamber. Soil temperature was measured at -10 cm, and air temperature at 30 cm, 100 cm, and 300 cm from the soil surface. No difference ($p < 0.10$) occurred in soil temperature (-10 cm) between chambered and unchambered plots. Likewise, the temperatures at 30 cm were similar in all plots. At 100 cm, the temperature inside chambers was slightly more than 2°C higher than that in unchambered plots on the hottest days for the period 1000 h to 1500 hr CST ($p < 0.10$). Also on the hottest days, the maximum temperature at 300 cm was some 5°C higher in chambered than in unchambered plots from 0800 hr to 1500 hr CST ($p < 0.10$). The air delivery system for chambered plots kept temperatures at plant canopy height approximately equal to ambient conditions. T_{dp} averaged 1°C higher in chambered plots than in unchambered plots from 1200 hr to 1500 hr CST ($p < 0.10$). Even though that difference was slight, higher humidities inside the chambers may have reduced evapotranspiration and indirectly affected soil water content (Owensby et al. 1993a). Soil water content, measured at 20, 30, 40, 50, 60, 80, 100, 120, 140, 160, 180, and 200 cm below the soil surface, was significantly higher in chambered plots than in unchambered plots from mid June to late August in both years ($p < 0.10$). **CEN** plots had higher soil moisture levels than **CAN** plots under drought conditions during the sampling period ($p < 0.10$).

Meteorological conditions

Precipitation was slightly below normal in 1990 except in August, and temperatures averaged slightly above normal during the growing season (Table 1). Precipitation in 1991 was much below normal during June through October, and temperatures were above normal.

Aboveground biomass sampling

In 1990, sampling began on 14 May and continued at 2-week intervals until 23 July and then at 4-week intervals until 15 October. Samples were clipped to ground level from two, 0.2 × 0.5 m plots randomly located in the ungrazed half of each plot. In 1991, peak live biomass was estimated by clipping two, 0.25 m^2 plots randomly located in the ungrazed half of each plot on 8 August. Peak biomass in tallgrass prairie normally occurs in early August. All biomass sample plots that had been previously clipped were excluded from the randomization for the following sample dates. Clipped samples were placed in an ice chest and transported to the laboratory where they were refrigerated until separation into the following components: *A. gerardii*, other C_4 grasses, *P. pratensis*, other C_3 grasses, grass-like plants (including *Carex* and *Cyperus*) and forbs. Immediately after separation of the samples, leaf area was estimated for each species or species group using a leaf area meter (LI-COR, Lincoln, NE, USA; Mod-

Table 1. Monthly precipitation (mm) and average daily maximum temperatures (C) and deviation from normal for the study site

Month	1990 Ppt. (mm)	1990 Dev. (mm)	1990 Temp (C)	1990 Dev. (C)	1991 Ppt. (mm)	1991 Dev. (mm)	1991 Temp (C)	1991 Dev. (C)
Jan	27	6	11.3	8.2	34	13	2.4	-0.7
Feb	22	-2	10.4	3.5	1	-23	14.3	7.4
Mar	105	52	15.5	3	35	-18	18.2	5.7
Apr	23	-48	19.3	-0.6	107	36	21.1	1.1
May	100	-14	22.8	-2.4	130	16	26.9	1.7
Jun	124	-10	32.1	1.9	51	-83	31.8	1.7
Jul	79	-22	33.2	0.1	47	-54	35.5	2.3
Aug	180	100	31.7	-0.8	56	-25	33.7	1.3
Sep	20	-83	30.7	3.1	44	-59	28.9	1.3
Oct	27	-46	23.1	1.4	33	-40	23.6	1.8
Nov	52	14	17.9	5.3	83	46	8.7	-3.9
Dec	26	3	5.4	-0.8	48	25	8.1	1.9
Total	785	-50			669	-166		

el LI-3100). After leaf area determination, samples were dried in a forced-air oven for 72 hr at 55°C, and weighed directly from the oven. In 1991, 20 tillers of *A. gerardii* and *P. pratensis* were collected and dried as above starting on 15 May and continuing at 2-week intervals until 24 July and then at 4-week intervals until 16 October for determination of N concentration.

Belowground biomass sampling

An estimate of relative belowground biomass production was obtained using buried root ingrowth bags (Lund et al., 1970). In early June 1990, four, 5-cm diameter soil cores were removed to a depth of 15 cm from the center of each plot. Fine-mesh nylon bags, filled with a mixture of fine and coarse sand to a volume equal to the soil core, were placed in the core holes. Eight bags were placed in new holes in each plot in late March, 1991. Root ingrowth bags were removed from the soil in early November of each year. Roots that had grown into the bags were removed, dried for 72 hr at 55°C, and weighed.

Chemical determinations

Root and shoot tissues from the biomass sampling were ground to pass a screen with 1 mm diameter holes and digested with sulfuric acid/hydrogen peroxide solution (Linder and Harley, 1942). N concentrations were colorimetrically determined (Technicon, 1977). Average total standing-crop N was determined by multiplying N concentration by standing biomass.

Data analysis

Data for each year were analyzed separately using ANOVA (SAS 6.06.01, SAS Inst., Cary, NC, USA) as a randomized complete block design. The model included replication and CO_2 treatment, and, when applicable, date and the date × CO_2 treatment interaction were included. Relative values (%) were analyzed following an arc sine transformation. Statistical significance for the F test was at $p < 0.10$. Means were separated using Duncan's Multiple Range Test ($p < 0.10$). Statistical analysis indicated no sampling date × CO_2 treatment interaction for aboveground biomass in 1990 and N concentration in 1990 and 1991; therefore, seasonal means are reported.

Results

Aboveground biomass and leaf area

In 1990, *A. gerardii* biomass (bm) and leaf area (la) averaged over all clipping dates were greater for **CEN** plots than for **CAN** and **AN** plots (bm, $p = 0.0001$; la, $p = 0.0001$) (Fig. 1). *P. pratensis* biomass and leaf

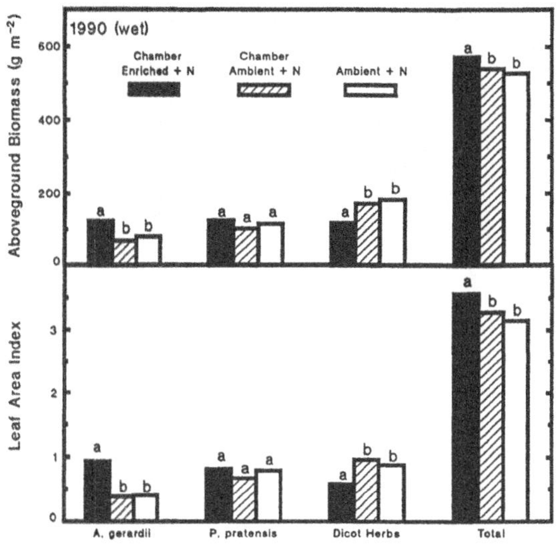

Fig. 1. Mean aboveground biomass (g m^{-2}) and leaf area index for indicated species and species groups averaged over nine growing season sampling dates in 1990 for native tallgrass prairie exposed to twice-ambient and ambient CO_2 concentrations and with 56 kg ha^{-1} added as ammonium nitrate. Means within species or species groups with a common letter do not differ [Duncan's Multiple Range Test, $p < 0.10$].

area averaged over all clipping dates did not differ among treatments (bm, $p = 0.3630$; la, $p = 0.1186$). Forb biomass and leaf area averaged over all clipping dates were significantly greater in **CAN** and **A** plots than in **CEN** plots (bm, $p = 0.1830$; la, $p = 0.1987$). Total biomass and leaf area for all species groups combined averaged over all clipping dates were significantly greater in **CEN** plots than in **CAN** and **A** plots (bm, $p = 0.0011$; la, $p = 0.0001$).

In 1991, biomass and leaf area were sampled only in early August. Biomass and leaf area of *A. gerardii* (bm, $p = 0.0717$; la, $p = 0.0130$) and all species combined (bm, $p = 0.0721$; la, $p = 0.0010$) were greater in **CEN** plots than in **CAN** and **AN** plots (Fig. 2). *P. pratensis* biomass and leaf area did not differ among the treatments (bm, $p = 0.2128$; la, $p = 0.2330$). *P. pratensis* peak biomass and leaf area occurred in early June, so the August sampling date does not reflect an estimate of peak biomass, but should indicate relative treatment responses. Forb biomass was greater in chambered than unchambered plots (bm, $p = 0.7398$; la, $p = 0.6138$).

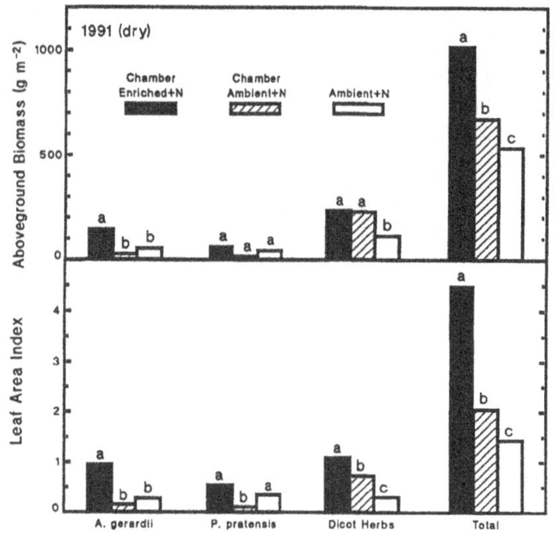

Fig. 2. Mean aboveground biomass (g m^{-2}) and leaf area index for indicated species and species groups in early August, 1991 for native tallgrass prairie exposed to twice-ambient and ambient CO_2 concentrations and with 56 kg ha^{-1} added as ammonium nitrate. Means within species or species groups with a common letter do not differ [Duncan's Multiple Range Test, $p < 0.10$].

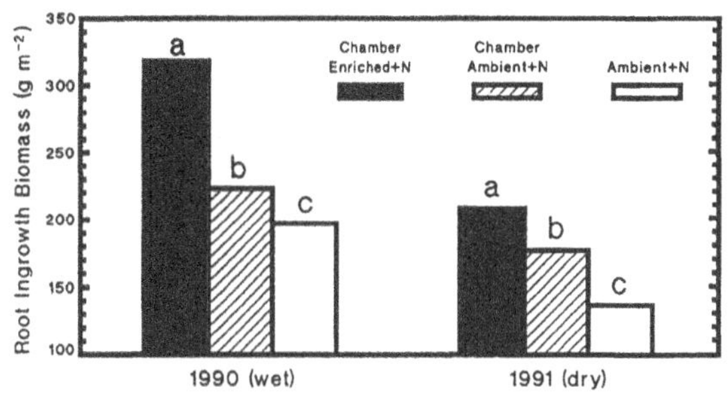

Fig. 3. Root biomass in ingrowth bags to a 15-cm depth in tallgrass prairie exposed to twice ambient and ambient CO_2 concentrations and with 56 kg ha^{-1} added as ammonium nitrate. Data are means of four bags per plot in 1990 and eight bags per plot in 1991 replicated three times. Means with a common letter within year do not differ [Duncan's Multiple Range Test, $p < 0.10$].

Root ingrowth biomass

In 1990, root ingrowth biomass, measured in nylon bags, was essentially doubled in **CEN** plots compared to **CAN** and **AN** plots ($p = 0.0001$) (Fig. 3). In 1991, **CEN** plots also had greater root ingrowth biomass than **CAN** plots, which had greater biomass than **AN** plots ($p = 0.0536$).

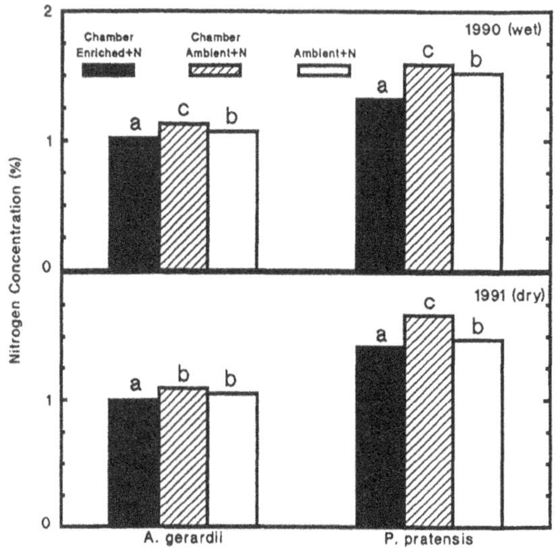

Fig. 4. Nitrogen concentration (%) for aboveground biomass of *Andropogon gerardii* and *Poa pratensis* averaged over nine sampling dates in 1990 and 1991 for native tallgrass prairie exposed to twice-ambient and ambient CO_2 concentrations with 56 kg ha^{-1} N added as ammonium nitrate. Means within species or species groups with a common letter do not differ [Duncan's Multiple Range Test, $p < 0.10$].

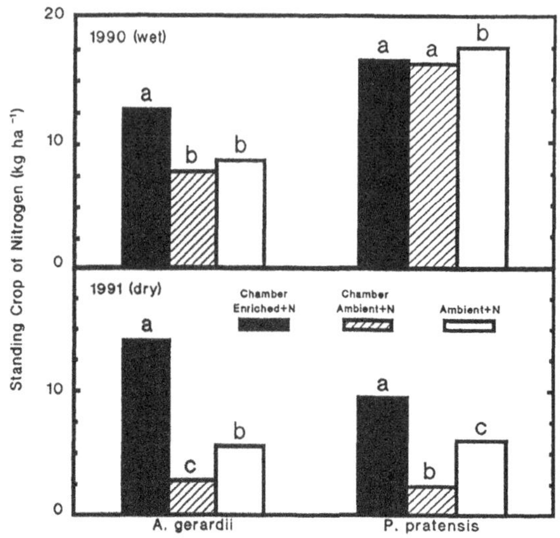

Fig. 5. Standing crop total N (kg ha^{-1}) of *Andropogon gerardii* and *Poa pratensis* averaged over nine sampling dates in 1990 and 1991 for native tallgrass prairie exposed to twice-ambient and ambient CO_2 concentrations with 56 kg ha^{-1} N added as ammonium nitrate. Means within species or species groups with a common letter do not differ [Duncan's Multiple Range Test, $p < 0.10$].

N concentration and standing crop N

The N concentration in aboveground tissues of *A. gerardii* and *P. pratensis*, averaged over nine sampling dates, was always lower in **CEN** plots than in **CAN** and **AN** plots in both 1990 and 1991 ($p = 0.0001$) (Fig. 4). In 1990, *P. pratensis* and *A. gerardii* biomass both had higher N concentrations in **CAN** plots than **AN** plots; in 1991, *P. pratensis* biomass showed the same response, but *A. gerardii* had similar N concentrations in **CAN** and **AN** plots ($p = 0.0001$). Average standing crop of N of *A. gerardii* was greater in **CEN** plots than in **CAN** and **AN** plots in both 1990 and 1991 ($p = 0.0001$) (Fig. 5). Average N standing crop in *A. gerardii* was similar in **CAN** and **AN** plots in 1990, but was greater in **AN** than **CAN** plots in 1991 (0.0001). Average standing crop N of *P. pratensis* during 1990 did not differ in CEN and CAN plots, but was higher in **AN** plots ($p = 0.0001$). However, in 1991, average N standing crop of *P. pratensis* was greater in **CEN** plots than in **CAN** and **AN** plots, with **AN** values being greater than **CAN** values ($p = 0.0001$).

Comparison of unfertilized and fertilized response to elevated CO_2

In order to determine whether N availability limited the response of tallgrass prairie to elevated CO_2, we compared relative biomass production response of N-fertilized plots to those from unfertilized plots in a study conducted concurrently on the same site (Owensby et al., 1993a) (Fig. 6). These data show that aboveground biomass production under elevated CO_2 was limited by N availability, particularly in 1991, a dry year. During 1990, the increased aboveground biomass production in CO_2-enriched plots compared to unchambered ambient plots was primarily from the C_4 perennial grasses. Owensby et al. (1993a) showed that the fumigation chamber affected water relations in a manner similar to that of elevated CO_2, and that the effects were not additive. Therefore, we compared the unchambered ambient plots to the CO_2-enriched plots. When fertilized, total biomass production was 24% greater on elevated CO_2 plots than ambient and 16% greater on unfertilized. The primary increase came from *A. gerardii*. However, in 1991 the increase in biomass production from elevated CO_2 with N fertilizer was 90% and 33% on the the unfertilized plots. *A. gerardii* biomass production in 1991 under elevated CO_2 compared to ambient was apparently greatly lim-

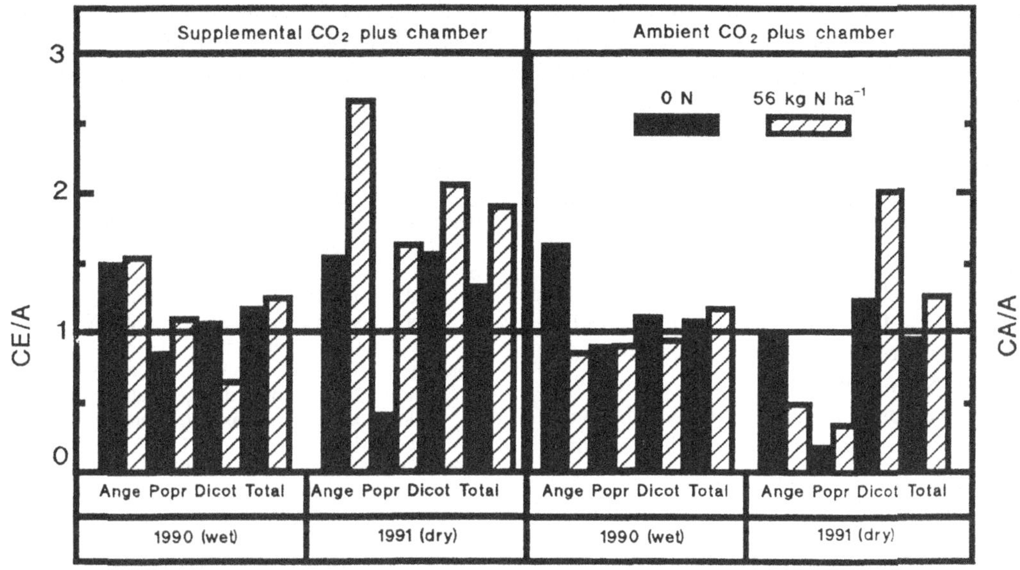

Fig. 6. Relative stimulation of aboveground biomass production by twice-ambient carbon dioxide concentrations compared to unchambered ambient plots with (treatment ratio, CEN/A) and without (CE/A) supplemental nitrogen (left) and by the open-top chamber effect with (CAN/A) and without (CA/A) supplemental nitrogen (right). Treatment abbreviations are defined in the text. Ange= *A. gerardii*, Popr= *P. pratensis*, Dicot= dicot herbs. Unfertilized data from Owensby et al. (1993a).

ited by N availability (a 166% increase in biomass on fertilized plots and a 19% increase on unfertilized). Aboveground biomass production enhancement by supplemental CO_2 in *P. pratensis*, and dicot herbs was amplified greatly by N fertilizer in 1991.

Discussion and conclusions

The primary purpose of this study was to test the hypothesis that the increased biomass production in a tallgrass prairie under elevated CO_2 would be limited by N availability, in both the short and long term. Short-term limitations of seasonal biomass production under elevated CO_2 would be associated with the inherent low availability of N in tallgrass prairie. This hypothesis assumes that other resources are not limiting. In the tallgrass prairie, both water and N have been shown to limit aboveground biomass production (Owensby et al., 1969). In a previous study without supplemental N, Owensby et al. (1993a,b) concluded that the primary impact of elevated CO_2 on biomass production was mediated through improved water-use efficiency. In addition, Knapp et al. (1993) reported that the photosynthetic capacity under optimal conditions of *A. gerardii* was greater in a dry year under elevate CO_2. Therefore, we predict that increased water-use efficiency and increased photosynthetic capacity of the dominant C_4 perennial grasses during years of suboptimal precipitation in a CO_2-enriched tallgrass prairie will result in increased above- and belowground production. Owensby et al. (1993a,b) also concluded that increased N uptake occurred because of increased root exploration of the soil mass and that N-use efficiency increased because the N requirement of the plant was reduced. Numerous studies have shown reductions in N concentration for C_3 and C_4 species under elevated CO_2 across a wide range of nitrogen availabilities (Coleman et al., 1991; Hoching and Meyer, 1985; Larigauderie et al., 1988).

If the response of aboveground biomass production to elevated CO_2 results primarily from increased water use efficiency (Owensby et al., 1993a), the N availability may more greatly limit the response of tallgrass prairie to elevated CO_2 in years when higher WUE is expressed as an increase in biomass production. This expectation is supported by data on aboveground biomass production in this study and by similar results for unfertilized prairie. Hunt et al. (1991) modeled ecosystem function with elevated CO_2 (2 × ambient) and their results predicted persistent increase in primary production in spite of what they concluded was N limitation, but nutrient cycles in ecosystems exposed to elevated CO_2 are essentially unstudied. However, root ingrowth biomass production was greater in 1990, a wet year, than in dry 1991. Since roots do not grow into dry soil, and the root ingrowth bags extended to only 15 cm, this result is not unexpected. Root growth may have been shifted to a greater soil depth in the dry year. Bazzazz (1990) and Newton (1991) reviewed the research dealing with natural ecosystems and concluded that generally root growth increased proportionately more with elevated CO_2 than shoot growth.

The chamber effects determined by plotting the ratios **CAN/A** and **CA/A** (Fig. 6) gave mixed results. Nevertheless, we concluded that presence of the chamber imposed N limitation on biomass productivity. The total aboveground biomass response was due to a large effect on dicot herbs. Owensby and Smith (1979) reported that dicot herb production and population density were increased by N fertilization in tallgrass prairie.

Long-term impacts of the apparent N limitation to growth can be inferred from the effects of elevated CO_2 on N concentration in biomass. Owensby et al. (1993b) reported a general reduction in N concentration of unfertilized tallgrass prairie species subjected to elevated CO_2. Those reductions in N concentration were greater than the reductions found in the current fertilized study. Nevertheless, CO_2 enrichment caused reductions in tissue N concentration even with supplemental N. Reductions in N concentration in above- and belowground biomass could potentially slow N cycling by limiting microbial decomposition rate. Indeed, Rice et al. (1993) measured direct stimulation of microbial activity by adding N to the same plots used in this study. Without added N, CO_2-enriched plots had a greater soil carbon content than ambient CO_2 plots, indicating reduced microbial activity or greater C input. They concluded that N limitation under elevated CO_2 would slow nutrient cycling and exacerbate N limitation.

Based on the data presented here and those presented earlier by Owensby et al. (1993a,b), we conclude that the future impact of elevated CO_2 on tallgrass prairie will entail increased biomass production in years with water stress and that N availability will limit the magnitude of the response.

Acknowledgements

Dale Strickler, Denise Garrett, Dean Larson, Garry Harter, and Neal Adam were extremely valuable in accomplishing.this research. Nasser Sionit helped in the initial setup of the study. This research was supported by the U.S. Department of Energy, Carbon Dioxide Research Division. Contribution No. 94-256-J from the Kansas Agricultural Experiment Station.

References

Bazzaz F A 1990 The response of natural ecosystems to the rising global CO_2 levels. Annu Rev. Ecol. Syst. 21,167–196

Boden T A, Kanciruk P and Farrell M P 1990 Trends '90, A compendium of data on global climate change. Carbon Dioxide Information Analysis Center, Oak Ridge National Laboratory, Oak Ridge TN. 257 p.

Coleman J S, Rochefort L, Bazzaz F A and Woodward F I 1991 Atmospheric CO_2, plant nitrogen status and the susceptibility of plants to an acute increase in temperature. Plant, Cell Environ 14, 667–674.

Curtis P S, Drake B G, Leadley P W, Arp W J and Whigham D F 1989a Growth and senescence in plant communities exposed to elevated CO_2 concentrations on an estuarine marsh. Oecologia 78, 20–29.

Curtis P S, Drake B G and Whigham D F 1989b Nitrogen and carbon dynamics in C_3 and C_4 estuarine marsh plants grown under elevated CO_2 in situ. Oecologia 78, 297–301.

Hocking P J and Meyer C P 1985 Responses of noogoora burr (*Xanthium occidentale* Bertol.) to nitrogen supply and carbon dioxide enrichment. Ann. Bot. 55, 835–844.

Hunt H W, Trlica M Z, Redente E F, Moore J C, Detling J K, Kittel T G F, Walter D E, Fowler M C, Klein D A and Elliott E T 1991 Simulation model for the effects of climate change on temperate grassland ecosystems. Ecol. Mod. 53, 205–246.

Kimball B A 1983 Carbon dioxide and agricultural yield, an assemblage and analysis of 430 prior observations. Agron. J. 75, 779–788.

Knapp A K, Hamerlynck E P and Owensby C E 1993 Photosynthesis and water relations responses to elevated CO_2 in the C_4 grass *Andropogon gerardii*. Int. J. Plant Sci. 154, 459–466.

Larigauderie A, Hilbert D W and Oechel W C 1988 Effect of CO_2 enrichment and nitrogen availability on resource acquisition and resource allocation in a grass, *Bromus mollis*. Oecologia 77, 544–549.

Linder R C and Harley C P 1942 A rapid method for the determination of nitrogen in plant tissue. Science 96, 565–566.

Lund Z F, Pearson R W and Buchanan G A 1970 An implanted soil mass technique to study herbicide effects on root growth. Weed Sci. 18, 279–281.

Mooney H A, Drake B G, Luxmore R J, Oechel W C and Pitelka L F 1991 Predicting ecosystem responses to elevated CO_2 concentrations. Bioscience 41, 96–104.

Newton P C D 1991 Direct effects of increasing carbon dioxide on pasture plants and communities. N. Z. J. Agric. Res. 34, 1–24.

Nijs I, Impens I and Behaeghe T 1988 Effects of rising atmospheric carbon dioxide concentration on gas exchange and growth of perennial ryegrass. Photosynthetica 22, 44–50.

Oechel W C and Strain B R 1985 Native species responses to increased carbon dioxide concentration. *In* Direct Effects of Increasing Carbon Dioxide on Vegetation. Eds. B R Strain and J D Cure. pp 117–154. U. S. D.O. E., NTIS, Springfield, VA.

Owensby C E and Anderson K L 1967 Yield responses to time of burning in the Kansas Flint Hills. J. Range Manage. 23, 12–16.

Owensby C E, Hyde R M and Anderson K L 1969 Effects of clipping and supplemental nitrogen and water on loamy upland bluestem range. J. Range Manage. 23, 341–346.

Owensby C E and Smith E F 1979 Fertilizing and burning Flint Hills bluestem range. J. Range Manage. 32, 254–258.

Owensby C E, Coyne P I, Ham J M, Auen L M and Knapp A K 1993a. Biomass production in a tallgrass prairie ecosystem exposed to ambient and elevated levels of CO_2. Ecol. Appl. 3, 644–653.

Owensby C E, Coyne P I and Auen L M 1993b Nitrogen and phosphorus dynamics of a tallgrass prairie ecosystem exposed to elevated carbon dioxide. Plant, Cell Environ. 16, 843–850.

Reichers G H and Strain B R 1988 Growth of blue grama (*Bouteloua gracilis*) in response to atmospheric CO_2 enrichment. Can. J. Bot. 66, 1570–1573.

Rice C W, Garcia F O, Hampton C O and Owensby C E 1994. Microbial response in tallgrass prairie to elevated CO_2. Plant and Soil (*In press*).

Strain B R and Cure J D 1985 Direct effect of increasing carbon dioxide on vegetation. Carbon Dioxide Research, State of the Art. U.S. D.O.E. Washington, DC.

Technicon Industrial Systems 1977 Individual/simultaneous determination of nitrogen and/or phosphorus in BD acid digests. Industrial Method NO. 334–74 WB+.

Tolley L C and Strain B R 1984 Effects of atmospheric CO_2 enrichment and water stress on growth of *Liquidamber styraciflua* L. and *Pinus taeda* L. seedlings. Can J. Bot. 10, 2135–2139.

Wray S M and Strain B R 1986 Response of two old field perennials to interactions of CO_2 enrichment and drought stress. Am. J. Bot. 73, 1486–1491.

Plant and Soil **165**: 115-127, 1994.

Effects of elevated CO_2 and nitrogen fertilization pretreatments on decomposition on tallgrass prairie leaf litter

Paul R. Kemp[1], Deborah G. Waldecker[2], Clenton E. Owensby[3], James F. Reynolds[1] and Ross A. Virginia[4]
[1]*Department of Botany, The Phytotron, Duke University, Durham, NC 27706, USA,* [2]*Department of Biology, San Diego State University, San Diego, CA 92182, USA,* [3]*Department of Agronomy, Kansas State University, Manhattan, KS 66506, USA and* [4]*Environmental Studies Program, Dartmouth College, Hanover, NH 03755, USA*

Key words: Andropogon gerardii, carbon chemistry, carbon dioxide, litter quality, *Poa pratensis, Sorghastrum nutans*

Abstract

Standing dead and green foliage litter was collected in early November 1990 from *Andropogon gerardii* (C_4), *Sorghastrum nutans* (C_4), and *Poa pratensis* (C_3) plants that were grown in large open-top chambers under ambient or twice ambient CO_2 and with or without nitrogen fertilization (45 kg N ha^{-1}). The litter was placed in mesh bags on the soil surface of pristine prairie adjacent to the growth treatment plots and allowed to decay under natural conditions. Litter bags were retrieved at fixed intervals and litter was analyzed for mass loss, carbon chemistry, and total Kjeldahl nitrogen and phosphorus. The results indicate that growth treatments had a relatively minor effect on the initial chemical composition of the litter and its subsequent rate of decay or chemical composition. This suggests that a large indirect effect of CO_2 on surface litter decomposition in the tallgrass prairie would not occur by way of changes in chemistry of leaf litter. However, there was a large difference in characteristics of leaf litter decomposition among the species. *Poa* leaf litter had a different initial chemistry and decayed more rapidly than C_4 grasses. We conclude that an indirect effect of CO_2 on decomposition and nutrient cycling could occur if CO_2 induces changes in the relative aboveground biomass of the prairie species.

Introduction

Increasing atmospheric CO_2 and associated potential climate changes are expected to have a number of direct and indirect effects on plants and ecosystems (Bazzaz, 1990; Field et al., 1992; Strain, 1987). The effects of elevated CO_2 on ecosystem processes, and decomposition of plant litter in particular, have received little attention (Lambers, 1993). The process of decomposition depends upon the chemical makeup of litter and its interactions with soil biota (Aber and Melillo, 1982; Berg and Staaf, 1980; Coûteaux et al., 1991; Meentmeyer, 1978; Swift et al., 1979). Increasing atmospheric CO_2 could directly influence the chemical composition of plant tissues and the resulting litter (Coûteaux et al., Field et al., 1991, 1992; Norby et al., 1986; Woodward et al., 1991), bringing about changes in patterns of nutrient cycling and carbon storage through changes in decomposition (Lambers, 1993).

Tallgrass prairie ecosystems may be relatively sensitive to changes in rates of surface litter decomposition and nutrient release, since accumulations of standing dead and surface litter have considerable influence on the amounts, productivity, and mixture of resident species (Knapp and Seastedt, 1986; Risser and Parton, 1982). Standing dead can also influence the frequency and intensity of fire, which, in turn, can affect prairie species diversity. Any changes in rates of litter decomposition and subsequent release of limiting nutrients could bring about important changes in tallgrass prairie structure and function.

There are few data from which to extrapolate the response of tallgrass prairies, especially decomposition, to increasing CO_2. Elevated atmospheric CO_2 might increase litter C/N ratios resulting in slower rates of decomposition (Lambers, 1993). On the other hand, reallocation of nonstructural N during senescence (Curtis et al., 1989) could result in little difference in litter C/N. Furthermore, changes in decomposition food webs (Coûteaux et al., 1991) could result in little difference in decomposition even if litter chemistry is altered by elevated CO_2. The objectives of this study were to determine: 1) rates of mass loss, 2) carbon chemistry, and 3) nitrogen (N) and phosphorous (P) concentrations of decomposing leaf litter from three principal grasses in a prairie near Manhattan, Kansas that had been subjected to ambient or twice ambient CO_2 fumigation using large open-top chambers, and with or without nitrogen fertilization. The decomposition experiment was carried out under ambient atmospheric CO_2 levels in a tract of natural prairie adjacent to the CO_2 treatment plots.

Materials and methods

Study site and CO_2 enrichment methods

The CO_2 enrichment study was conducted in a pristine tallgrass prairie just north of Kansas State University in Manhattan, Kansas (39.12 °N, 96.35 °W, 324 m elevation). The dominant plant at this site is the C_4 grass, big bluestem (*Andropogon gerardii* Vitman). Other important species include Indiangrass (*Sorghastrum nutans* (L.) Nash.), Kentucky bluegrass (*Poa pratensis* L.), tall dropseed (*Sporobolus asper* (Michx.) Kunth.), western ragweed (*Ambrosia psilostachya* DC.), Louisiana sagewort (*Artemesia ludoviciana* Nutt.), and ironweed (*Vernonia baldwinii* (Small) Schub.).

Average annual precipitation is 835 mm, with 500 mm occurring from April through August. Long-term mean minimum and maximum temperatures for January are -8.8 and 2.4 °C and for July are 19.0 and 38.1 °C. Average minimum temperature is below freezing from about mid-November through mid-March. For the period of experimental treatment, precipitation was much below normal (-156 mm) from January through June 1989, and much wetter than normal (+202 mm) from August through October 1989. Precipitation for 1990 was near normal. For the period of the decomposition study, precipitation was near normal for winter/spring 1991, below normal (-125 mm) from June through October, and above normal (+94 mm) for winter 1991/1992. More details of precipitation and temperature during the study period are included in Owensby et al. (1993a, b).

The study consisted of 15 experimental plots and six treatments. Ten of the plots were enclosed with open-top chambers (4.5 m diameter and 4 m height) in which the CO_2 concentration could be maintained at two levels: ambient (330 to 450 ppm) or twice ambient (660 to 900 ppm) levels. The CO_2 concentration in the chambers were monitored with an infrared gas analyzer 24 hours a day, and the doubled CO_2 concentration in the enriched chambers was achieved by adding pure CO_2 using transducers actuated by a computer monitoring system. In addition to CO_2 treatment, some plots received N fertilization (NH_4NO_3 applied at 45 kg N ha^{-1}) each spring. Five of the plots were not covered by chambers, of which three served as controls and two received N fertilization. Thus, the six experimental treatments were: 1) chamber, ambient CO_2, no N fertilization (3 plots); 2) chamber, ambient CO_2, N fertilization (2 plots); 3) chamber, elevated CO_2, no N fertilization (3 plots); 4) chamber, elevated CO_2, N fertilization (2 plots); 5) no chamber, no N fertilization (3 plots); and 6) no chamber, N fertilization (2 plots). Chambers and elevated CO_2 fumigation were maintained during the growing season (April through October), but removed during winter. The plots without fertilization were begun in May 1989 and the fertilized plots were begun in April 1990. See Owensby et al. (1993b) for further details of the experimental design.

Litter placement and retrieval

At the end of the growing season (November 1, 1990), standing dead and green foliage was harvested from three grass species (*Andropogon gerardii*, *Sorghastrum nutans*, and *Poa pratensis*) from every plot. This material proved to be a good representation of the actual litterfall, since three days following the collection, subfreezing temperatures and a large snowfall killed most of the remaining green leaves and crushed the standing dead to the surface. The foliage litter was separated by species and that from replicate plots was combined to produce a single, homogeneous sample for the treatment (see *Statistical analyses*). The litter was air dried for 5 to 7 days and then placed in 10 × 10 cm, mesh (1 mm^2 opening), polypropylene bags. A total of 630 bags (210 per species) were prepared, each containing ca. 3.5 g of grass litter. Ninety bags were randomly chosen for immediate chemical analyses (5 bags for each of 6 treatments and 3 species), and the remaining bags (540) were placed into the field for subsequent retrieval. The field site was an area of undisturbed prairie adjacent to the CO_2 enrichment plots. Bags were positioned in contact with the mineral soil surface, by carefully removing the natural litter from

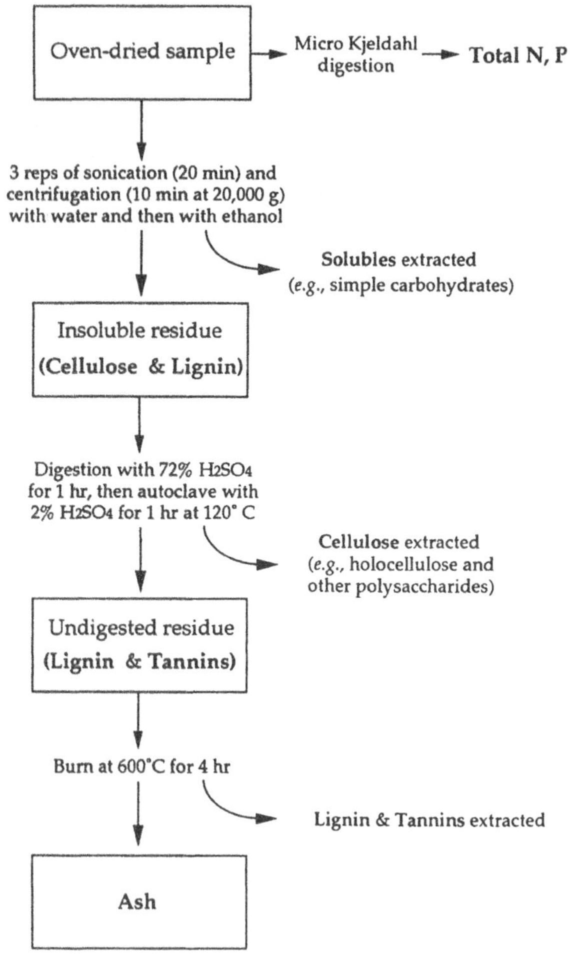

Fig. 1. Flowchart of analyses used to obtain chemical fractions of litter samples.

an area the size of the bag. Bags were placed about 15 cm apart and the area between the bags was not disturbed (benches were used to place and retrieve bags). Ninety bags were retrieved from the field on each of the following sampling dates: March 1, May 1, August 1, and November 1, 1991, and May 1 and November 1, 1992. The bags retrieved in November 1992 were highly contaminated with mineral soil, apparently as a result of heavy rainfalls during the summer of 1992, and were thus unusable in the analyses.

Chemical analyses

Each set of litter samples was analyzed for the fraction of principal carbon constituents (solubles, cellulose, and acid resistant aromatics [lignin and tannins]) and total Kjeldahl N and P. The litter samples were oven-dried (55 °C for 48 h), weighed, and ground through a 40-mesh screen Wiley mill. The total N and P concentrations were measured on a subsample (0.1 g) of the ground material using a modified micro Kjeldahl method and a Technicon Autoanalyzer (Bremner and Mulvaney, 1982). Carbon fractions were determined using the procedure summarized in Figure 1. The soluble component of the litter was determined by subjecting subsamples (ca. 0.3 g) of the ground material to three repetitions of sonication in 20 mL water (20 min) followed by centrifugation (10 min at 20,000 *g*) and removal of solvent. This procedure was then repeated using 20 mL ethanol as the solvent. The soluble fraction was calculated as the loss in weight of the remaining oven-dried residue (Berg et al., 1982). The residue, consisting primarily of cellulose and lignin, was digested with 72% H_2SO_4 for 1 h at 30 °C to hydrolyze the cellulose. This was followed immediately by dilution to 2% H_2SO_4 and autoclaving at 120 °C to achieve secondary hydrolysis (Effland, 1977). The resulting solution was filtered (Whatman #42), and the remaining residue oven-dried for 24 h, and weighed to obtain the Klason lignin fraction. The difference between the starting residue weight and the lignin residue weight was taken to be the cellulose fraction. Ash content was determined by burning the lignin residue at 600 °C for 4 h in a muffle furnace. Mass loss and chemical fractions were calculated on an ash-free basis (Schlesinger, 1985).

Statistical analyses

At the start of the decomposition experiment it was assumed that the major source of experimental variation would be due to differences in microbial colonization and weathering of litter bags. In order to minimize the starting variability among bags, the litter from different plots of the same treatment were pooled and homogenized. This procedure was justified based on earlier analyses of plant tissue and litter from different plots, which indicated that the plots within the same treatment were not significantly different from one another. Two sets of analyses of variance ANOVA) were performed: one with species as a factor and one without species as a factor (i.e., separate ANOVA for each species' data). Prior to carrying out the ANOVAs, percent mass loss and chemical fraction data were arcsin transformed in order to approximate a more normal distribution. Tukey-Kramer *post-hoc* tests were carried

Table 1. Initial chemical composition of leaf litter and summary of decomposition over 18 months. The annual mass loss coefficient, k, was determined by least squares fit to the equation: $Mass_t = Mass_{initial} * e^{kt}$. The k values were compared using Tukey-Kramer test. Turnover time = 1/k. Values within a column that are not followed by the same letter are significantly different ($p < 0.05$)

	Initial chemical composition						Decomposition		
Treatment	% Soluble	% Cellulose	% Lignin	% N	C/N	% P	% Total mass loss	k	Turnover time (yr)
Andropogon									
2 × CO_2	14.6a	70.2a	15.1bc	0.42c	95.36c	0.052ab	19.78a	-0.171a	5.85
1 × CO_2	20.0bcd	65.2c	14.8bc	0.39c	102.15c	0.099efg	22.18ab	-0.197ab	5.08
No Chamber	16.8abc	65.7bc	17.5a	0.55d	72.74d	0.088ef	22.56ab	-0.190ab	5.26
2 × CO_2 + N	19.3bcd	65.8bc	14.8bc	0.42c	95.74c	0.077cde	19.53a	-0.172a	5.81
1 × CO_2 + N	17.2abc	67.4abc	15.1bc	0.39c	103.28c	0.059abc	19.62a	-0.165a	6.06
No Cham + N	19.1bcd	64.3c	16.5ab	0.53d	75.27d	0.066bcd	24.58abc	-0.215b	4.65
Sorghastrum									
2 × CO_2	22.9d	64.4c	12.5ef	0.40c	99.03c	0.139h	24.75abc	-0.216b	4.63
1 × CO_2	20.6cd	66.7abc	12.7ef	0.39c	103.12c	0.108fgh	22.59ab	-0.193ab	5.18
No Chamber	16.2ab	70.0a	13.8cd	0.38c	104.89c	0.063bcd	22.28ab	-0.212b	4.72
2 × CO_2 + N	18.5abc	68.4abc	12.9e	0.24a	167.09a	0.085def	19.74a	-0.179ab	5.71
1 × CO_2 + N	19.1bcd	68.4abc	12.4ef	0.34b	119.26b	0.054ab	21.61ab	-0.180ab	5.56
No Cham + N	17.2abc	69.1a	13.6cd	0.34b	117.83b	0.046a	23.14ab	-0.205ab	4.88
Poa									
2 × CO_2	34.9f	54.1e	10.9f	1.04e	38.65e	0.123gh	33.96d	-0.303d	3.30
1 × CO_2	37.1f	51.1e	11.6ef	1.21f	33.02f	0.134gh	35.41d	-0.313d	3.19
No Chamber	33.2ef	54.6de	12.3def	1.43g	28.04g	0.178i	29.74bcd	-0.255c	3.92
2 × CO_2 + N	34.0ef	54.6de	11.4ef	0.94e	42.62e	0.118gh	28.68bcd	-0.244c	4.10
1 × CO_2 + N	34.9f	53.3e	11.8ef	1.23f	32.42f	0.140h	33.40d	-0.283cd	3.53
No Cham + N	29.3e	58.9d	11.8ef	1.34fg	29.79fg	0.144h	31.35cd	-0.272cd	3.68

out to test for specific differences among means when F values from ANOVAs were significant ($p < 0.05$).

Results

By far the most significant factor affecting decomposition was differences among species (see overall ANOVA for mass loss and chemical constituents in Tables A1–A6 of the Appendix). Mass loss was significantly greater for *Poa* litter than for the two C_4 grasses (Table 1; Fig. 2). The initial carbon chemistry and N and P content of the litter was significantly different among species. *Poa* litter had higher soluble, N and P fractions and lower cellulose than *Andropogon* or *Sorghastrum* (Table 1; Figs. 3 and 4). There were also a number of significant effects of treatment factors (CO_2 and N fertilization) upon mass loss and litter chemistry (Tables A1–A6). However, there were no treatment effects on chemical composition that were consistent for all species (Table 1). Because of the overriding differences between species, and because the species responded differently to treatments (as evidenced by numerous species × treatment interactions in the overall ANOVA), a separate set of ANOVAs was performed for each species, to identify significant responses of the individual species to CO_2 and N fertilization treatments. The results of these ANOVAs are presented symbolically within Figures 2–4.

The pattern of organic mass loss was similar for all three species. Most of the loss of material occurred during the growing season from May 1991 (181 days from start) to November 1991 (365 days from start; Fig. 2). The growth treatments (elevated CO_2, N fertilization, and the chamber itself) had significant effects that varied depending upon the species and time. The chamber

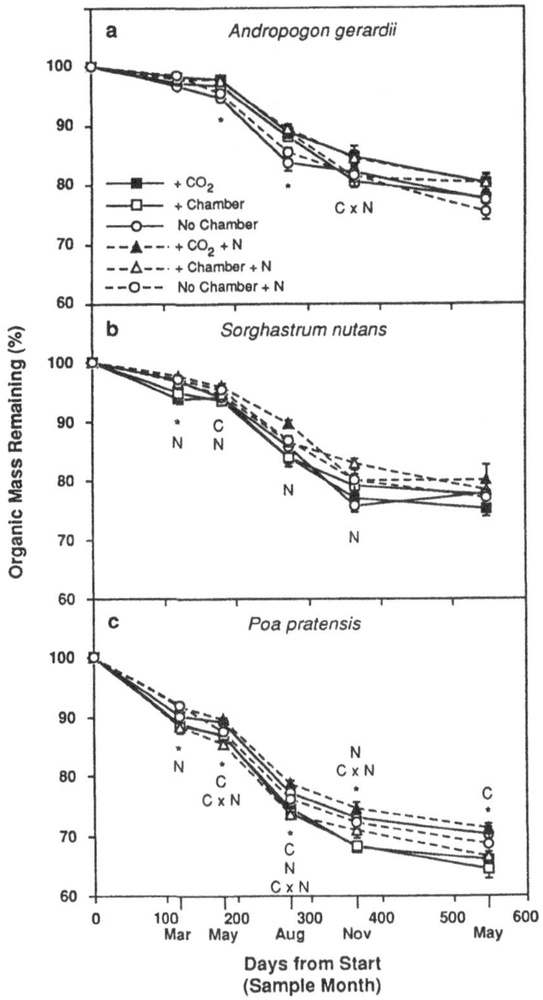

Fig. 2. Mass loss through time of leaf litter from *Andropogon gerardii* (C_4), *Sorghastrum nutans* (C_4), and *Poa pratensis* (C_3) that were grown with two levels of CO_2 and with or without N fertilization. Growth treatments were: (+ CO_2) — plants in chambers with elevated CO_2 (720 ppm), (+ Chamber) — plants in chambers with ambient CO_2 (360 ppm), (No Chamber) — plants in open plots with no pretreatment, (+ CO_2 + N) — plants in chambers with elevated CO_2 (720 ppm) and N fertilization, (No Chamber + N) — plants in chambers with ambient CO_2 (360 ppm) and N fertilization, and (No Chamber + N) — plants in open plots with no chamber and N fertilization. Values are means of 5 samples ± S.E. Significant ($p < 0.05$) treatment effects at a sample date are identified on the graphs as follows: * = chamber effect, C = CO_2 effect, N = N fertilization effect, C × N = CO_2 × N fertilization interaction.

affected loss of organic matter in both *Andropogon* and *Poa*. Nitrogen fertilization significantly affected mass loss of litter in *Sorghastrum* and *Poa* during the four middle sample times (Fig. 2b, c). For *Poa* litter, there was an effect of CO_2 and an interaction between CO_2 and N fertilization. However, by the end of the 18 month study period, total mass loss was similar among treatments within a species (Table 1), but different between *Poa* and the C_4 grasses. Total mass loss of *Poa* litter (averaged over all treatments) was 32%, whereas mass loss of both C_4 grasses averaged 22%.

Other investigators have found that mass loss through time can be described by an exponential curve. In the present study mass loss was linear during the winter and then declined exponentially during the summer. However, in order to compare our results with other studies of a year or more, the data was fit to a single exponential function to obtain the annual mass loss coefficient, *k*. All exponential regressions were significant (average $r^2 = 0.92$). The *k* values, which reflect the whole period of decay, emphasize the difference between *Poa* and the C_4 grasses (Table 1). Litter turnover (1/k) was significantly faster in *Poa* (Table 1).

Carbon chemistry of the litter changed significantly through time (Fig. 3) and there were significant differences among species in the various carbon fractions (Tables A2–A4). *Poa* litter initially had a larger fraction of soluble material than the C_4 grass litter (Figs. 3a, b, c). The soluble fraction in the *Poa* litter diminished during the first half of the study, becoming similar to the soluble fraction of the C_4 grass litter by August 1991. The soluble fraction in the C_4 grass litter declined slightly by the first sample date (March 1991) and was relatively unchanged thereafter. There were several effects of growth treatment conditions upon the soluble fraction that varied with species and time. N fertilization had the largest impact on the soluble fraction of the decomposing litter. However, the effect was inconsistent through time, causing decreased solubles in the May 1991 samples from *Andropogon* and *Poa*, the November 1991 sample from *Sorghastrum*, and the May 1992 sample from *Andropogon*, but increased solubles in the August 1991 sample from *Poa*. Elevated CO_2 produced small and inconsistent differences in the initial soluble fraction of the C_4 grass litter (Figs. 3a, b). By the end of the experiment, the soluble fraction was similar among all treatments and species.

Along with a higher initial soluble fraction, *Poa* litter had a lower initial fraction of cellulose than the C_4 grasses (Figs. 3g, h, i). The fraction of cellulose in litter differed among the C_4 grasses, with *Sorghastrum* having a greater fraction of cellulose than *Andropogon* after the first sample date. The cellulose fraction of the litter followed a similar course through time for all species, increasing (significantly in *Poa* and slightly in the C_4 species) during the first 180 days (winter), then

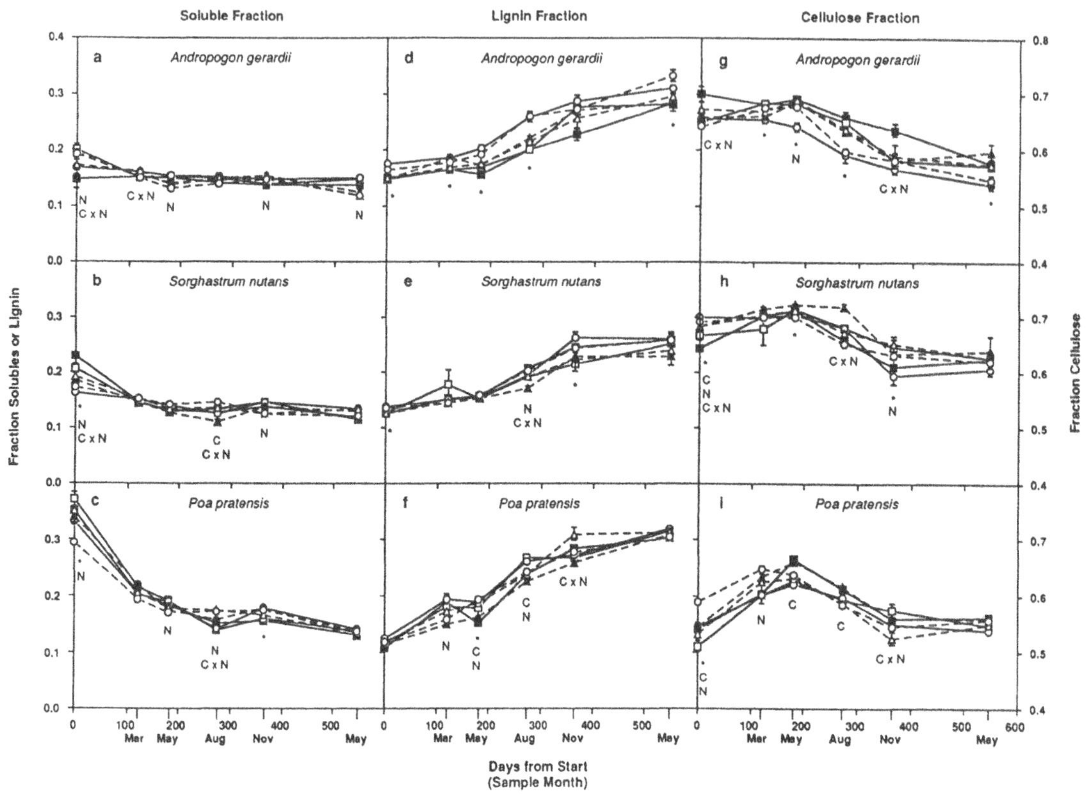

Fig. 3. Fraction of ash-free solubles, lignin, and cellulose in litter from 3 species of grasses grown with two levels of CO_2 and with or without N fertilization. Legend and description of growth treatments are as in Figure 2. Values are means of 5 samples ± S.E. Note that the axes differ for cellulose compared to solubles and lignin.

declining during the next 180 days (growing season). There were small, significant effects of the chamber, CO_2, and N during the middle of the decomposition period. The effect of the chamber was to increase the cellulose fraction in *Andropogon* and *Sorghastrum* litter. Nitrogen fertilization increased cellulose fraction at some isolated sample dates for each species. Elevated CO_2 caused a significant increase in *Poa* litter cellulose at the middle sample times. However, by the end of the experiment only species and chamber differences were significant; *Sorghastrum* litter generally had a higher cellulose fraction than *Andropogon* or *Poa*. *Andropogon* litter from inside chambers had higher cellulose than *Poa* litter, but litter from outside chambers had similar cellulose contents.

The lignin fraction of the litter increased steadily through the course of the experiment in all species and treatments (Figs. 3d, e, f). As with the other carbon fractions, the most significant source of variation in decomposition response was due to species differences (Table A4). There was a significant effect of the chamber on the lignin fraction of *Andropogon* and *Sorghastrum* litter (greatest lignin in litter from unchambered plots; Figs. 3d, e, f; Table 1). There were significant effects of elevated CO_2 and N fertilization on the lignin fraction of *Poa* litter; elevated CO_2 lowered the lignin fraction and fertilization had a variable effect.

The total initial N concentration of litter was significantly affected by all treatment factors (Table A5; Table 1). Total N concentration was similar for litter of both C_4 grasses, and increased from ca. 0.3% to 1.1% over the 18 month period (Figs. 4a, b). Total N concentration of *Poa* litter was significantly greater than that of the C_4 grasses, and increased from ca. 1.1% to 2.7% over the 18 month period (Fig. 4c). *Andropogon* litter collected from inside the chambers generally had lower N concentration than did litter collected from outside the chambers, and this difference was maintained during decomposition. *Sorghastrum* litter from N fertilized plots had lower N concentration than litter from unfertilized plots. Fertilization also had a significant effect on N concentration in *Poa* litter, but the

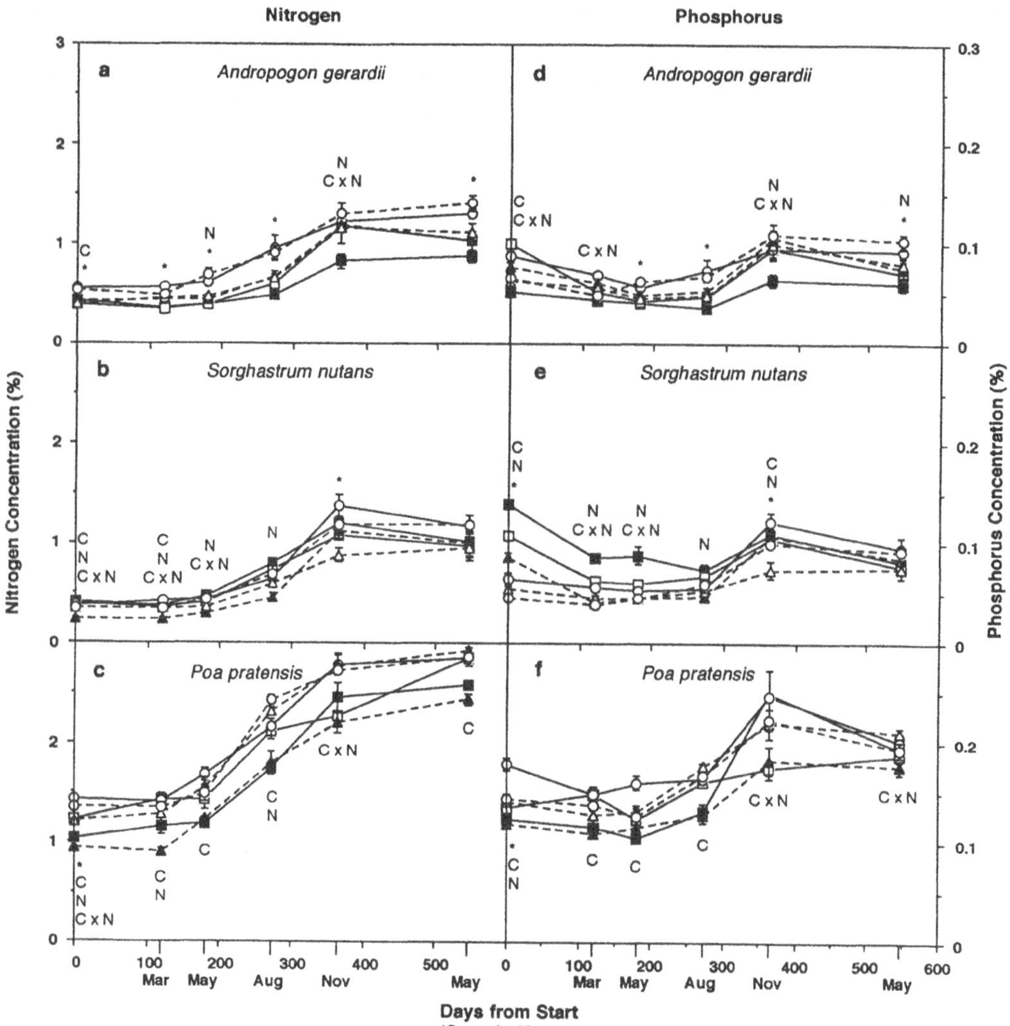

Fig. 4. Total Kjeldahl nitrogen and total phosphorus concentrations in litter from 3 species of grasses grown with two levels of CO_2 and with or without N fertilization. Legend and description of growth treatments are as in Figure 2. Values are means of 5 samples ± S.E.

effect varied through time. Elevated CO_2 caused *Poa* litter to have consistently lower N throughout decomposition.

Total P concentrations of the litter behaved similarly among the species during decomposition (Figs. 4d, e, f). The P concentration tended to decline slightly during the first half of decomposition, then increase slightly. There was little net change in P of C_4 grass litter over the course of the study; averaging about 0.08 % over all treatments. Total P concentration of *Poa* litter was greater than for the C_4 grasses, and increased more from May 1991 to November 1991; P averaged about 2.0 % in *Poa* litter at the end of the study (Fig. 4f). All of the treatment factors had a significant effect on P concentration at some point during the experiment, but none were consistent throughout the study nor for different species. There was a significant effect of CO_2 and an interaction of CO_2 with N fertilization on the initial P concentration of litter in both C_4 grasses, this resulted from *Andropogon* litter having lowest P in high CO_2, unfertilized plots, whereas *Sorghastrum* litter had highest P in high CO_2, unfertilized plots (Table 1). Elevated CO_2 caused reduced P concentration of *Poa* litter except at the November 1991 sample date.

Discussion

The principal objective of this study was to ascertain how decomposition might be indirectly effected by elevated CO_2 through its impact on plant litter chemistry. Although there were significant effects of elevated CO_2 on litter decomposition rate and/or litter chemistry, these differences were small in comparison to the differences among species. Elevated CO_2 failed to induce major changes in the chemistry of fresh litter. There was no overall significant effect induced by CO_2 in soluble, cellulose, lignin, or phosphorus fractions of the initial litter (Tables A2–A6). There was, however, an overall significant difference in N content of the litter between CO_2 treatments. But the effect was different among species. Elevated CO_2 reduced N in *Poa* litter, slightly increased N in *Andropogon* litter, and decreased N in *Sorghastrum* litter in fertilized plots (Table 1). However, except for *Poa*, the differences in N concentration disappeared soon after the onset of decomposition.

The generally small changes observed in litter chemistry of C_4 grasses grown at high CO_2 are consistent with small or lacking changes in chemistry of mature leaves of C_4 grasses exposed to elevated CO_2 (Curtis et al., 1989; Wong, 1979). Owensby (1993a) reported that elevated CO_2 did not change N concentrations in *Andropogon* leaves during 1990, the year that we collected litter from these same plots. However, elevated CO_2 reduced N concentration slightly in the previous and following years. Curtis et al. (1989) found that elevated CO_2 did not effect N concentration of mature leaves or subsequent litter of the C_4 grass *Spartina patens*. Owensby (1993a) found that elevated CO_2 induced a slight decrease in P in mature leaves of *Andropogon* in 1989, but had no effect in 1990 and 1991. Our study found that elevated CO_2 increased P in litter of *Sorghastrum*.

The effect of elevated CO_2 on litter chemistry of the C_3 grass *Poa* is consistent with the effects of elevated CO_2 on leaves of C_3 plants (Overdieck et al., 1988; Wong, 1990). However, it contrasts with the findings for senescing leaves of C_3 graminoids. Larigauderie et al. (1988) found that the large initial differences in leaf N induced by elevated CO_2 in *Bromus mollis* disappeared as the annual plants approached maturity. Curtis et al. (1989) found that although elevated CO_2 induced decreased N concentration in mature leaves of *Scirpus olneyi*, reallocation of N at senescence led to similar residual N levels in litter from different CO_2 treatments. The foliage that we collected at the end of the season showed no reduction of N, compared to mature leaves. Owensby et al. (1993a) found that during the growing season, N concentrations for mature leaves of *Poa* averaged 1.15% at ambient CO_2 and 1.05% at 2 × ambient, compared to the 1.24% and 1.04%, respectively, that we found at the end of the season. The relatively high N values for *Poa* at the end of the season may be due to the fact that *Poa* regrows in late autumn after the senescence of the taller C_4 grass canopy. Thus, *Poa* leaves apparently have relatively high N even though they contribute to the litter soon after being killed by freezing temperatures (normally early to mid November). Also, our N concentrations may be high compared to Owensby et al. (1993a) since we collected only leaves, whereas they collected all aboveground tissue (including stems and culms with lower N).

The results for C_4 grass litter and some C_3 graminoid litter (e.g., Curtis et al., 1989) contrast with those for deciduous trees, which exhibit a very large effect of CO_2 on N concentrations of abscised leaves (Coûteaux et al., 1991). The lack of a strong effect of CO_2 on litter N concentration of grasses (especially the C_4 grasses of this study) should act to limit potential changes in surface litter decomposition and nutrient cycling in the prairie as atmospheric CO_2 increases. However, the situation for surface litter of the prairie can be contrasted with that for root litter (Rice et al., this volume), in which elevated CO_2 significantly increases soil organic carbon, nitrogen, and microbial activity, especially with the addition of N fertilization. Furthermore, if increased CO_2 leads to greater aboveground productivity, then surface litter mass of the prairie could be expected to increase.

Some of the observations regarding CO_2 response may be confounded by the effect of the chamber. After species differences, the chamber was the most significant factor influencing litter chemistry (Tables A2–A6). Owensby et al. (1993b) indicated that the elevated CO_2 effect and the chamber effect were similar, mediated through changes in water use efficiency. However, there may be additional chamber effects different from elevated CO_2. For example, both the chamber and elevated CO_2 caused a decrease in initial N concentration of *Poa* litter, but in *Andropogon* litter the chamber caused a decrease in N, whereas elevated CO_2 caused a slight increase in N. The chamber caused a significant increase in lignin in both C_4 grasses, whereas CO_2 had no effect. In any case, the litter chemistry changes induced by CO_2 × chamber interactions were very few and small compared to other differences.

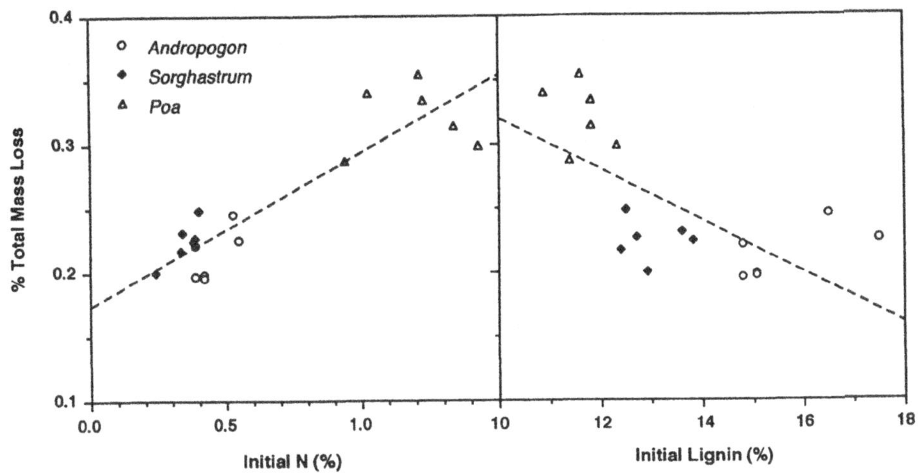

Fig. 5. Relationship between mass loss and initial N and lignin in litter. Each data point is the mean for a treatment (n = 5). Lines are least squares regression lines (for mass loss vs. N: $y = 0.175 + 0.118\ x$, $r^2 = 0.81$; for mass loss vs. lignin: $y = 0.509 - 1.907\ x$, $r^2 = 0.44$).

In general, mass loss through decomposition in this experiment was similar to that of forest (Staaf and Berg, 1982) and chaparral (Schlesinger, 1985) foliage litter, but somewhat less than desert grasses (MacKay et al., 1987), where physical weathering may be a more important factor (Moorhead and Reynolds, 1989). Mass loss in our study (32% for *Poa* litter and 22% for C_4 grass litter) bracketed the 25% mass loss over 18 months reported by Seastedt (1988) for mixed foliage at a similar tallgrass prairie site. Other research has emphasized either climate or litter chemistry as controlling rates of decomposition (Anderson, 1992). Our results show that both factors control decomposition in the prairie. During the period from November to March, low temperatures limit microbial activity and mass loss is small (Fig. 2) and related to leaching of soluble materials (Figs. 3a, b, c). In spring and summer, mass loss increases (Fig. 2), cellulose fraction declines, and N and P concentrations increase. Others (Aber and Melillo, 1982; Schlesinger, 1985; Seastedt, 1988) have found that N and P concentrations of foliage litter generally increase during the first one or two years of decomposition. These nutrients are immobilized by the decomposers for extended periods and the litter mass thus acts as a sink for nutrients. Our study found that there was also a difference in the extent to which the litter acted as a sink. *Poa* litter showed a greater increase in N, and particularly P, than did the C_4 grass litter (Fig. 4). Mass loss among all species and treatments was strongly correlated ($r^2 = 0.81$) with initial N content of the litter and to a lesser extent negatively correlated ($r^2 = 0.44$) with initial lignin (Fig. 5), supporting the conclusions of Seastedt (1988) and Taylor et al. (1989) that, at leaf for herbaceous litter, N is a better indicator of potential decomposition rate than lignin.

The differences in rate of decomposition between C_3 and C_4 surface litter has important implications with regard to decomposition and nutrient availability in the prairie. These differences are much larger than the effects of elevated CO_2 on decomposition within a particular species. Changes in prairie surface litter decomposition and nutrient cycling in response to CO_2 are more likely to arise from shifts in species composition (especially C_3 versus C_4), rather than through changes in chemistry of a particular litter type. Knapp and Seastedt (1986) concluded that decomposition is slow in the tallgrass prairie and could be expected to control nutrient availability. The results of our study suggest that this is due, in part, to the dominance of C_4 grasses, which decay considerably slower than the C_3 grass litter (and presumably C_3 forbs). If elevated CO_2 and/or climate change alter species composition of the community, there could be a subsequent change in decomposition rates and release of nutrients. Owensby (1993b) found that *Poa* biomass declined under elevated CO_2. This may be related to the improved water relations of the C_4 grasses at elevated CO_2, in combination with their height dominance. A shift to greater C_4 biomass would lead to litter with higher C/N ratio, higher lignin and cellulose, and slower turnover, resulting in greater accumulation, and further growth limitation, by surface litter (Knapp and Seastedt, 1986). On the other hand, if C_3 forbs increase or, in the case of wetter prairies and marshlands, C_3 graminoids increase (Arp et al., 1993), then litter decomposition would involve an increased immobilization of inorganic nutrients, but a faster overall decay, faster turnover of surface litter, and subsequently greater total nutrient content in the surface soil.

Acknowledgements

We thank Lisa Auen for assistance in the field and Robert MacAller for assistance with the chemical analyses. The research was supported by grants DE-FG05-92ER61493 and DE-FG02-84ER60253 from the Department of Energy.

References

Aber J D and Melillo J M 1982 Nitrogen immobilization in decaying hardwood leaf litter as a function of initial nitrogen and lignin content. Can. J. Bot. 60, 2263–2269.

Anderson J M 1992 Responses of soils to climate change. Adv. Ecol. Res. 22, 163–210.

Arp W J, Drake B G, Pockman W T, Curtis P S and Whighan D F 1993 Interactions between C_3 and C_4 salt marsh plant species during four years of exposure to elevated atmospheric CO_2. Vegetatio 104/105, 133–143.

Bazzaz F A 1990 The response of natural ecosystems to the rising global CO_2 levels. Annu. Rev. Ecol. Syst. 21, 167–196.

Berg B and Staaf H 1980 Decomposition rate and chemical changes of Scots pine needle litter. II. Influence of chemical composition. *In* Structure and Function of Northern Coniferous Forests An Ecosystem Study. Ed. T Persson. pp 373–390. Ecological Bulletins, Stockholm, Sweden.

Berg B, Wessen B and Ekbohm G 1982 Nitrogen level and decomposition in Scots pine needle litter. Oikos 38, 291–296.

Bremner J M and Mulvaney C S 1982 Nitrogen — total. *In* Methods of Soil Analysis. Part 2. Chemical and Microbial Properties. Eds. A L Page, R H Miller and D R Keeney. pp 595–624. Am. Soc. Agron. and Soil Sci. Soc. Am., Madison, WI.

Coûteaux M M, Mousseau M, Célérier M L and Bottner P 1991 Increased atmospheric CO_2 and litter quality: decomposition of sweet chestnut leaf litter with animal food webs of different complexities. Oikos 61, 54–64.

Curtis P S, Drake B G and Whigham D F 1989 Nitrogen and carbon dynamics in C_3 and C_4 estuarine marsh plants grown under elevated CO_2 in situ. Oecologia 78, 297–301.

Juliano Effland M J 1977 Modified procedure to determine acid soluble lignin in wood and pulp. TAPPI 60, 143–144.

Field C B, Chapin F S III, Matson P A and Mooney H A 1992 Response of terrestrial ecosystems to the changing atmosphere: a resource-based approach. Annu. Rev. Ecol. Syst. 23, 201–235.

Knapp A K and Seastedt T R 1986 Detritus accumulation limits productivity of tallgrass prairie. BioScience 36, 662–668.

Lambers H 1993 Rising CO_2, secondary plant metabolism, plant-herbivore interactions and litter decomposition. Vegetatio 104/105, 263–271.

Larigauderie A, Hilbert D W and Oechel W C 1988 Effect of CO_2 enrichment and nitrogen availability on resource acquisition and resource allocation in a grass, *Bromus mollis*. Oecologia 77, 544–549.

MacKay W P, Fisher F M, Silva S and Whitford W G 1987 The effects of nitrogen, water and sulfur amendments on surface litter decomposition in the Chihuahuan Desert. J. Arid Environ. 12, 223–232.

Meentmeyer V 1978 Macroclimate and lignin control of litter decomposition rates. Ecology 59, 465–472.

Moorhead D L and Reynolds J F 1989 Mechanisms of surface litter mass loss in the northern Chihuahuan Desert: a re-interpretation. J. Arid Environ. 16, 157–163.

Norby R J, Pastor J and Melillo J M 1986 Carbon-nitrogen interactions in CO_2 enriched white oak: physiological and long-term perspectives. Tree Physiol. 2, 223–241.

Overdieck D, Reid C and Strain B R 1988 The effects of preindustrial and future CO_2 concentrations on growth, dry matter production and the C/N relationship in plants at low nutrient supply: *Vigna unguiculata* (cowpea), *Abelmoschus esculentus* (okra) and *Raphanus sativus* (radish). Angew. Botanik 62, 119–134.

Owensby C E, Coyne P I and Auen L M 1993a Nitrogen and phosphorus dynamics of a tallgrass prairie ecosystem exposed to elevated carbon dioxide. Plant Cell Environ. 16, 843–850.

Owensby C E, Coyne P I, Ham J M and Auen L M 1993b Biomass production in a tallgrass prairie ecosystem exposed to ambient and elevated CO_2. Ecol. Appl. 3, 644–653.

Rice C W, Garcia F, Hampton C O and Owensby C E 1994 Soil microbial biomass and respiration in a tallgrass prairie exposed to increased levels of atmospheric CO_2. Plant and Soil (This volume).

Risser P G and Parton W J 1982 Ecosystem analysis of the tallgrass prairie: nitrogen cycle. Ecology 63, 1342–1351.

Seastedt T R 1988 Mass, nitrogen and phosphorus dynamics in foliage and root detritus of tallgrass prairie. Ecology 69, 59–65.

Schlesinger W H 1985 Decomposition of chaparral shrub foliage. Ecology 66, 1353–1359.

Staaf H and Berg B 1982 Accumulation and release of plant nutrients in decomposing Scots pine needle litter. II. Long-term decomposition in a Scots pine forest. Can. J. Bot. 60,1561–1568.

Strain B R 1987 Direct effect of increasing atmospheric CO_2 on plants and ecosystems. Trend. Ecol. Evol. 2,18–21.

Swift M J, Heal O W and Anderson J M 1979 Decomposition in Terrestrial Ecosystems. Studies in Ecology, Volume 5. Blackwell, Oxford, England.

Taylor B R, Parkinson D and Parsons W J F 1989 Nitrogen and lignin content as predictors of litter decay rates: a microcosm test. Ecology 70, 97–104.

Wong S-C 1979 Elevated atmospheric partial pressure of CO_2 and plant growth. I. Interactions of nitrogen nutrition and photosynthetic capacity in C_3 and C_4 plants. Oecologia 44, 68–74.

Wong S-C 1990 Elevated atmospheric partial pressure of CO_2 and plant growth. II. Nonstructural carbohydrate content in cotton plants and its effect on growth parameters. Photosyn. Res. 23, 171–180.

Woodward F I, Thompson G B and McKee I F 1991 The effects of elevated concentrations of carbon dioxide on individual plants, populations, communities and ecosystems. Ann Bot. (Supplement 1) 67, 23–38.

Appendix

Table A1. F and p values from ANOVA of percent dry matter remaining in litter bags at various sample dates

Factor	March 1991	May 1991	Aug 1991	Nov 1991	May 1992
	F	F	F	F	F
Species	219.33***	418.84***	277.49***	98.35***	76.84***
Chamber	10.99**	0.04	0.35	0.22	0.00
CO_2	4.12*	20.75***	10.67**	0.00	2.39
Nitrogen	23.77***	10.01**	21.27***	8.82**	2.54
Sp × Ch	3.04	19.71***	20.11***	4.85*	4.62*
Sp × CO_2	0.89	0.12	1.93	2.55	0.95
Sp × N	2.01	3.26*	4.14*	5.16**	0.83
Ch × N	0.21	0.01	0.52	2.60	3.72
CO_2 × N	1.10	2.87	4.02	1.34	0.83
Sp × Ch × N	2.13	0.07	0.38	1.11	0.23
Sp × CO_2 × N	4.29	5.50**	2.94	5.74**	1.34

Table A2. F and p values from ANOVA of soluble matter in litter at various sample dates

Factor	Nov 1990	March 1991	May 1991	Aug 1991	Nov 1991	May 1992
	F	F	F	F	F	F
Species	495.24***	294.36***	324.07***	46.57***	67.45***	9.98***
Chamber	28.66***	1.55	0.08	0.01	3.55	0.03
CO_2	2.29	0.65	2.53	4.17*	1.74	6.38*
Nitrogen	1.03	0.00	38.77***	6.33*	0.01	16.27***
Sp × Ch	4.09*	3.20*	4.61*	0.62	1.89	0.53
Sp × CO_2	2.54	0.33	1.66	0.72	1.66	0.44
Sp × N	10.89***	0.09	9.45***	10.19***	10.57***	18.13***
Ch × N	4.39*	3.81	2.06	0.02	4.19*	9.02**
CO_2 × N	4.34*	0.04	0.11	11.04**	0.85	0.01
Sp × Ch × N	3.95*	1.87	0.24	1.77	0.10	2.91
Sp × CO_2 × N	9.98***	5.34**	1.09	4.79*	0.81	0.02

Table A3. F and p values from ANOVA of cellulose in litter at various sample dates

Factor	Nov 1990	March 1991	May 1991	Aug 1991	Nov 1991	May 1992
	F	F	F	F	F	F
Species	390.29***	107.17***	365.04***	206.83***	45.58***	67.81***
Chamber	12.86***	0.32	25.06***	26.41***	5.29*	5.77*
CO_2	3.30	1.43	29.68***	11.11**	0.08	3.09
Nitrogen	2.35	8.59**	10.27**	0.20	0.00	3.66
Sp × Ch	11.53***	1.86	8.23***	16.63***	2.15	2.03
Sp × CO_2	4.44*	1.31	14.81***	2.87	4.49*	0.17
Sp × N	10.16***	5.43**	2.41	3.76*	4.80*	0.08
Ch × N	2.31	1.28	6.21*	0.07	3.73	0.84
CO_2 × N	4.42*	0.11	0.33	9.69**	0.02	0.56
Sp × Ch × N	3.25*	2.74	2.41	2.30	0.48	0.10
Sp × CO_2 × N	7.29**	0.33	3.03	11.73***	5.34**	0.29

Table A4. F and p values from ANOVA of lignin in litter at various sample dates

Factor	Nov 1990	March 1991	May 1991	Aug 1991	Nov 1991	May 1992
	F	F	F	F	F	F
Species	193.87***	16.34***	127.86***	115.62***	16.56***	66.10***
Chamber	32.28***	0.00	54.09***	34.42***	3.86	7.75**
CO_2	0.52	0.77	44.96***	4.49*	0.56	0.89
Nitrogen	0.30	9.40**	7.09**	4.48*	0.04	0.39
Sp × Ch	4.60*	2.52	19.96***	25.34***	4.89*	3.61*
Sp × CO_2	1.16	1.63	23.42***	4.64*	3.81*	0.09
Sp × N	0.81	6.70**	2.38	10.57***	1.27	3.32*
Ch × N	2.15	0.19	4.59*	0.13	2.18	0.02
CO_2 × N	0.15	0.07	0.26	0.90	0.38	0.66
Sp × Ch × N	0.91	2.36	2.78	1.17	0.95	0.35
Sp × CO_2 × N	1.03	2.80	2.60	6.36**	9.63***	0.51

Table A5. F and *p* values from ANOVA of total Kjeldahl N content of litter at various sample dates

Factor	Nov 1990	March 1991	May 1991	August 1991	Nov 1991	May 1992
	F	F	F	F	F	F
Species	3645.54***	654.83***	621.29***	447.97***	213.04***	276.85***
Chamber	101.18***	13.39***	30.77***	11.99***	12.05***	10.88**
CO_2	43.57***	16.69***	5.54*	11.49 **	0.26	2.88
Nitrogen	95.65***	14.53***	0.81	0.07	0.09	0.72
Sp × Ch	42.40***	4.72*	7.92***	8.71***	1.79	3.15
Sp × CO_2	34.12***	4.79*	2.40	1.00	4.84*	1.49
Sp × N	44.00***	11.98***	9.75***	10.20***	5.79**	1.25
Ch × N	1.10	4.53*	0.45	0.37	0.03	0.00
CO_2 × N	32.01***	7.65**	4.96*	0.62	0.81	0.05
Sp × Ch × N	1.09	3.90*	1.40	1.80	1.57	0.10
Sp × CO_2 × N	14.92***	3.02	1.73	4.06*	5.97**	0.31

Table A6. F and *p* values from ANOVA of total Kjeldahl P content of litter at various sample dates

Factor	Nov 1990	March 1991	May 1991	Aug 1991	Nov 1991	May 1992
	F	F	F	F	F	F
Species	208.41***	367.22***	385.50***	311.37***	135.26***	233.48***
Chamber	3.51	0.00	10.87**	5.54*	8.27**	12.54***
CO_2	0.06	1.26	0.35	6.72*	0.68	0.67
Nitrogen	47.80***	38.18***	10.25**	0.14	0.01	1.09
Sp × Ch	18.04***	2.79	9.52***	9.80***	1.03	3.44*
Sp × CO_2	25.08***	6.98	6.54**	3.14	4.42*	1.99
Sp × N	19.50***	26.63***	23.55***	6.11**	8.87***	2.64
Ch × N	3.19	2.80	0.46	0.12	0.00	0.36
CO_2 × N	25.51***	0.00	1.05	0.01	1.17	0.07
Sp × Ch × N	7.49**	3.96*	3.59*	1.80	2.17	0.34
Sp × CO_2 × N	18.50***	8.77***	4.78*	3.95*	8.33***	1.12

Plant and Soil **165**: 129-138, 1994.

Soil pCO_2, soil respiration, and root activity in CO_2-fumigated and nitrogen-fertilized ponderosa pine

Dale Johnson[1], Donn Geisinger[2], Roger Walker[2], John Newman[1], James Vose[3], Katherine Elliot[3] and Timothy Ball[1]
[1]*Desert Research Institute and Environmental and Resource Sciences, University of Nevada, Reno, NV 09506-60220, USA,* [2]*Environmental and Resource Sciences, University of Nevada, Reno, NV, USA and* [3]*Coweeta Hydrologic Lab, U.S. Forest Service, Otto, NC, USA*

Key words: carbon dioxide, nitrogen, ponderosa pine, soil respiration, soil carbon

Abstract

The purpose of this paper is to describe the effects of CO_2 and N treatments on soil pCO_2, calculated CO_2 efflux, root biomass and soil carbon in open-top chambers planted with *Pinus ponderosa* seedlings. Based upon the literature, it was hypothesized that both elevated CO_2 and N would cause increased root biomass which would in turn cause increases in both total soil CO_2 efflux and microbial respiration. This hypothesis was only supported in part: both CO_2 and N treatments caused significant increases in root biomass, soil pCO_2, and calculated CO_2 efflux, but there were no differences in soil microbial respiration measured in the laboratory. Both correlative and quantitative comparisons of CO_2 efflux rates indicated that microbial respiration contributes little to total soil CO_2 efflux in the field. Measurements of soil pCO_2 and calculated CO_2 efflux provided inexpensive, non-invasive, and relatively sensitive indices of belowground response to CO_2 and N treatments.

Introduction

There is increasing evidence that belowground processes can be strongly affected by increases in atmospheric CO_2. Several studies have shown that root growth responds disproportionately to increases in atmospheric CO_2 (Norby et al., 1986, 1987, 1992; Rogers et al., 1992; Walker et al., 1994). Norby et al. (1987) found increased root exudation from *Pinus echinata* under elevated atmospheric CO_2, and suggested that this may provide a mechanism for enhancing rhizosphere nutrient availability. Körner and Arnone (1992) noted increases in fine root biomass and soil respiration with elevated CO_2 in an artificial tropical ecosystem. The authors also found a reduction in soil C, which they attributed to stimulation of decomposition in the rhizosphere though root exudation. Zak et al. (1993) found that elevated CO_2 caused increases in labile C and N in soil from *Populus grandidentata* seedlings grown under elevated CO_2. The authors suggested that elevated CO_2 may create a positive feedback on soil C and N dynamics and tree growth because of increased N availability in the rhizosphere caused by root exudation.

It is obvious from the literature cited above that belowground effects of elevated atmospheric CO_2 are important and must be monitored in experiments involving CO_2 treatments. There are some significant methodological problems in monitoring belowground activity, however. Destructive methods such as soil coring and root harvesting give the best quantitative estimates of belowground response, but cannot be conducted on a routine basis in small plots. Non-destructive measurements such as soil respiration and nutrient leaching provide sensitive indices of overall belowground activity, but generally do not provide information on the responses of individual belowground components (i.e., roots vs microbes).

The Forest Response to CO_2 project (Ball et al., 1992) is studying the responses of ponderosa and loblolly pine (*Pinus ponderosa* and *P. taeda*) to CO_2 and N treatments in open top chamber facilities. A major goal of this research is to gain a comprehensive picture of belowground response to treatments by combining a variety of non-destructive belowground monitoring techniques (video imaging with mini-rhizotrons, soil solution sampling, soil atmosphere CO_2 concentration [pCO_2] and CO_2 efflux) — with periodic destructive sampling of soils and vegetation. Previous papers have reported the initial results of soil respiration measurements using dynamic chambers (Vose et al., 1994) and root phenology using mini-rhizotrons (Johnson et al., 1994) in the ponderosa pine site. This paper will focus upon responses of soil pCO_2 and calculated CO_2 efflux to treatments, comparing these responses to changes in root biomass and soil carbon content. Of all of the methods listed, measurements of soil pCO_2 and calculation of CO_2 efflux are the least invasive, require the least equipment and are most easily made on an extensive, routine basis (de Jong and Schappert, 1972). Based upon the literature, we hypothesized that elevated CO_2 and N treatments would cause increased root biomass which would in turn cause increased total soil CO_2 efflux, root and microbial respiration.

Site and methods

Site

The open-top chamber site was located at the Institute of Forest Genetics in Placerville, California. The soil is Aiken clay loam, a Xeric Haplohumult derived from andesite. Soils were intensively sampled prior to chamber establishment, and were found to be very uniform. Some average chemical and physical properties of the soils from the site are shown in Table 1.

Treatments

During February–April 1991, 24 hexagonal open-top chambers (3.6 m in diameter) were established on the site. The basic experimental design consisted of three levels of nitrogen (0, 10, and 20 g m^{-2} yr^{-1} of N as ammonium sulfate, applied in early spring), and four CO_2 treatments (ambient, no chamber; ambient, chambered; 525 $\mu L\ L^{-1}$ CO_2; and 700 $\mu L\ L^{-1}$ CO_2). Water was delivered to each plot via a timed stand pipe to a looped one inch diameter manifold, and low pressure spray heads. Each of the chambered treatments was replicated three times, and each of the unchambered treatments was replicated twice. Only the results from the chambered measurements will be reported here. Due to cost limitations, the 10 g m^{-2} yr^{-1} N, 525 $\mu L\ L^{-1}$ CO_2 treatment was excluded. Treatments were begun in May, 1991. A full description of chamber operation is given by Ball et al. (1992). In May of 1991, Ponderosa pine (*Pinus ponderosa*) was planted in each chamber. Seedlings were grown from seed (21 planting locations per chamber) and seedlings (21 per chamber), the latter being a backup in the event of excessive mortality. Seed-grown seedling survival was very good, and the seedling-grown stock was removed in October 1991. Weeds were controlled by laying weedcloth around seedlings in each chamber. Weedcloth was found to have no effect on CO_2 retention in the soil: CO_2 concentrations beneath the weedcloth and above the soil surface were at ambient levels for the chamber being sampled, and there were no discernible effects of weedcloth presence or absence upon pCO_2.

Biomass harvesting

In October 1991, three trees from each chamber were harvested, including complete root systems. In 1992, three trees from each chamber were harvested again, but only one complete root system per chamber was obtained because of the increased size of the seedlings and concern for excessive plot disturbance. Root biomass by size class and mycorrhizal infection were analyzed in each case and will be reported in later papers (R.F. Walker, unpubl. data). Only total root biomass will be reported here.

Soil pCO_2, temperature and moisture monitoring

Gas wells were established at 15 and 30 cm depths in each chamber. The gas wells consisted of 4 mm tubing inserted to the proper depths in the soil and fitted with a stoppered, female end of a plastic union at the surface. During gas collections, stoppers were removed and 15 mL of gas was withdrawn from the well (enough to completely evacuate the tubing and obtain soil gas) using a 50 mL syringe fitted with tygon tubing and the male half of the union. Samples for CO_2 analyses were obtained with Hamilton gas syringes from the section of tygon tubing between the large syringe and the union. CO_2 analyses were performed on a LiCOR 6250 CO_2 analyzer using peak heights compared to a standard gas of 0.877% CO_2. Soil moisture was measured by various methods during the early part of the study. From July–August 1992 portable tensiometers (Soil Moisture Corp.) were used. In that soil moisture tension was normally kept well below 50 kPa by irrigation, there was little concern that tensiometers would become inoperable. The portable tensiometers provided adequate estimates of soil moisture content, as evidenced by comparisons with gravimetric analyses, but they were abandoned in October 1992 in

Table 1. Some chemical and physical properties of the Placerville site soils (Aiken clay loam, Xeric Haplohumult derived from andesite)

Horizon and depth (cm)	Db ($g\ cm^{-3}$)	%>2mm (%)	C[a] ($mg\ g^{-1}$)	N[b] ($mg\ g^{-1}$)	C/N	Bray P[b] ($mg\ kg^{-1}$)	pH[c]	CEC[d]	Exchangeable cations Ca	Mg	K	Na	Al	%BS[e] (%)
								($cmol_c\ kg^{-1}$)						
Ap (0–18)	1.14	1	22.0	0.9	24	12.1	5.1	11.24	4.37	0.62	0.74	0.04	0.68	51
Bw (18–30)	1.24	1	18.0	0.9	21	10.9	5.1	9.39	4.26	0.62	0.74	0.03	0.78	65
Bt (30+)			7.1	0.4	16	1.6	5.5	14.89	6.11	1.18	0.90	0.04	0.02	57

[a] Perkin-Elmer 2400 CHN Analyzer.
[b] 0.5 *M* HCl + 1 *M* NH_4F (Olson and Sommers, 1982).
[c] 0.1 *M* $CaCl_2$.
[d] Cation exchange capacity and exchangeable cations by 1 *M* NH_4Cl extraction followed by 1 *M* KCl.
[e] Percent base saturation.

favor of gravimetric samples because the time necessary to obtain tensiometer measurements was greater than that needed to take soil samples for gravimetric analyses. Between October 1992 and May 1993 gravimetric analyses were used for estimation of soil water content. After that time, gypsum blocks were calibrated and used for estimations of water content, because of concern over the repeated effects of destructive soil sampling.

Measurement and calculation of soil CO_2 efflux

Respiration by the chamber method was measured with a continuous flow infrared gas analyzer (IRGA) system. Soil CO_2 flux was measured from soil chambers (10 cm diameter, 10 cm high, 785 cm^3 volume) inserted 1.25 cm into the ground. During measurement, caps with inlet and outlet ports were placed on each core to measure CO_2 evolution. Flow rate was 800 to 1200 mL min^{-1} to minimize turbulence within the sample core and to ensure measurable levels of CO_2 evolution. In addition, inlet and outlet flow were carefully monitored to ensure that no suction was created in the outlet sample. Carbon dioxide concentrations of air entering and leaving the chambers were measured and logged electronically with an ADC LCA3 IRGA and a Campbell 21 X data logger. Continuous measurements were taken across treatments (i.e., all factorial combinations) within a randomly selected replication (chamber) at each sample period (October 1992, April 1993, and June 1993.)

Soil CO_2 efflux was calculated according to the CO_2 profile method outlined by de Jong and Schappert (1972). This method is based upon the assumption that CO_2 efflux is dominated by diffusion and therefore controlled by the partial pressure of CO_2 in the soil atmosphere (pCO_2) and the diffusivity of CO_2 in the soil (de Jong and Schappert, 1972; Rolston, 1986). At steady-state,

$$q = D\frac{dC}{dz} \tag{1}$$

where q= CO_2 efflux (g CO_2 -C m^{-2} day^{-1}) = root and microbial respiration at steady-state, C = soil CO_2 -C concentration (g m^{-3}), z = depth (m), D = diffusion coefficient (m^2 day^{-1}).

Because CO_2 diffusion through water is much lower than in air, D is strongly affected by soil water content. There are several formulations for D (Collin and Rasmuson, 1988), all of which take soil moisture content into account. In this paper, we used a modification of the equation given by Millington (1959) (as quoted by Rolston, 1986):

$$D = (\partial)(D_a)(P_{eff}^{10/3})/E^2 \tag{2}$$

where D = diffusion coefficient of CO_2 in air (cm^2 sec^{-1}), (E = voids ratio, or total soil porosity, P_{eff} = effective porosity = total porosity (E) minus volumetric water content (V_w), and ∂ = a coefficient to account for non-ideal pore shape and dead-end pores (Collin and Rasmuson, 1988). For the Placerville soil, the value of

∂ was determined to be 0.1 based upon comparisons with measured CO_2 efflux using dynamic chambers (Vose et al., 1994).

Soil sampling and analysis

In March of 1991 and 1993, three replicate soil samples were taken in each chamber from the Ap (0–18 cm) and Bw (18–30 cm) horizons for bulk density, percent gravel, and chemical analyses. Samples were dried, sieved (< 2 mm), bulked by chamber, and analyzed for total C and total N on a Perkin-Elmer 2400 CHN analyzer.

On July 8, 1993, additional samples were taken as above from the Ap horizon for laboratory incubation. The samples were sieved and bulked by chamber in a field-moist condition, taking great care to remove root fragments. 100 g of field-moist soil was placed in 255 mL fruit jars fitted with septa for gas sampling and incubated at 25°C for 21 days. Gas samples from the headspace were taken at the initiation and on a daily basis and analyzed for CO_2 as described above. When headspace pCO_2 levels exceeded 1.5% (the maximum levels found at 15 cm in the field), the chambers were flushed with ambient air, resampled to establish baseline values again, and allowed to incubate further. Flushings took place on days 6 and 14.

Statistical analyses

Statistical analyses included two-way analysis of variance, with treatment effects considered significant only at $p \lesseqgtr 0.05$ (SYSTAT software). Treatment means were compared using Tukey's HSD procedure, $p \leq 0.05$.

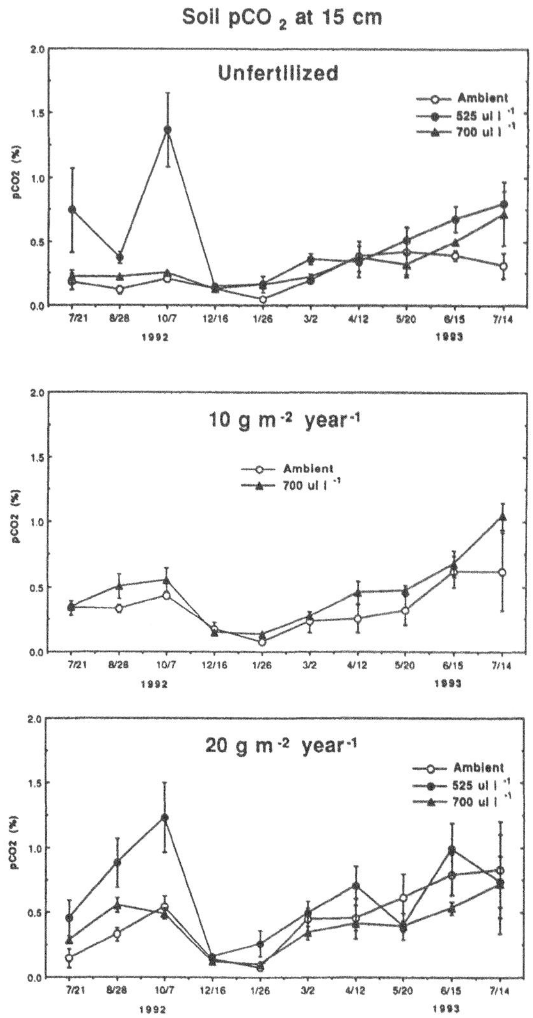

Fig. 1. Seasonal trends in soil pCO_2 at the 15 cm depth at the Placerville field site.

Results and discussion

Effects of treatments on root biomass, soil pCO_2, soil CO_2 efflux, and soil C pools

In the summer and autumn of 1992, there was a significant positive effect of CO_2 treatment on soil pCO_2 and calculated CO_2 efflux, especially in the 525 μL L^{-1} CO_2-treated chambers (Figs. 1–3). There was also a smaller but significant N fertilizer effect upon soil pCO_2 and calculated CO_2 efflux during the spring and summer of 1992. Soil pCO_2 in all treatments decreased substantially during the winter of 1992–1993, probably in response to the precipitous decrease in soil temperature (Fig. 4). The removal of 20% of the biomass in late October 1992 may also have caused some reduction in soil pCO_2 and calculated CO_2 efflux. Both soil pCO_2 and temperature rose again in the spring and summer of 1993, and by June of 1993, there were significant effects of both CO_2 and N treatments. However, the predominance of the 525 μL L^{-1} CO_2 treatments did not re-emerge after the winter of 1992.

Between October 1991 and October 1992, root biomass increased by approximately 2 orders of magnitude, and CO_2 treatment effects began to predominate over the initial N treatment effects (Fig. 5). In October 1991, there was a significant N treatment effect upon root biomass but no significant CO_2 treatment effect. In October 1992, there was a significant N treatment

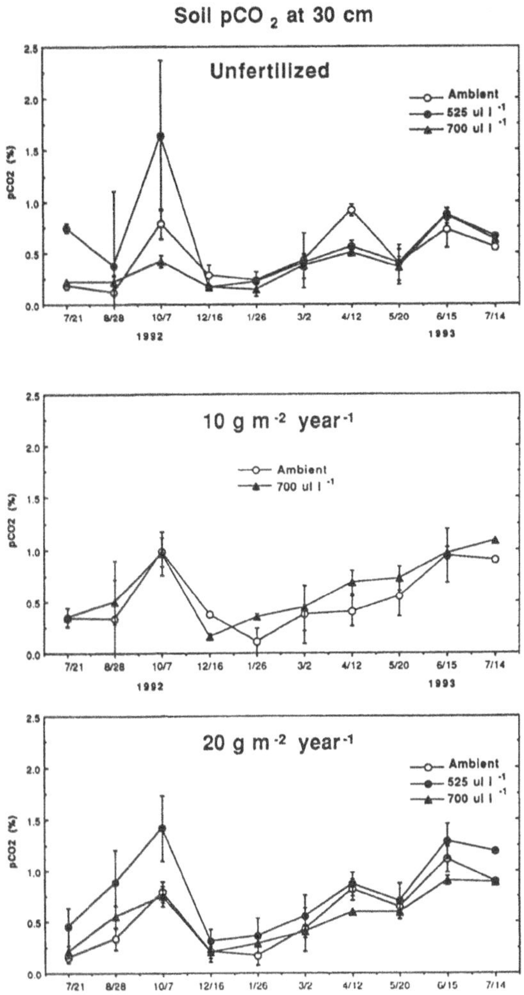

Fig. 2. Seasonal trends in soil pCO_2 at the 30 cm depth at the Placerville field site.

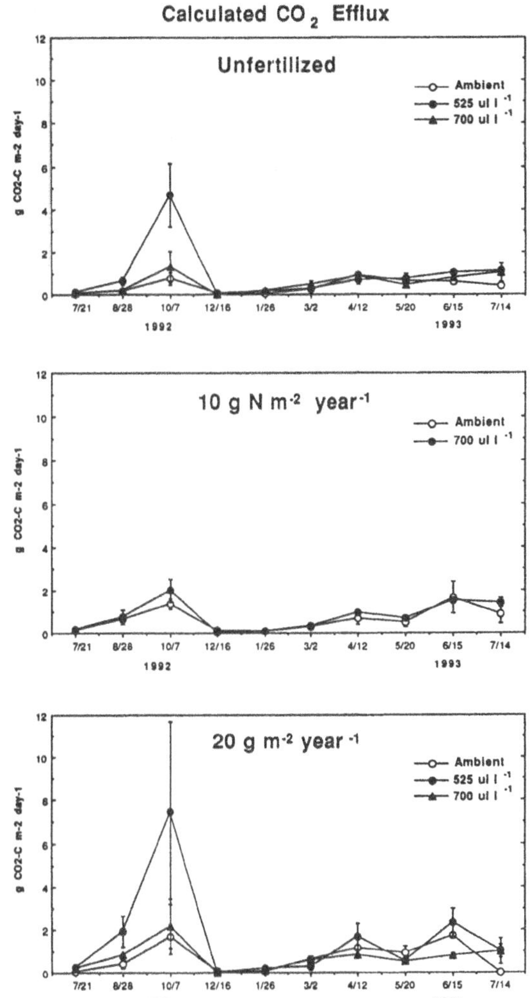

Fig. 3. Seasonal trends in calculated CO_2 efflux at the Placerville field site.

effect in the ambient and 700 $\mu L\ L^{-1}$ CO_2 chambers, but CO_2 effects were larger and statistically significant at all fertilization levels. There was a tendency for greater root biomass in the 525 $\mu L\ L^{-1}$ than in the 700 $\mu L\ L^{-1}$ CO_2 treatments, but the differences were statistically significant only in the unfertilized chambers.

There were statistically significant correlations among root biomass, measured and calculated soil CO_2 efflux, pCO_2 at 15 and 30 cm in October 1992 (Table 2). These correlations suggest that roots make a major contribution to total soil respiration. The literature suggests that from 1/3 to 2/3 of total soil respiration can be attributed to roots in mature forest ecosystems with well-developed litter layers (Edwards and Harris, 1977; Johnson et al., 1975; Ewell et al., 1987b; Raich and Nadelhoffer, 1989). Given the fact that no litter layer was present in this study, it is probable that the role of roots in total soil respiration was even more significant than in mature forests.

Results from laboratory incubations suggest that microbial respiration from bulk soils contributed relatively little to total soil respiration. There were no significant treatment effects on CO_2 efflux from laboratory-incubated soils samples, whereas there were statistically significant effects of both CO_2 and N treatment upon soil pCO_2 and calculated CO_2 efflux in the field on the sampling date (Fig. 6). Furthermore, the rates of CO_2 efflux in the laboratory, when

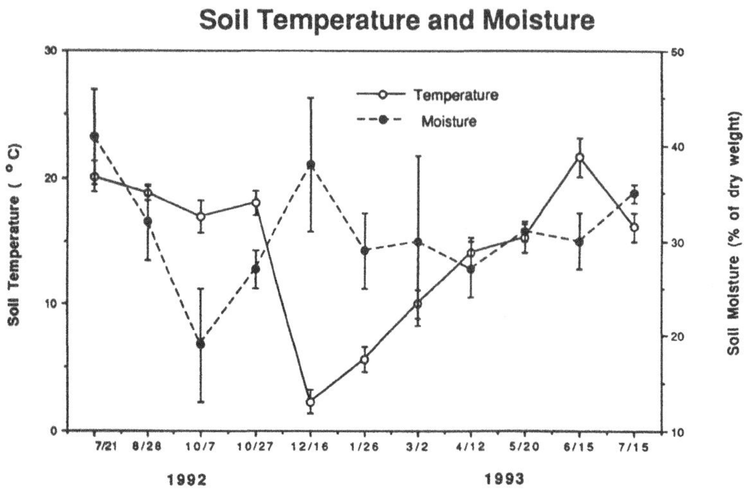

Fig. 4. Seasonal trends in soil temperature and moisture at the Placerville field site.

Table 2. Correlation coefficients (r^2) among root biomass, pCO_2, and CO_2 flux in the Placerville field site in October 1992

Variable	Root Biomass (g tree^{-1})	pCO_2 at 15 cm (%)	pCO_2 at 30 cm (%)	Measured CO_2 flux (g m^{-2}day^{-1})
pCO_2 at 15 cm (%)	0.71	-	-	-
pCO_2 at 30 cm (%)	0.50	0.83	-	-
Measured CO_2 flux (g m^{-2} day^{-1})	0.45	0.69	0.70	-
Calculated CO_2 flux (g m^{-2} day^{-1})	0.58	0.83	0.54	0.50

scaled up to a g m^{-2} day^{-1} level (using Ap horizon soil weights) equalled about 10% of calculated CO_2 efflux in the field for the sampling date (Fig. 6). Actual microbial CO_2 efflux rates in the field were probably considerably less than those determined in the laboratory under ideal temperature and moisture conditions after substantial soil disturbance due to sampling and processing.

In contrast to the results of Körner and Arnone (1992) we found no reduction in soil C with CO_2 treatments. With one exception, there were no statistically significant differences in soil C content between 1991 and 1993 (Table 3). The exception was in the Bw horizon of the 700 μL L^{-1}, 20 g N m^{-2} year^{-1} treatment, where 1993 soil C was significantly greater than 1991 soil C. However, the 1991 soil C in this particular case was unusually low, suggesting the possibility of a Type II statistical error.

One of the probable reasons for the differences in our results and those of Körner and Arnone (1992) is that we used a natural soil with relatively large, stable soil C pools (approximately 6,000 to 8,000 g C m^{-2}), whereas the artificial soil used by Körner and Arnone (1992) (a mineral mixture of silicate sand and vermiculite) contained only 15 to 20% as much total C (1280 g C mm^{-2}) which was derived from overlying compost material. Also, soil C losses of the magnitude reported by Körner and Arnone (1992) (75 to 300 g C m^{-2}) could have gone undetected in the Placerville soil. The differences in soil C between 1991 and 1993 at Placerville (from -725 to +1239 g C m^{-2}), while mostly non-significant, were much greater than those measured by Körner and Arnone (1992) (Table 3).

The absolute values of soil C change calculated as a daily loss can be compared to soil CO_2 efflux rates determined in the laboratory and in the field in order to gain additional insight into the relative roles of roots and microbes in affecting total soil respiration. These daily values were similar in magnitude to soil CO_2 efflux rates measured in the laboratory (Table 3 and

Table 3. Changes in soil C at Placerville, 1991–1993

Treatment		Horizon	Soil C content (g m^{-2})			Percent change	Daily soil C loss	Average annual CO_2-C Efflux
CO_2	N		1991	1993	1993-1991	(%)		(g m^{-2} day $^{-1}$)
Ambient	Unfertilized	Ap	4557±604	4205±264	-352	-8	0.48	
Ambient	Unfertilized	Bw	2857±434	2484±295	-373	-13	0.51	
Ambient	Unfertilized	Total	7415±695	6689±296	-725	-10	0.99	1.4±0.5
Ambient	10 g m^{-2} yr^{-1}	Ap	4150±470	3832±416	-318	-8	0.44	
Ambient	10 g m^{-2} yr^{-1}	Bw	2651±230	2779±374	128	5	-0.17	
Ambient	10 g m^{-2} yr^{-1}	Total	6802±524	6612±560	-191	-3	0.26	3.4±1.4
Ambient	20 g m^{-2} yr^{-1}	Ap	4448±669	4069±411	-379	-9	0.52	
Ambient	20 g m^{-2} yr^{-1}	Bw	2636±269	2469±123	-167	-6	0.23	
Ambient	20 g m^{-2} yr^{-1}	Total	7085±749	6539±429	-546	-8	0.75	1.8±0.8
525 $\mu L\ L^{-1}$	Unfertilized	Ap	4564±328	5105±520	542	12	-0.74	
525 $\mu L\ L^{-1}$	Unfertilized	Bw	2970±537	2627±806	-344	-12	0.47	
525 $\mu L\ L^{-1}$	Unfertilized	Total	7534±630	7732±959	198	3	-0.27	2.2±0.7
525 $\mu L\ L^{-1}$	20 g m^{-2} yr^{-1}	Ap	5329±981	4780±692	-548	-10	0.75	
525 $\mu L\ L^{-1}$	20 g m^{-2} yr^{-1}	Bw	2789±245	2563±317	-226	-8	0.31	
525 $\mu L\ L^{-1}$	20 g m^{-2} yr^{-1}	Total	8118±1011	7343±761	-774	-10	1.06	2.6+0.7
700 $\mu L\ L^{-1}$	Unfertilized	Ap	4543±809	4015±96	-528	-12	0.72	
700 $\mu L\ L^{-1}$	Unfertilized	Bw	2484±274	2622±549	137	6	-0.19	
700 $\mu L\ L^{-1}$	Unfertilized	Total	7028±854	6637±557	-391	-6	0.54	2.7±0.8
700 $\mu L\ L^{-1}$	10 g m^{-2} yr^{-1}	Ap	3974±438	4631±1002	657	17	-0.90	
700 $\mu L\ L^{-1}$	10 g m^{-2} yr^{-1}	Bw	2877±450	2293±264	-584	-20	0.80	
700 $\mu L\ L^{-1}$	10 g m^{-2} yr^{-1}	Total	6852±628	6924±1036	73	1	-0.10	6.0±4.1
0.00								
700 $\mu L\ L^{-1}$	20 g m^{-2} yr^{-1}	Ap	4225±583	4536±393	311	7	0.43	
700 $\mu L\ L^{-1}$	20 g m^{-2} yr^{-1}	Bw	1934±356	2872±362***	938	48	-1.28	
700 $\mu L\ L^{-1}$	20 g m^{-2} yr^{-1}	Total	6160±683	7409±534*	1249	20	-1.71	2.4±1.1

**Statistically significant, 95% level.
***Statistically significant, 99% level.

Fig. 6), but were 30–90% lower than average annual calculated CO_2 efflux (Table 3).

It seems clear from these comparisons and the correlations among root biomass, soil pCO_2, and CO_2 efflux that microbial respiration contributes little to total CO_2 efflux. This does not necessarily imply that autotrophic root respiration completely dominates total soil CO_2 efflux, however: there is the distinct possibility that rhizosphere microbial activity is also a significant CO_2 source (Zak et al., 1993). The possibility of treatment effects on rhizosphere soil C and microbial communities is currently under investigation.

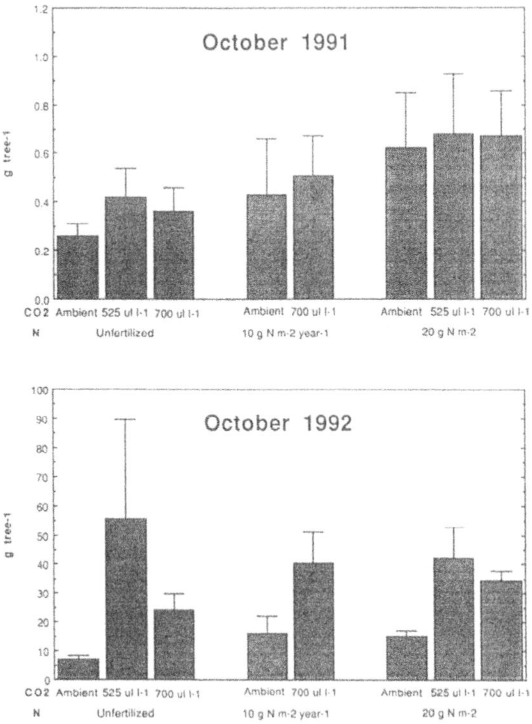

Fig. 5. Root biomass in October 1991 and October 1992.

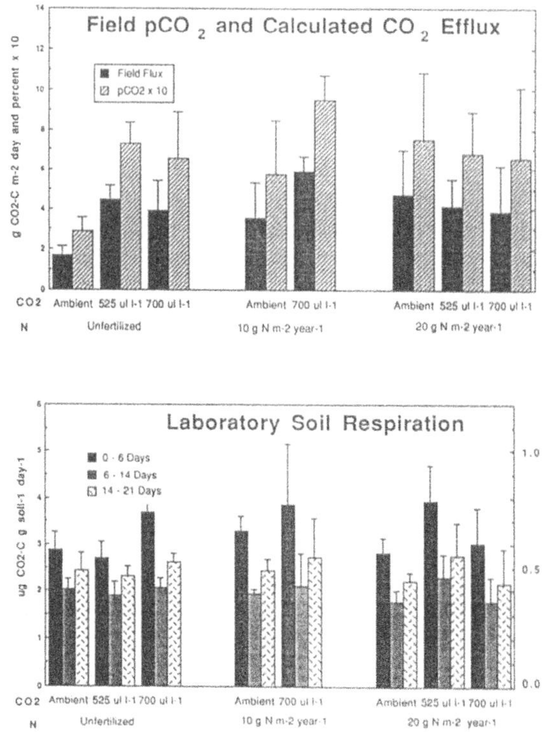

Fig. 6. Soil pCO_2 and calculated CO_2 efflux in the field (top) and soil CO_2 efflux during laboratory incubations of samples taken in July 1993.

Physical factors affecting soil pCO_2 and calculated CO_2 efflux

The data presented above suggest that soil pCO_2 and calculated CO_2 efflux can be used as indices of below-ground activity at the Placerville site. There remains the question as to how soil pCO_2 might respond to changes in the diffusion coefficient (D) caused by changes in soil moisture, however (de Jong and Schappert, 1972; Solomon and Cerling, 1987). The extent to which soil pCO_2 is sensitive to total soil respiration (q, or total soil CO_2 efflux), soil depth (z), and moisture content (Vw) can be seen by combining equations 1 and 2, integrating, and solving for C:

$$C_z = \frac{(q)(z)}{J(Da)[(E - V_w)^{10/3}/E^2]} + C_o \qquad (3)$$

where C_z = pCO_2 at depth z and C_o = pCO_2 in the atmosphere above the soil.

Equation 3 shows that soil pCO_2 (C_z) increases with depth (z) and total respiration (q) for a given (constant) soil moisture content (V_w). Soil pCO_2 at 30 cm was nearly always (89% of the time) greater than pCO_2 at 15 cm, as predicted by equation 3 (as well as by the more sophisticated equations of Wesseling [1962]). Equation 3 also shows that soil pCO_2 at a given depth is more sensitive to changes in soil moisture (V_w) than total respiration (q): C_z is directly proportional to q and inversely proportional to the 10/3 power of the quantity [E-V_w] containing the soil moisture term.

Despite the potential importance of soil moisture on soil pCO_2, there was no correlation between soil pCO_2 and soil moisture over the sampling period (r^2 = 0.04) or among chambers on any specific sampling date. There was, however, a weak but statistically significant correlation between soil pCO_2 and soil temperature (r^2 = 0.29). In contrast, calculated CO_2 efflux was less correlated with temperature (r^2 = 0.12) than with moisture (r^2 = 0.31).

The effect of soil moisture on pCO_2 and calculated CO_2 efflux can also be seen clearly from the temporal patterns in these parameters in the spring and summer of 1992. For experimental reasons, the soil was allowed to dry in October 1992 (Fig. 4), causing D to decrease by two orders of magnitude (from 5.4×10^{-8} cm^2 sec^{-1} in July to 1.9×10^{-6} cm^2 sec^{-1}

in October). As the soil dried, calculated CO_2 efflux increased, as would be expected from equation 1, but soil pCO_2 increased as well (Figs. 1–3). Thus, the temporal variations in pCO_2 in this soil appear to be driven primarily by variations in root and soil microbial respiration rather than by variations in soil moisture and D, whereas D is a major factor in calculated CO_2 efflux. This contrasts with the results of de Jong et al. (1974), where wetting and drying cycles had a major effect upon soil respiration in native grasslands and cultivated soils. Solomon and Cerling (1987) found that the presence of a snowpack created a diffusion barrier and produced elevated soil pCO_2 even with low respiration rates in a montane meadow in Utah. Thus, variations in soil moisture may significantly affect soil pCO_2 under other circumstances and will nearly always significantly affect calculated CO_2 efflux.

Conclusions

The hypothesis that elevated CO_2 and N fertilization would cause increased root biomass and total soil CO_2 efflux was supported by the results of these studies: both CO_2 and N treatments had significant, positive effects upon root biomass, soil pCO_2 and calculated CO_2 efflux. The intermediate (525 $\mu L\ L^{-1}$) CO_2 treatment produced the highest soil pCO_2 and root biomass. The hypothesis that treatments would cause increases in microbial respiration were not supported, however: there were no differences in laboratory-determined soil CO_2 efflux. It appears that microbial respiration in bulk soils contributed little to total soil respiration and that the patterns in both pCO_2 and CO_2 efflux are due mainly to differences in root biomass. It remains to be seen as to whether rhizosphere microbial respiration is significant or not, however.

It appears that soil pCO_2 and calculated CO_2 efflux provide sensitive and relatively cheap indices of belowground response to treatments. Treatment effects on soil CO_2 efflux measured by dynamic chambers, although similar in overall pattern and significantly correlated with root biomass, soil pCO_2, and calculated CO_2, were not statistically significant because of high variability (Vose et al., 1994). Soil pCO_2 measurements offer the decided advantage of being easily and, therefore, cheaply obtained, allowing greater flexibility in number of samples and frequency of sampling. The relative times required for soil pCO_2 and chamber CO_2 efflux measurements (four hours as opposed to one week) also give the pCO_2 method the advantage of avoiding temporal variations during the sampling period. Finally, the soil pCO_2 method offers the advantage of observing pCO_2 responses and calculating CO_2 effluxes at different depths. The major disadvantage of the soil pCO_2 method is that CO_2 efflux calculations are highly sensitive to assumptions about the diffusion coefficient (D), a parameter that cannot be directly monitored on a routine basis. Future papers will compare D values calculated from various models in the literature to D values calculated from measurements of CO_2 efflux and soil pCO_2 in the field over a range of soil moisture and temperature conditions.

Acknowledgements

Research Supported by Southern California Edison, the Electric Power Research Institute (RP3041-02) and the Nevada Agricultural Experiment Station, University of Nevada, Reno. Technical assistance by Valerie Yturiaga and Carol Johnson is greatly appreciated.

References

Amundson R G and Davidson E A 1990 Carbon dioxide and nitrogenous gases in the soil atmosphere. J. Geochem. Expl. 38, 13–41.

Ball J T, Johnson D W, Strain B R, Thomas R and Walker R F 1992 Effects of CO_2 on Forests. Second Annual Report. Desert Research Institute, Reno, NV.

Castelle A J and Galloway J N 1990 Carbon dioxide dynamics in acid forest soils in Shenandoah National Park, Virginia. Soil Sci. Soc. Am. J. 54, 252–257.

Collin M and Rasmuson A 1988 A comparison of gas diffusivity models for unsaturated porous media. Soil Sci. Soc. Am. J. 52, 1559–1656.

De Jong E, Redman R E and Ripley E A 1979 A comparison of methods to measure soil respiration. Soil Sci 127, 300–306.

De Jong E and Schappert H J V 1972 Calculation of soil respiration and activity from CO_2 profiles in the soil. Soil Sci. 119, 328–333.

De Jong E, Schappert H J V and MacDonald K B 1974 Carbon dioxide evolution from virgin and cultivated soil as affected by management practices and climate. Can. J. Soil Sci. 54, 299–307.

Edwards N T 1975 Effects of temperature and moisture on carbon dioxide evolution in a mixed deciduous forest floor. Soil Sci. Soc. Am. Proc. 39, 361–365.

Edwards N T and Harris W F 1977 Carbon cycling in a mixed deciduous forest floor. Ecology 58, 433–437.

Ewell K C, Cropper W P and Gholz H L 1987a Soil CO_2 evolution in Florida slash pine plantations 1. Changes through time. Can. J. For. Res. 17, 325–329.

Ewell K C, Cropper W P and Gholz H L 1987b Soil CO_2 evolution in Florida slash pine plantations 11. Importance of root respiration. Can. J. For. Res. 17, 330–333.

Gordon A M, Sclentner R E and Van Cleve K 1987 Seasonal patterns of soil respiration and CO_2 evolution following harvesting in the white spruce forests of interior Alaska. Can. J. For. Res. 17, 304–310

Johnson D W, Cole D W and Gessel S P 1975 Processes of nutrient transfer in a tropical rain forest. Biotropica 7(3), 208–215.

Johnson M G, Tingey D T, Storm M J and Phillips D L 1994 Patterns of ponderosa pine fine root growth and turnover as affected by elevated CO_2: Initial results. Plant and Soil (*In press*).

Körner C and Arnone J A 1992 Responses to elevated carbon dioxide in artificial tropical ecosystems. Science 257, 1672–1675.

Millington R J 1959 Gas diffusion in porous media. Science 130, 100–102.

Norby R J, Gunderson C A, Wullschleger S D, O'Neill E G and McCracken M K 1992 Productivity and compensatory responses of yellow-poplar trees in elevated CO_2. Nature 357, 322–324.

Norby R J, O'Neill E G and Luxmoore R J 1986a Effects of atmospheric CO_2 enrichment on the growth and mineral nutrition of *Quercus alba* seedlings in nutrient-poor soil. Plant Physiol. 82, 83–89.

Norby R J, O'Neill E G, Hood W G and Luxmoore R J 1987 Carbon allocation, root exudation, and mycorrhizal colonization of *Pinus echinata* seedlings grown under CO_2 enrichment.Tree Physiol. 3, 203–210.

Olsen S R and Sommers L E 1982 Phosphorus. *In* Methods of Soil Analysis. Part 2. Chemical and Microbiological Properties. Second Edition. Number 9 in the Series, Agronomy. Eds. A L Page, R H Miller and D R Keeney. pp 403–430. Am. Soc. Agron. Madison, WI.

Raich J W and Nadelhoffer K J 1989 Belowground carbon allocation in forest ecosystems: global trends. Ecology 70, 1346–1354.

Rogers H H, Peterson C M, McCrimmon J N and Cure J D 1992 Response of plant roots to elevated atmospheric carbon dioxide. Plant Cell Environ. 15, 749–752.

Rolston D E 1986 Gas diffusivity. Chapter 4. *In* Methods of Soil Analysis, Part 1. Physical and Mineralogical Methods. Eds. Am. Soc. Agron. — Soil Sci. Soc. Am.

Solomon D K and Cerling T E 1987 The annual carbon dioxide cycle in a montane soil: Observations, modeling, and implications for weathering. Water Resour. Res. 23, 2257–2265.

Vose J M, Elliot K J and Johnson D W 1994 Diurnal and daily patterns in soil CO_2 flux in response to elevated atmospheric CO_2 and nitrogen fertilization. Adv. Soil Sci. (*In press*).

Walker R F, Geisinger D R, Ball J T and Johnson D W 1994 Enriched atmospheric CO_2 and soil N effect on growth and ectomyrrohizal colonization of juvenile ponderosa pine. For. Sci. (*In press*).

Wesseling J 1962 Some solutions of the steady state diffusion of carbon dioxide through soils. Neth. J. Agric. Sci. 10, 109–117.

Zak D R, Pregitzer K S, Curtis P S, Teeri J A, Fogel R and Randlett D L 1993 Elevated atmospheric CO_2 and feedback between carbon and nitrogen cycles. Plant and Soil 151, 105–117.

Plant and Soil **165**: 139-146, 1994.

Enhanced root system C-sink activity, water relations and aspects of nutrient acquisition in mycotrophic *Bouteloua gracilis* subjected to CO_2 enrichment*

J. A. Morgan[1], W. G. Knight[2], L. M. Dudley[3] and H. W. Hunt[4]
[1]*Rangeland Resources Research Unit, USDA-ARS, Ft. Collins, CO, USA,* [2]*Science Department, Front Range Community College, Ft. Collins, CO, USA,* [3]*Department of Plant, Soils and Biometeorology, Utah State University, Logan, UT, USA and* [4]*Natural Resource Ecology Laboratory and Department of Rangeland Ecosystem Science, Colorado State University, Fort Collins, CO, USA*

Key words: blue grama, *Bouteloua gracilis*, C_4 grass, CO_2 enrichment, mycorrhizae, nitrogen, phosphorus, VAM, water relations

Abstract

In order to better elucidate fixed-C partitioning, nutrient acquisition and water relations of prairie grasses under elevated [CO_2], we grew the C_4 grass *Bouteloua gracilis* (H.B.K.) lag ex Steud. from seed in soil-packed, column-lysimeters in two growth chambers maintained at current ambient [CO_2] (350 μL L^{-1}) and twice enriched [CO_2] (700 μL L^{-1}). Once established, plants were deficit irrigated; growth chamber conditions were maintained at day/night temperatures of 25/16°C, relative humidities of 35%/90% and a 14-hour photoperiod to simulate summer conditions on the shortgrass steppe in eastern Colorado. After 11 weeks of growth, plants grown under CO_2 enrichment had produced 35% and 65% greater total and root biomass, respectively, and had twice the level of vesicular-arbuscular mycorrhizal (VAM) infection (19.8% versus 10.8%) as plants grown under current ambient [CO_2]. The CO_2-enriched plants also exhibited greater leaf water potentials and higher plant water use efficiencies. Plant N uptake was reduced by CO_2 enrichment, while P uptake appeared little influenced by CO_2 regime. Under the conditions of the experiment, CO_2 enrichment increased root biomass and VAM infection via stimulated growth and adjustments in C partitioning below-ground.

Abbreviations: [ ] concentration, DAP-Days after planting, VAM-vesicular-arbuscular mycorrhizae

Introduction

Although some work has been done documenting the effects of elevated atmospheric CO_2 on the growth and physiology of prairie grasses (Kirkham et al., 1991; Morgan et al., 1994; Nie et al., 1992; Owensby et al., 1993a, 1993b; Riechers and Strain, 1988), little has been published on the importance of fixed-C allocation below-ground and its implications in enhancing plant response to CO_2 on the prairie. Enriched CO_2 environments are known to stimulate biomass (Curtis et al., 1990; Owensby et al., 1993b) and the C-sink capacity of root systems (Rogers et al., 1992). Increased root development promoted by the allocation of fixed-C below-ground may enhance the potential for water and nutrient acquisition via the production of additional absorptive surface area (Norby et al., 1986, 1987; Owensby et al., 1993a; Rogers et al., 1992). However, the reduction in stomatal aperture often reported in plants grown at elevated CO_2 can reduce transpiration and increase plant water use efficiency (Bazzaz, 1990; Newton, 1991).

Native grasses in low moisture/fertility settings have evolved nearly ubiquitous, mutualistic associations with vesicular-arbuscular mycorrhizal (VAM) fungi (Trappe, 1981). VAM serve ecologically important nutrient acquisition and perhaps drought resistance functions under sub-optimal plant growth conditions (Abbott and Robson, 1984; Stribley, 1987). The degree of plant infection by and dependence on mycorrhizae normally increases with decreasing soil fertility and/or increasing drought stress, in some cases approaching

* The U.S. Government right to retain a non-exclusive, royalty free licence in and to any copyright is acknowledged.

obligatrophy (Miller et al., 1987). VAM are reported to improve the productivity and the water relations of drought-stressed prairie grasses (Allen, 1982; Allen and Allen, 1986). Because the growth and activities of VAM fungi are dependent upon plant-fixed and translocated carbon (Lewis, 1975), they may be expected to respond positively to the increased photosynthesis in plants grown in CO_2-enriched atmospheres. Previous studies have established that CO_2 enrichment can stimulate ectomycorrhizal infection in various species (Conroy et al., 1990; Norby et al., 1986, 1987; O'Neill et al., 1987); we are not aware of any studies in which the response of grasses and associated VAM to CO_2 enrichment have been explored, with the exception of the report by Monz et al. (1994).

The influence of elevated CO_2 on plant nutrient uptake has been little explored, except for N. Plants grown at enriched [CO_2] typically have reduced concentrations of plant N and various photosynthetic enzymes (Conroy, 1992; Curtis et al., 1989; Larigauderie et al., 1988; Newton, 1991; Norby et al., 1986; Owensby et al., 1993a). This appears to occur independent of plant N availability, and may result from a decreased demand for N in enriched CO_2-grown plants (Conroy, 1992). On the other hand, the increased root mass and mycorrhizae of CO_2-enriched plants (Conroy et al., 1990; Norby et al., 1986, 1987; Owensby et al., 1993a; Rogers et al., 1992) may facilitate nutrient uptake of immobile nutrients like P.

B. gracilis is an important, dominant C_4 grass species in the shortgrass steppe of the USA whose photosynthesis and growth are enhanced by CO_2 enrichment (Morgan et al., 1994; Riechers and Strain, 1988). In the following study growth, plant water relations, tissue [N] and [P], water use efficiency and VAM infection were determined on growth chamber-grown *Bouteloua gracilis* plants submitted to near-ambient and twice-ambient atmospheric [CO_2]. Once established, plants were grown at low humidity and were deficit irrigated so that water potentials dropped to levels typical of those experienced by grasses growing on the semi-arid shortgrass steppe of Colorado in summer. We hypothesized that CO_2 enrichment would stimulate plant growth, increase C partitioning below-ground, enhance root VAM colonization, and improve plant water use efficiency and water relations. We further hypothesized that based on previous reports, tissue [P] would be little affected or slightly enhanced by CO_2 enrichment, but that plant [N] would be reduced.

Materials and methods

Plant culture

An Ascalon sandy loam soil characteristic of those sites occupied by blue grama (*Bouteloua gracilis* L.) was obtained from the Central Plains Experimental Range, which is located approximately 12 km north of Nunn, Colorado, USA (40° 42′ north, 104° 42′ west). The soil was shredded, air-dried, and 10.3 kg used to fill each of eight soil columns, measuring 45-cm in length by 15-cm in diameter. An initial air-dry sample of this soil was analyzed for NO_3-N, NH_4-N and HPO_4-P (below).

Twenty caryopses of *B. gracilis* were planted into each of the soil columns, placed in a greenhouse, and the top several cm of soil kept moist for the week prior to emergence. Six days after planting (DAP), half of the columns were placed in each of two growth chambers, one maintained at near-ambient [CO_2] (350 μL L^{-1}) and the other at twice elevated [CO_2] (700 μL L^{-1}). Both EGC[1] high light intensity growth chambers (equipped with sodium vapor and metal halide lamps, supplying about 900 μmol m^{-2} S^{-1} photosynthetic photon flux density at plant height) were maintained at day/night temperatures of 25/16 °C, relative humidities of 35%/90% and a 14-hour photoperiod. For the first 12 days after columns were placed in the chambers, 100 mL of water was added every day. At the end of this time, columns were thinned to 8 plants per column, and the watering regime was curtailed to 100 mL every other day through 24 DAP. At 32 and 63 DAP, columns were saturated with water to simulate two major rainfall events that would encourage deep rooting (Fig. 1). Smaller amounts of water were applied to all columns between 45 and 60 DAP (Fig. 1) only when stress-induced leaf curling was evident in the ambient CO_2-grown columns. Leaf curling was never observed in the enriched CO_2-grown columns. Fifteen hundred ml of water was added to all columns at 75 DAP to collect soil solution samples for NO_3-N analysis (below). At 76 DAP, plants were destructively harvested and soil samples taken (below).

Plant and soil measurements

Periodic determinations of leaf xylem water potential (ψ) were obtained from plants in each column on 47, 50, 53, 59 and 69 DAP (Fig. 1). On each measurement

* Mention of a trademark or manufacturer by the USDA does not imply its approval to the exclusion of other products.

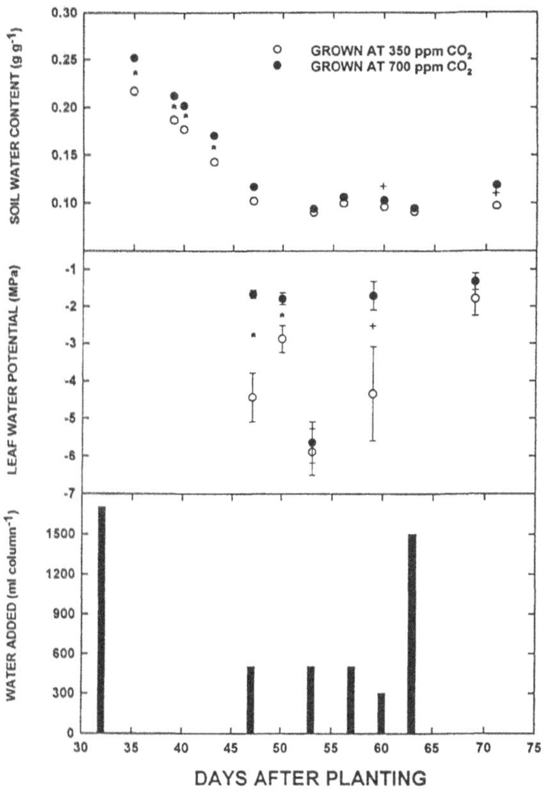

Fig. 1. Soil water contents, leaf water potentials and water additions to columns containing eight *B. gracilis* plants grown at 350 μL L^{-1} [CO_2] and 700 μL L^{-1} [CO_2]. $*$ and + indicate significant treatment differences at $p \leq 0.05$ and $p \leq 0.10$ levels, respectively.

day at six hours after the beginning of the daily photoperiod, a representative, most recent fully-expanded leaf was excised near the leaf collar, placed in a PMS Instrument Company (Corvallis, Oregon, USA) pressure chamber, and pressurized to determine ψ.

At 32 DAP and 75 DAP, soil solution samples were obtained via syringe from ceramic probes (5.1 cm length, 0.6 cm outside diameter) placed in the center of the soil columns 23 cm below the soil surface, and attached with tubing to ports located on the column surface. These solutions were gathered as the wetting front from irrigation water was passing through the soil surrounding the probes. Soil solutions were placed in a glass bottle and microbial activity was minimized by adding two drops of CCl_4 and refrigerating until analysis at 4 °C. Prior to analysis each sample was passed through a 0.45 μm filter. Nitrate concentrations were determined by high-pressure liquid chromatography (Dionex Series 4500) with a conductivity detector, an anion-exchange column (Ionpac-AS4A column) and a background suppressor column (Anion Micromembrane AMMS). The eluant was 1.80 m*M* Na_2CO_3 – 1.70 m*N* $NaHCO_3$ and the regenerant was 50 m*M* H_2SO_4 (Dionex, 1987). Standards were prepared from $NaNO_3$ that had been dried overnight at 100 °C. The position of the NO_3 peak was verified by known addition and the area under the NO_3 peak in the chromatogram was linearly related to concentration ($r^2 = 0.99$) over a range of 0.05 to 10.00 mg NO_3-N L^{-1}.

Gravimetric measurements of soil columns were performed periodically from 35 DAP through the end of the study. These measurements, along with initial column weights and oven-dry soil weights, were used to determine consumptive water use and to track soil water content during the latter stages of the study.

At the termination of the experiment, plant shoots were excised at the crown and oven-dried at 65 °C to constant weight. The intact soil was removed from each column, and spatulas used to extract 10 representative soil samples throughout the profile. These were combined within each column for nutrient analysis (below). The soil was washed away from intact crown-root systems, and the latter cut into three 2-cm sections at depths of 5–7 cm, 10–12 cm and 25–27 cm for analysis of VAM infection. The remaining below-ground biomass was oven-dried at 65°C to constant weight.

VAM infection was determined on each root sample using the grid-line intersect method of Phillips and Hayman (1970), sampling 100 intersects per sample. We stained root samples with Chlorozal Black E Stain instead of the Trypan Blue described in Phillips and Hayman (1970). Dry weights of these sections were added to total below-ground weights. VAM infection per column was averaged across the three depth subsamples.

Soil samples were air-dried, crushed and thoroughly mixed. Ammonium and ortho-P were extracted with 0.5*M* $NaHCO_3$ (Olsen and Sommers, 1982) and concentrations of the two species were determined colorimetrically using a flow analysis system. Nitrate was determined by reduction to nitrite in a copperized cadmium column followed by diazotizing with sulfanilamide and coupling with N-(1-naphthyl) ethylenediamine to produce a colored complex. The concentration of the complex was quantified by measuring absorbance at 520 nm (Robinson, 1991b). Ammonium was reacted with alkaline phenol then sodium hypochlorite to form indophenol blue that was quantified by measuring absorbance at 630 nm (Robinson, 1991a). Phosphate was determined by reaction with ammonium molybdate and antimony potassium tartrate followed by reduction with ascorbic acid and quantification by measuring absorbance at 880 nm (Stieg, 1991). Soil sample data gathered on the final day of the study (76 DAP) indicated abnormally high [NH_4] relative to [NO_3] for the prairie soil used in this study (Schimel and Parton, 1986), and therefore are not included in this presentation. We suspect that the relatively high [NH_4] was an artifact, caused by sampling the soils more than 24 hours after they had incubated near field capacity, and then drying them prior to analysis. Such wetting-drying sequences can enhance mineralization, thereby altering soil N concentrations and forms (Haynes, 1986).

Plant and root samples were digested in HNO_3 and $HClO_4$ and P was determined by inductively coupled plasma spectroscopy. Plant tissue and soil [N] were determined via the Kjeldahl method.

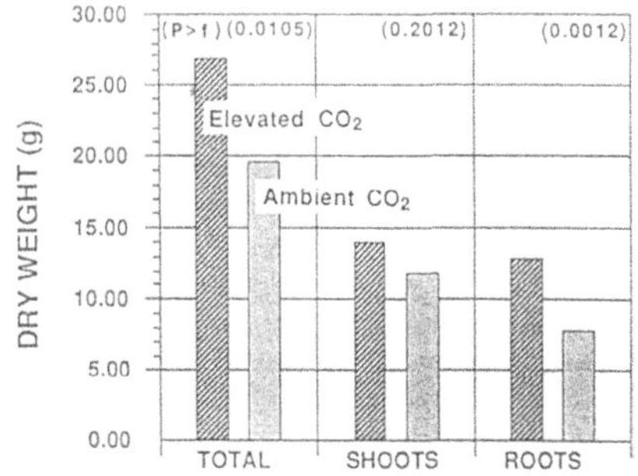

Fig. 2. Terminal dry weights (76 days after planting) of eight *B. gracilis* plants (total), shoots, and their roots grown at 350 μL L^{-1} [CO_2] and 700 μL L^{-1} [CO_2].

Statistics

Analysis of variance (ANOVA) was performed on all plant and soil traits assuming a completely randomized design, with four replications and two CO_2 treatments. Unless otherwise indicated, treatment differences mentioned in the text are assumed to be significant at the $p \leq 0.05$ level of significance.

Results and discussion

Growth, VAM infection and carbon partitioning

Plants grown under elevated atmospheric [CO_2] produced 37% more total biomass (at-harvest dry weight) and 65% more dry root matter (roots plus crowns) over the 11 wk growing period than plants grown under current ambient atmospheric [CO_2] (Fig. 2). The root systems of plants from both CO_2 treatments extended to the bottoms of the columns, although neither treatment appeared root-bound. These findings are consistent with previous studies which have indicated significant photosynthetic and growth responses of this C_4 grass to CO_2 enrichment (Morgan et al., 1994; Riechers and Strain, 1988).

There were no significant at-harvest differences in aboveground biomass (shoots) between treatments. Consequently, root:shoot ratios were 44% greater for plants grown at 700 μL L^{-1} [CO_2] than for the current ambient CO_2-treatment plants. CO_2 enrichment has commonly been observed to increase root:shoot ratios (Bazzaz, 1990; Larigauderie et al., 1988; Newton, 1991; Rogers et al., 1992; Wilson, 1988). Since the allocation of recently fixed carbon below-ground increases with the severity of water stress (Wilson, 1988), CO_2 and soil water supply likely interact from their combined influence on carbon partitioning. Thus, the exact proportion of carbon diverted below-ground as a function of [CO_2] will depend on plant water status. Our results and those of Monz et al. (1994), both conducted on periodically water-stressed plants, suggest that much of the potential growth responses of native grasses of this region to CO_2 will be below-ground.

VAM activity has also been correlated with increased root:shoot ratios in infected rangeland grasses (Knight et al., 1989; Miller et al., 1987), presumably as a consequence of sub-optimal nutrient and/or water availabilities. In the present study, nearly twice the root length of CO_2-enriched plants was infected by VAM compared to plants grown under current ambient [CO_2] (19.8% versus 10.9% respectively). While these infection rates may appear low, they are similar to those reported for *B. gracilis* plants growing in the field at CPER where soil used for this study was obtained (Monz et al., 1994). Differences in total VAM infection were even more pronounced between the CO_2 treatments since below-ground biomass of plants grown at elevated [CO_2] was greater than for plants grown at present ambient [CO_2]. These observations of enhanced VAM infection by CO_2 enrichment are consistent with the findings reported for *B. gracilis* in this volume (Monz et al., 1994) as well as for ectomycorrhizal species (Conroy et al., 1990; Norby et al., 1986, 1987).

Water relations

The influence of CO_2 enrichment on plant growth was at least partly mediated by soil/plant water relations. Despite identical irrigation regimes for both treatments, soil water content of CO_2-enriched grown plant columns was significantly greater when columns were first weighed at 35 DAP (Fig. 1). This treatment difference diminished with time as soil water became more limiting in all plants. CO_2 enrichment causes leaf stomatal closure in *B. gracilis* (Morgan et al., 1994) and numerous other species (Bazzaz, 1990; Larigauderie et al., 1988; Newton, 1991), and can lead to transpirational water savings (Rogers et al., 1984; Wray and Strain, 1986). In our study, consumptive water use must have been reduced in the enriched CO_2-treatment prior to our first column weight measurement on DAP 35. However, treatment differences in soil water content could not be maintained in the face of a dwindling water supply, so soil water contents eventually dropped to similar values of approximately 0.1 g g^{-1}. A 1400-mL irrigation at 63 DAP elevated soil water content sufficiently that treatment differences were again apparent nine days later, when column measurements revealed higher soil water contents in CO_2-enriched columns (Fig. 1, $p = 0.07$).

In addition to curtailing plant water consumption, CO_2-enrichment appeared to have a direct influence on plant water relations. From 47 through 72 DAP when soil water contents were similarly low in both treatments, periodic measurements revealed lower ψ in the ambient-grown plants on three of five measurement dates (Fig. 1). With the exception of measurements obtained at 53 DAP when ψ of both treatments dropped below - 5 MPa, the enriched CO_2-grown leaves maintained a relatively stable ψ of approximately - 2 MPa. Values of ψ were generally - 3 MPa or lower and more variable in the ambient-grown plants. Leaf curling was evident in plants grown at current ambient [CO_2] a day or two before irrigating at 32, 57, 60 and 63 DAP; CO_2-enriched plants never exhibited visible signs of water stress. This alleviation of plant water stress via CO_2-enrichment has been observed previously, and has been attributed to maintenance of either higher plant water potentials or enhanced leaf turgor (Nie et al., 1992; Sionit and Patterson, 1985; Tolley and Strain, 1984; Wray and Strain, 1986). In our study, the occurrence of treatment ψ differences of over 2 MPa on two dates when soil water contents were similarly low (Fig. 1) suggests that the benefit to water relations was not simply due to turgor maintenance via osmotic adjustment, but rather to the capability of the CO_2-enriched plants to maintain a higher ψ. Reductions in stomatal aperture and transpiration from CO_2 enrichment (above) likely limited the degree of midday desiccation, thereby improving water relations of plants grown at 700 $\mu L\,L^{-1}$ [CO_2]. The more substantial root/VAM system in the CO_2-enriched plants may have contributed to their improved plant water relations by enabling a more thorough soil exploration and intimate contact of the below-ground biomass with the soil water compared to the ambient CO_2-grown plants (Allen and Allen, 1986).

Growth at elevated [CO_2] can improve water use efficiency through transpirational water savings, and also directly through enhanced photosynthesis and growth (Bazzaz, 1990; Larigauderie et al., 1988; Morgan et al., 1994; Newton, 1991; Nie et al., 1992; Owensby et al., 1993b; Rogers et al., 1984). With some C_4 grasses, CO_2 enrichment has been reported to influence photosynthesis and growth only indirectly through improved water relations via reductions in transpiration, and not via a direct photosynthetic/growth response (Kirkham et al., 1991; Knapp et al., 1994, Nie et al., 1992). However, CO_2-induced increases in water use efficiency of *B. gracilis* may result directly from both increases in plant photosynthesis and decreases in transpiration (Morgan et al., 1994). In the present study, water use efficiency was greater for CO_2-enriched plants at the final harvest (3.8 g L^{-1} versus 2.9 g L^{-1} for plants grown at ambient [CO_2]). However, soil water became so limiting during the last 30 days of the study that treatment differences in soil water use converged. The increased biomass of CO_2-enriched plants likely resulted from a direct stimulation of photosynthesis in addition to an indirect growth benefit due to improved plant water relations.

Nutrient uptake

Plant tissue [N] was reduced in CO_2-enriched plants (Table 1). This is a common feature in CO_2-enrichment studies and appears independent of fertility (Conroy, 1992). Part of this can be explained by the accumulation of soluble carbohydrates in CO_2-enriched plants and their greater growth which tends to dilute plant tissue [N]. However, this phenomenon does not entirely account for the reduced plant tissue [N] in CO_2-enriched plants (Conroy, 1992).

A somewhat surprising finding was that shoot and, to a lesser extent, total plant N (p = 0.07) were reduced under CO_2-enrichment, while total root N

Table 1. Total and plant-part tissue N and P concentrations and total and plant-part N and P uptake in plants grown under ambient (350 $\mu L\ L^{-1}$) and elevated (700 $\mu L\ L^{-1}$) CO_2

	PLANT NUTRIENT CONCENTRATION					
	N			*P*		
	Total	Shoot	Root	Total	Shoot	Root
	g kg^{-1} dry weight					
Elevated CO_2	7.85	7.40	8.35	1.17	1.38	0.95
Ambient CO_2	12.24	13.6	10.2	1.87	2.37	1.12
P > F	0.0001	0.0001	0.014	0.0001	0.008	0.223
	PLANT NUTRIENT CONTENT					
	mg/column					
Elevated CO_2	210.3	103.1	107.2	31.4	19.2	12.2
Ambient CO_2	238.9	160.2	78.7	36.7	28.1	8.6
P > F	0.068	0.0013	0.0072	0.181	0.031	0.0085

increased (Table 1). In other recent field (Owensby et al., 1993a) and growth chamber (Rogers et al., 1992) studies where CO_2 enrichment was found to stimulate root growth, total plant N uptake increased, presumably because the increased root mass allowed for better exploration of the soil. Soil samples taken at the beginning of the study indicated the soil initially contained 7.0 mg NO_3-N kg^{-1} dry soil (72 mg NO_3-N $column^{-1}$) and 1.6 mg NH_4-N kg^{-1} dry soil (16.5 mg NH_4-N $column^{-1}$), for a total available N of 88.5 mg $column^{-1}$. This indicates that the more than 200 mg N assimilated by plants $column^{-1}$ over the course of the study (Table 1) was derived mostly from N mineralized during the experiment. Because the dynamics of N availability and plant uptake and release throughout the experiment are complex, a strict accounting of how and when N availability influenced plant uptake is not possible. Nevertheless, periodic measurements of soil [NO_3] revealed that CO_2-induced alterations in plant N uptake may have been influenced at times by soils-based processes. At 32 DAP, CO_2 regime had no apparent influence on plant N-uptake, as solution samples obtained from the soil columns revealed similar [NO_3-N] averaging 6.5 mg NO_3-N L^{-1} across CO_2 treatments (Table 2). However, on the final day of the study (DAP 75), concentrations of NO_3-N in columns exposed to 700 $\mu L\ L^{-1}$ [CO_2] were only 0.14 mg L^{-1} soil solution, considerably lower than the 3.30 mg NO_3-N L^{-1} detected in columns exposed to 350 $\mu L\ L^{-1}$ [CO_2]. Inasmuch as NO_3-N is considered to be the dominant form of available N in these soils (Schimel and Parton, 1986), our results suggest that available N was more limiting at 75 DAP for plants grown at 700 $\mu L\ L^{-1}$ [CO_2]. The increased allocation and release of organic carbon belowground in plants grown at elevated CO_2 may have led to immobilization of a significant fraction of the available N (Hunt et al., 1986). Whether this was a dominant process regulating plant N-uptake throughout the study is unknown. If it was, then it would explain the reduced N-uptake by plants grown at elevated [CO_2]. Reduced N-uptake in the plants grown at 700 $\mu L\ L^{-1}$ CO_2 may also have been due to a plant metabolically-regulated phenomenon arising from a reduced requirement for N (Conroy, 1992). However, this apparent reduced requirement for N in CO_2-enriched plants appears most evident in C_3 species, not C_4 species, presumably because of former's high N-investment in carboxylating protein (Conroy, 1992). Further, a reduced N-requirement would lead to increased soil available N, not a decrease as was indicated at 75 DAP (Table 2).

Similar amounts of P were assimilated by ambient and CO_2-enriched plants, although concentrations of this element were reduced in the CO_2-enriched treatment, probably because of their greater growth. As with N, CO_2 enrichment caused reductions in this nutrient in shoots, but increases in the root fraction. No treatment differences were found in soil [HPO_4]; concentrations remained stable throughout the study at 18.9 mg HPO_4-P kg^{-1} dry soil. Our findings are distinctly different from those cited by Conroy (1992) and in general, those reported by Owensby et al. (1993) where plant P-uptake was proportional to growth increases under CO_2 enrichment. While there

Table 2. Soil solution NO_3-N concentrations extracted from middle of soil columns containing blue grama plants at 32 days after planting (DAP), and 75 DAP

Days After Planting	CO_2 Treatment 350 $\mu L\ L^{-1}$ [CO_2]	CO_2 Treatment 700 $\mu L\ L^{-1}$ [CO_2]
	mg NO_3-N L^{-1} Soil Solution	
32	6.37	6.68
75	3.30	0.14**

** indicates significant treatment differences at $p \leq 0.01$.
Data Log-transformed before analysis of variance.

is evidence that greater rooting density and VAM activity can facilitate nutrient uptake of immobile nutrients like P (Stribley, 1987) and in some cases, even mobile nutrients like N, the higher infection and larger root systems of the CO_2-enriched plants appeared not to influence the results. VAM-facilitated P uptake is most important when soil HPO_4 levels are low, and therefore may not have been a major factor in our study as soil HPO_4 levels remained high throughout the study (Table 2). Maintenance of low soil water contents the last 30 days of the study may have resulted in lower [P] in CO_2-enriched plants as well. Owensby et al. (1993a) found that plant P-uptake in CO_2-enriched *Andropogon gerardii* plants was not sufficient to maintain nominal tissue [P] under extreme drought, and suggested that drought may have impaired the role of mycorrhizae in P uptake.

Conclusion

CO_2-enrichment increased growth, VAM infection, root:shoot ratios, and improved water relations in the C_4 grass *B. gracilis*. Yet despite the large increase in C partitioning belowground in CO_2-enriched plants, N uptake appeared reduced and P uptake little influenced by [CO_2]. These results indicate that the assimilation of soil nutrients relative to C obtained in photosynthesis may be less in plants grown at elevated [CO_2]. Results from the literature and this study suggest that such alterations in plant nutrient uptake may be caused by both plant-and soils-based phenomena. An understanding of the relative importance of these different phenomena and their interactions with the environment will be required before accurate predictions of plant responses to CO_2-enrichment will be possible, particularly in ecosystems in which water and nutrient supplies are dominant limiting growth factors.

Acknowledgements

We would like to thank Mark Angrick and Chris Monz for their suggestions and diligence with the mycorrhizal work, along with John Read, Dan LeCain and Barry Weaver for their valuable assistance throughout the experiment. This research was supported by award number 92–37100–7670 of the National Research Initiative Competitive Grants Program of the USDA and award BSR–8818269 of the National Science Foundation.

References

Abott L K and Robson A D 1984 The effect of mycorrhizae on plant growth. *In* V A Mycorrhiza. Eds. D J Bagyaraj and C L Powell. CRC Press, Inc. Boca Raton, FL.

Allen M F 1982 Influence of vesicular-arbuscular mycorrhizae on water movement through *Boutelous gracilis* (H.B.K.) Lag Ex Steud. New Phytol. 91, 191–196.

Allen E B and Allen M F 1986 Water relations of xeric grasses in the field: Interaction of mycorrhizae and competition. New Phytol. 104, 559–571.

Bazzaz F A 1990 The response of natural ecosystems to the rising global CO_2 levels. Ann. Rev. Ecol. Syst. 21, 167–196.

Conroy J P, Milham P J and Barlow E W R 1990 Increases in phosphorus requirement for CO_2-enriched pine species. Plant Physiol. 92, 977–982.

Conroy J P 1992 Influence of elevated atmospheric CO_2 concentrations on plant nutrition. Aust. J. Bot. 40, 445–456.

Curtis P S, Drake B G and Whigham D F 1989 Nitrogen and carbon dynamics in C_3 and C_4 estuarine marsh plants grown under elevated CO_2 in situ. Oecologia 78, 297–301.

Curtis P S, Balduman L M, Drake B G and Whigham D F 1990 Elevated atmospheric CO_2 effects on belowground processes in C_3 and C_4 estuarine marsh communities. Ecology 71, 2001–2006.

Dionex Coporation 1987 Determination of Anions Using the HPIC-AS4A Seperator, Document No. 034035, Dionex Corp., Sunnyvale, CA, USA.

Haynes R J 1986 Mineral nutrition in the plant-soil system. Academic Press, Inc., New York, p 68 and p 108.

Hunt H W, Stewart J W B and Cole C V 1986 Concepts of sulfur, carbon, and nitrogen transformations in soil: Evaluation by simulation modeling. Biogeochemistry 2, 163–178.

Kirkham M B, He H, Bolger T P, Lawlor D J and Kanemasu E T 1991 Leaf photosynthesis and water use of big bluestem under elevated carbon dioxide. Crop Sci. 31, 1589–1594.

Knapp A K, Hamerlynck E P and Owensby C E 1994 Photosynthetic and water relations responses to elevated CO_2 in the C_4 grass *Andropogon gerardii*. Int. J. Plant Sci. 154 (*In press*).

Knight W G, Allen M F, Jurinak J J and Dudley L M 1989 Elevated carbon dioxide and solution phosphorus in soil with vesicular-arbuscular mycorrhizal western wheatgrass. Soil Sci. Soc. Am. J. 53, 1075–1082.

Larigauderie A, Hilbert D W and Oechel W C 1988 Effect of CO_2 enrichment and nitrogen availability on resource acquisition and resource allocation in a grass *Bromus mollis*. Oecologia 77, 544–549.

Lewis D H 1975 Comparative aspects of the carbon nutrition of mycorrhizas. *In* Endomycorrhizas. Eds. F E Sanders, B Mosse and P B Tinker Academic Press, London.

Miller R M, Jarstfer A G and Pillai J K 1987 Biomass allocation in an *Agropyron smithii-Glomus symbiosis*. Am. J. Bot. 74, 114–122.

Monz C A, Hunt H W, Reeves F B and Elliott E T 1994 The response of mycorrhizal colonization to elevated CO_2 and climate change in *Pascopyrum smithii* and *Boutelous gracilis*. Plant and Soil (*In press*).

Morgan J A, Hunt H W, Monz C A and LeCain D R 1994 Consequences of long-term growth at two carbon dioxide concentrations and temperatures for leaf gas exchange of *Pascopyrum smithii* (C_3) and *Bouteloua gracilis* (C_4). Plant Cell Environ. (*In press*).

Newton P C D 1991 Direct effects of increasing carbon dioxide on pasture plants and communities. N. Z. J. Agric. Res. 34, 1–24.

Nie D, He H, Mo G, Kirkham M B and Kanemasu E T 1992 Canopy photosynthesis and evapotranspiration of rangeland plants under doubled carbon dioxide in closed-top chambers. Agric. For. Meteorol. 61, 205–217.

Norby R J, O'Neill E G and Luxmoore R J 1986 Effects of atmospheric CO_2 enrichment on the growth and mineral nutrition of *Quercus alba* seedlings in nutrient-poor soil. Plant Physiol. 82, 83–89.

Norby R J, O'Neill E G, Hood W G and Luxmoore R J 1987 Carbon allocation, root exudation and mycorrhizal colonization of *Pinus echinata* seedlings grown under CO_2 enrichment. Tree Physiol. 3, 203–210.

Olsen D W and Sommers L E 1982 Total carbon, organic carbon, and organic matter. *In* Methods of Soil Analysis, Part 2, Chemical and Microbiological Properties, 2nd ed. Eds. A L Page, R H Miller D R Keeney. pp 921–922. Am. Soc. Agronomy, Madison, WI.

O'Neill E G, Luxmoore R J and Norby R J 1987 Increases in mycorrhizal colonization and seedling growth in *Pinus echinata* and *Quercus alba* in an enriched CO_2 atmosphere. Can. J. For. Res. 17, 878–883.

Owensby C E, Coyne P I and Auen L M 1993a Nitrogen and phosphorus dynamics of a tallgrass prairie ecosystem exposed to elevated carbon dioxide. Plant Cell Environ. 16, 843–850.

Owensby C E, Coyne P I, Ham J M, Auen L M and Knapp A K 1993b Biomass production in a tallgrass prairie ecosystem exposed to ambient and elevated CO_2. Ecol. App. 3, 644–653.

Phillips J M and Hayman D S 1970 Improved procedure for clearing roots and staining parasitic and vesicular-arbuscular mycorrhizal fungi for rapid assessment of infection. Trans. Br. Mycol. Soc., 55, 158–161.

Riechers G H and Strain B R 1988 Growth of blue grama (*Boutelous gracilis*) in response to atmospheric CO_2 enrichment. Can. J. Bot. 66, 1570–1573.

Robinson B 1991a Ammonia (Phenolate) Potable and Surface Waters. Lachat Instruments, Milwaukee, WI.

Robinson B 1991b Nitrate/Nitrite, Nitrite in Surface Water, Wastewater. Lachat Instrument, WI.

Rogers H H, Peterson C M, McCrimmon J N and Cure J D 1992 Response of plant roots to elevated atmospheric carbon dioxide. Plant Cell Environ. 15, 749–752.

Rogers H H, Sionit N, Cure J D, Smith J M and Bingham G E 1984 Influence of elevated carbon dioxide on water relations of soybeans. Plant Physiol. 74, 233–238.

Schimel D S and Parton W J 1986 Microclimatic controls of nitrogen mineralization and nitrification in shortgrass steppe soils. Plant and Soil 93, 347–357.

Sionit N and Patterson D T 1985 Responses of C_4 grasses to atmospheric CO_2 enrichment. II. Effect of water stress. Crop Sci. 25, 533–537.

Stieg S 1991 Orthophosphate in Waters. Lachat Instruments, Milwaukee, WI.

Stribley D P 1987 Mineral nutrition. *In* The Ecophysiology of Vesicular-Arbuscular mycorrhizae. Ed. G Safir. CRC Press, Inc. Boca Raton, FL.

Tolley L C and Strain B R 1984 Effects of CO_2 enrichment and water stress on growth of *Liquidambar straciflua* and *Pinus taeda* seedlings. Can. J. Bot. 62, 2135–2139.

Trappe J M 1981 Mycorrhizae and productivity of arid and semi-arid rangelands. *In* Advances in food producing systems for arid and semi-arid lands. Eds. J T Manassah and E J Briskey. pp 581–599. Academic Press, N.Y.

Wilson J B 1988 A review of evidence on the control of shoot: root ratio, in relation to models. Ann. Bot. 61, 433–449.

Wray S M and Strain B R 1986 Response of two old field perennial to interactions of CO_2 enrichment and drought stress. Am. J. Bot. 73, 1486–1491.

MODELLING

Plant and Soil **165:** 149-160, 1994.

Modeling the belowground response of plants and soil biota to edaphic and climatic change — What can we expect to gain?

Stan D. Wullschleger[1], Jonathan P. Lynch[2] and Glenn M Berntson[3]
[1]*Environmental Sciences Division, P.O. Box 2008, Oak Ridge National Laboratory, Oak Ridge, TN 37831–6034, USA,* [2]*Department of Horticulture, 101 Tyson Building, The Pennsylvania State University, University Park, PA 16802, USA and* [3]*Harvard University, Biological Laboratories, 16 Divinity Avenue, Cambridge, MA 02138, USA*

Key words: climate change, CO_2 enrichment, modeling, root biology, soil biota

Abstract

As atmospheric CO_2 concentrations continue to increase, so too will the emphasis placed on understanding the belowground response of plants to edaphic and climatic change. Controlled-exposure studies that address the significance of an increased supply of carbon to roots and soil biota, and the consequences of this to nutrient cycling will play a prominent role in this process. Models will also contribute to understanding the response of plants and ecosystems to changes in the earth's climate by incorporating experimental results into conceptual or quantitative frameworks from which potential feedbacks within the plant-soil system can be identified. Here we present five examples of how models can be used in this analysis and how they can contribute to the development of new hypotheses in the areas of root biology, soil biota, and ecosystem processes. Two examples illustrate the role of coarse and fine roots in nitrogen and phosphorus uptake from soils, the respiratory costs associated with this acquisition of nutrients, and the significance of root architecture in these relationships. Another example focuses on a conceptual model that has helped raise new ideas about the effects of elevated CO_2 on root and microbial biomass, and on nutrient dynamics in the rhizosphere. Difficulties associated with modeling the contribution of mycorrhizal fungi to whole-plant growth are also discussed. Finally, several broad-scale models are used to illustrate the importance of root turnover, litter decomposition, and nitrogen mineralization in determining an ecosystem's response to atmospheric CO_2 enrichment. We conclude that models are appropriate tools for use both in guiding existing studies and in identifying new hypotheses for future research. Development of models that address the complexities of belowground processes and their role in determining plant and ecosystem function within the context of rising CO_2 concentrations and associated climate change should be encouraged.

Introduction

Faced with the probability that human activities are contributing to a change in the earth's atmospheric CO_2 concentration, researchers in the plant and soil sciences are increasingly being asked to address questions of global proportions. By their very nature, answers to these questions will be complex and will likely include feedbacks between poorly understood components of the plant-soil system. Global increases in CO_2 concentration, for example, may yield primary effects on leaf and whole-plant photosynthesis and C allocation which in turn may spawn secondary consequences for root growth, fine-root turnover, microbial dynamics in the rhizosphere, litter quality, decomposition, and nutrient cycling. Although critical to our understanding of how plant, soil, and microbial processes can potentially constrain ecosystem responses to edaphic and climatic change, many of these effects will be difficult to study. Answers will come, however, from CO_2 enrichment studies that can successfully identify and focus on key processes that, while measured on individuals, can be related to broader questions of how these effects influence ecosystem behavior and function.

As a complement to existing CO_2 enrichment studies, models will also be important tools for both understanding plant function and predicting the response of plants, plant communities, and ecosystems to changing edaphic and climatic conditions. By incorporating experimental results into either conceptual or quantitative frameworks, models can be used effectively to identify gaps in our understanding of organismal and ecosystem processes, facilitate the development of well-defined hypotheses and focused objectives, and provide a platform from which to conduct studies with these ideas explicitly in mind. In addition, as our understanding of a plant-soil system matures, models will assume predictive attributes and will therefore avail themselves to answering questions beyond that which can be accomplished through controlled-exposure studies alone. To illustrate how models can be applied to the study of root biology, soil biota, and ecosystem processes, we present several brief examples that will hopefully demonstrate the utility of models in and beyond the experimental process. Where possible these examples focus on questions of special relevance to atmospheric CO_2 enrichment studies.

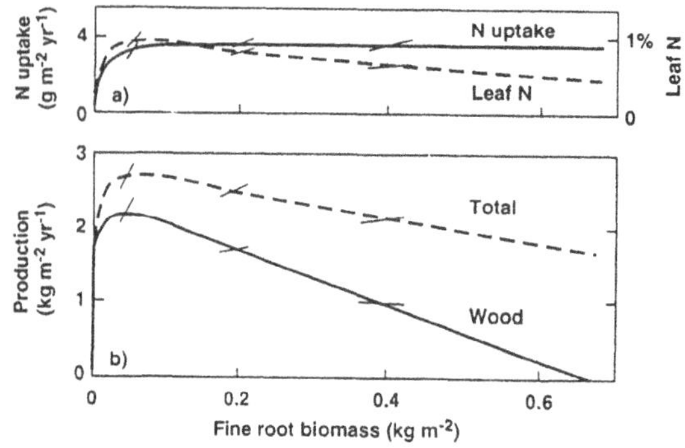

Fig. 1. Simulated effects of increasing fine root biomass on (**a**) stand nitrogen uptake and leaf nitrogen concentrations, and (**b**) total productivity and on the production of wood for an even-aged conifer forest. The line segments indicate how individual nitrogen uptake, nitrogen concentration, and wood and total production vary as an individual changes its root biomass relative to the surrounding stand. Reproduced from King (1993) with permission.

Root biology

Successful acquisition of water and nutrients by plants requires that functional roots be present within the soil profile and distributed in such a manner so as to facilitate the gathering of limited soil resources. Carbon allocated to coarse roots may increase the lateral development of root systems, while the increased allocation of C to fine roots may facilitate nutrient and water uptake. Changes in the relative amount of C allocated to different root types and in the temporal patterns of root production, elongation, and overall growth rate will affect root architecture, altering both the intensity and the amount of soil that is explored. These changes may be significant to the productivity of plants in competitive environments. There is, however, an energetic cost associated with root growth and nutrient uptake, against which the benefits of increased root proliferation must be evaluated (Johnson, 1990; Van der Werf et al., 1993). It is imperative that empirical and theoretical studies of root biology begin to address these concerns.

Carbon allocation to fine root biomass

Atmospheric CO_2 enrichment studies which document an increased allocation of C to roots, particularly fine roots, often imply that this conveys some advantage in terms of increased water or nutrient availability (O'Neill et al., 1987b; Stulen and den Hertog, 1993; Tolley and Strain, 1984). Greater fine-root proliferation in white oak (*Quercus alba* L.) seedlings grown for 40 weeks at elevated CO_2 concentrations was suggested to increase the extent of soil exploration and to favor phosphorus, but not nitrogen, uptake from nutrient-limited soils (Norby et al., 1986a). This increased acquisition of phosphorus, if mediated through greater fine-root growth, may assist in overcoming nutrient limitations that would otherwise constrain plant and ecosystem response to elevated CO_2 concentrations. Unfortunately, causal relationships between an increased growth of fine roots and enhanced nutrient uptake are difficult to establish in controlled-exposure studies. More difficult is the ability to link observations on belowground C allocation to ecosystem-level questions of nutrient cycling. These can, however, be addressed through simulation modeling by considering how the allocation of C between foliage and fine roots affects whole-plant productivity and resource acquisition.

Wanting to better understand how above- and belowground C allocation influences competition for limiting resources, King (1993) constructed an optimization model that relates forest growth both to nitrogen uptake and to the allocation of biomass among leaves, fine roots, and woody tissues. Simple relationships between C and nitrogen were used to describe growth in terms of canopy light interception, tissue nitrogen concentrations, and nitrogen uptake by roots. This model builds upon the earlier efforts of McMurtrie (1985, 1991) and was used by King (1993) to study the consequences of changing C allocation on nutrient uptake, wood production, and individual tree growth for even-aged conifer stands receiving adequate water, but suboptimum nitrogen. In addition to estimating whole-stand productivity, competition within a stand was simulated by considering the costs and benefits associated with an individual tree allocating slightly more or less biomass to leaves or roots than its neighbor. The influence of C allocation strategies on nutrient cycling and long-term site fertility were also examined.

Simulated results indicated that nitrogen uptake for a stand initially increased with increasing fine root biomass and reached a plateau at low to moderate investments of C in fine roots (Fig. la). Likewise, leaf nitrogen concentrations also increased in association with greater fine root biomass, but unlike nitrogen uptake, declined as C allocation to fine roots increased beyond a certain point. This decline in leaf nitrogen occurred because less nitrogen was available for incorporation into foliage as fine root biomass increased. As leaf nitrogen concentrations declined, so too did the light utilization coefficient (defined as dry mass produced per unit photosynthetically active radiation intercepted by the canopy), which through its effects on C gain led to a decrease in stand productivity and, in particular, the production of wood (Fig. 1b). Proportionately greater reductions in wood production were observed because an increasing fraction of C must be diverted from stem biomass and allocated instead to roots. Note that stand productivity was maximized at relatively low investments of C in fine root biomass.

These results provide insights into what might occur if all individuals within a stand functioned in a cooperative manner, sharing resources equally among all members. If an individual gains a competitive advantage over its neighbor, however, and allocates more of its resources to leaf or root biomass, it may acquire more than its share of limiting resources (e.g., light or nutrients), thereby increasing its own productivity. The line segments crossing the curves in Figure 1 illustrate the change in productivity that might be expected from an individual that allocates slightly more or less growth to roots than the entire stand. Trees that allocate more of their resources to the production of root biomass acquire more nitrogen, and in turn maintain higher leaf nitrogen concentrations, use irradiance more effectively, and produce greater biomass than that of the surrounding stand.

An obvious conclusion from this modeling exercise is that an increased allocation of C to fine root biomass represents a response that can elicit a multitude of secondary effects at the stand level. For individuals that share resources equally among their neighbors, greater productivity will not necessarily result from a greater investment of C in fine root biomass, nor will an increase in C allocation to fine roots guarantee higher leaf nitrogen concentrations. If a tree can gain a competitive advantage over its neighbor, however, an increased allocation of C to fine roots may benefit that individual relative to less competitive members of the stand. Given this distinction between sharing and competing for above- and belowground resources, a meaningful question to ask is whether the technique of growing plants in isolation (i.e., typical of pot studies) can provide information about how plant communities will likely respond to edaphic and climatic change. Reekie and Bazzaz (1989) studied competition and patterns of resource use in five tropical tree species and noted that the response of individuals to elevated CO_2 was not a good predictor by itself of performance when seedlings were grown in competitive arrays. In contrast to the individually grown plants, which showed no response to atmospheric CO_2 enrichment, elevated CO_2 had a stimulatory effect upon biomass when these species were grown in competition with one another. Competitive performance among the five species was attributed to canopy height, photosynthetic rates at low light, and biomass allocation to roots early in the experiment. Depending on one's specific objective, it is clear that predicting the outcome of plant-to-plant interactions in a higher-CO_2 world, either between different species or between ecotypes of the same species, will require a better understanding of how elevated CO_2 alters the above- and belowground acquisition of resources by competing individuals (Bazzaz and McConnaughay, 1992). This remains a challenge for future research.

King (1993) also discussed briefly the potential role of fine roots in nutrient cycling and concluded that an increased allocation of C to roots for trees competing within a stand may enhance long-term productivity by increasing nitrogen uptake, and thus decreasing nutrient losses from forest ecosystems. Post et al. (1992) recognized the importance of fine roots in determining nutrient availability for temperate hardwoods when they used modeled results to suggest that greater fine root productivity in response to rising CO_2 could increase ecosystem C storage through the effects of fine root biomass on nitrogen mineralization. Given the potential significance of fine roots as contributors to long-term site fertility, an important question to ask is whether CO_2 enrichment studies can be designed to confirm these simulated links between C allocation to fine roots, nutrient retention, and cycling efficiency. Artificial ecosystems, such as those constructed by

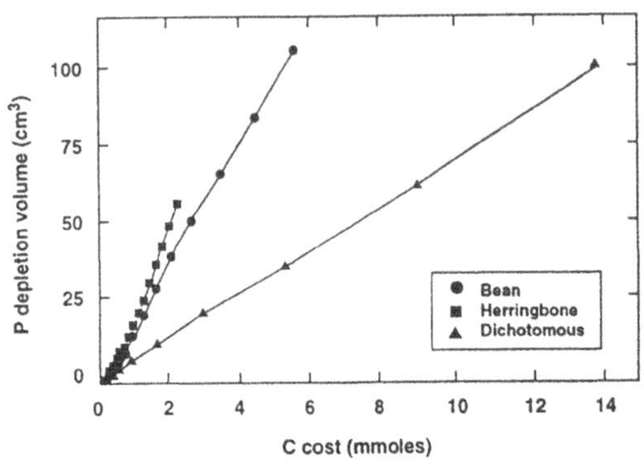

Fig. 2. Simulated C costs associated with three contrasting root architectures and the benefits derived in terms of phosphorus acquisition. Adapted from Nielsen et al. (1994) with permission.

Körner and Arnone (1992) are beginning to provide useful information on the growth of competitive forest communities in elevated CO_2 and nutrient cycling. Additional studies will be required, however, before a clear picture emerges.

Root architecture and nutrient acquisition

Root systems vary greatly in size, lateral extent, depth, rooting intensity, proportion of fine and coarse roots, and other characteristics that might affect the relationship of C utilization to soil resource acquisition, soil biota, and rhizodeposition. Although important, this 3-dimensional nature of the root system, or root architecture, is difficult to analyze. Root systems are, for example, highly complex geometric objects, within which C utilization and resource acquisition are nonuniform, and which interact dynamically with a spatially and temporally heterogeneous soil environment. These complexities are further magnified by an absence of conceptual models for describing root architecture and the lack of reasonable methods for quantifying the architecture of living root systems. However, the 3-dimensional arrangement of roots within the soil profile and the unique interaction of roots with their edaphic surroundings can be approached through geometric simulation modeling, a technique in which root growth, C utilization, and resource acquisition are simulated in a spatially explicit manner.

Nielsen et al. (1994) used the geometric simulation model, *SimRoot*, to compare three contrasting root architectures in terms of the C required to support root growth and maintenance processes, and the "benefit" derived from this investment of C as evidenced by soil phosphorus acquisition. *SimRoot* parallels the architectural analysis approach of Fitter et al. (1991), but includes an explicit treatment of C costs. The root architectures employed were i) herringbone, a pattern formed by simple taproot growth with files of lateral branches, ii) dichotomous, formed by regular bifurcation of the root apical meristem, and iii) the root architecture of bean (*Phaseolus vulgaris* L.) which tends to be intermediate in configuration. In order to parameterize this model, the spatial distribution of C utilization for biomass, root respiration, and exudation were measured on individual *Phaseolus* root axes, and then these results overlaid onto the root systems as simulated by the model. Phosphorus acquisition was chosen as an appropriate function to consider in terms of root-system architecture since the relative mobility of phosphorus in soils is low, thus making its acquisition highly dependent on the spatial and temporal patterns of soil exploration by the root system. Like nitrogen, phosphorus is also a limiting soil resource in many terrestrial ecosystems, including much of the humid tropics, volcanic highlands, and temperate forests.

Simulated results indicate that root systems of contrasting architecture consume substantially different amounts of C in acquiring inorganic phosphorus (Fig. 2). A dichotomous architecture, for example, expends approximately four times as much C in acquiring phosphorus than a herringbone architecture. Bean root architecture was intermediate between the two topological extremes. Differences in the acquisition of phosphorus by these three architectures were largely due to two factors, one related to respiratory losses of C by apical meristems and another related to the increased competition between roots within the soil profile. First, the bifurcation of apical meristems for dichotomous root systems necessitates, by definition, greater meristems per root system. Because apical meristems have higher rates of respiration compared to other root tissues, the C required to support this type of architecture is greater than that of the other two architectural designs. Coupled with these costs, there is also more intense inter-root competition within dichotomous root systems, and over time this leads to the growth of new roots into soils already depleted of phosphorus by older roots. Together, these two factors result in lower phosphorus acquisition efficiencies for root systems of a dichotomous architecture.

These results indicate that root architecture is an important variable to consider when determining relationships between the amount of C allocated to roots and soil resource acquisition. One might conclude from this simulation exercise that while quantifying root biomass in response to edaphic and climatic change will continue to be a significant measure, steps must also be taken to understand how this biomass is spatially distributed within the soil profile. There is some evidence that root proliferation does change with depth for plants grown at elevated CO_2 compared to ambient-grown controls (Berntson and Woodward, 1992; Rogers et al., 1992) and that these differences might be important for nutrient and water uptake. Berntson and Woodward (1992) recently concluded that the growth of *Senecio vulgaris* at elevated CO_2 resulted in more branched, longer root systems that foraged through larger soil volumes, presumably increasing the exploitation potential of these plants compared to ambient-grown controls. If such a change in the 3-dimensional arrangement of roots represents a universal response to atmospheric CO_2 enrichment, it is clear that this could have interesting and potentially unforeseen implications for resource acquisition by plants grown at elevated CO_2 concentrations. We note, however, that while an increased exploration of soils is advantageous for nutrient acquisition when considered apart from the development of other root systems, in a natural setting it may lead to higher interplant competition due to the increasing overlap of active rooting zones between individual plants. Competition between roots of differing architectures for growth-limiting nutrients like phosphorus would be amenable to geometric modeling using much the same framework as was used by Nielsen et al. (1994) for individual root systems. Geometric simulation models could also be used to estimate the spatial distribution and amount of C deposited in the soil with varying rates of root C utilization, which may be useful in predicting C utilization by soil biota and the incorporation of this C into soil organic matter. Attempts to merge descriptions of C allocation to roots, rates of C exudation to the rhizosphere, and the consequences of these processes to soil biota would vastly improve our ability to distinguish the mechanisms by which plants and ecosystems respond to edaphic and climatic change.

Soil biota

Plant roots do not exist in isolation within the soil profile, but are instead accompanied by a host of organisms associated with the rhizosphere. This root-soil interface, alive with bacteria, fungi, and assorted soil fauna, is greatly influenced by the root and supports microbial populations that are greater in number and often different in composition than the soil at large (O'Neill et al., 1987a). Most of these soil organisms share some degree of dependency on the root as a source of readily available C either from root exudates, mucigel, sloughed-off root cap cells, or intact root cells. It is this reliance on photosynthetic products combined with an ability of soil biota to increase nutrient availability that highlights the need to better understand microbial responses to rising CO_2 and the capacity of these organisms to facilitate plant nutrient uptake.

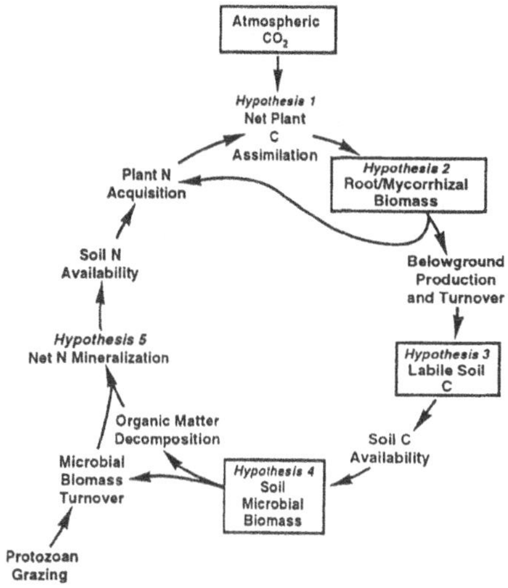

Fig. 3. Conceptual model depicting the influence of atmospheric CO_2 enrichment on whole-plant carbon assimilation, root proliferation, microbial activity, and the availability of soil nitrogen in the rhizosphere. This model served as a platform from which specific hypotheses were developed and later tested in a CO_2 enrichment study using seedlings of *Populus grandidentata* Michx. grown for one year in a nutrient-poor soil (Zak et al., 1993). Reproduced with permission.

Microbial activity and nutrient cycling in the rhizosphere

Since atmospheric CO_2 enrichment presumably increases the flux of C to roots, it is often suggested that the exudation of organic compounds from roots may also increase, thus stimulating microbial activity and supposedly nutrient cycling in the rhizosphere (Norby et al., 1986a). Few studies, however, have attempted to quantify the effect of elevated CO_2 on rhizospheric populations of bacteria and mycorrhizal fungi (Norby et al., 1987; O'Neill et al., 1987a; Rogers et al., 1992) and there has been no attempt to establish cause-and-effect relationships between increased root growth at elevated CO_2 and greater nutrient availability. Phosphorus uptake was shown to increase in white oak seedlings grown at elevated CO_2 (Norby et al., 1986a), although these authors could only *infer* that this resulted from either greater fine root proliferation or from mycorrhizae and bacteria stimulating phosphorus mineralization in the rhizosphere. Despite this lack of study, it is still generally felt that soil biota will play an important, although at present an ill-defined, role in determining plant and ecosystem productivity in a higher-CO_2 world (O'Neill et al., 1991). There exists a considerable challenge to develop studies that will clarify the effect of elevated CO_2 on microbial activity and the contribution of this to nutrient cycling.

To help visualize these relationships, Zak et al. (1993) recently developed a conceptual model depicting the influence of elevated CO_2 on above- and belowground plant productivity, soil C availability, microbial activity, and the rate at which nutrients are cycled within the plant-soil system (Fig. 3). Using this model, a series of hypotheses were identified, each of which would need to be studied in order to quantify the primary feedbacks between rising CO_2 and effects on the cycling of C and nitrogen in the rhizosphere. Specifically, the model was designed to examine the following five hypotheses — i) plants grown in an elevated CO_2 atmosphere should have greater rates of photosynthesis, although the magnitude of this response should depend on adequate soil nitrogen availability, ii) if nitrogen does not limit the photosynthetic response, total and belowground biomass should be greater for plants grown at elevated CO_2 than for ambient-grown controls, iii) atmospheric CO_2 enrichment should increase soil C availability through a greater input of root-derived C to the rhizosphere, iv) if soil biota are C limited, then greater soil C availability at elevated CO_2 should elicit an increase in microbial biomass within the rhizosphere, and v) net nitrogen mineralization should increase due to larger and more active microbial populations and greater rates of organic matter decomposition.

Once these hypotheses were formalized into a conceptual framework and specific objectives identified, an open-top chamber study was conducted to examine whether atmospheric CO_2 enrichment would stimulate greater C allocation to roots, in turn leading to an increase in soil microbial biomass, to increased rates of organic matter turnover, and to enhanced nitrogen availability (Zak et al., 1993). Sprouts of bigtooth aspen (*Populus grandidentata* Michx.) were planted in a nitrogen-deficient soil and exposed to either ambient or twice-ambient CO_2 for most of one growing season. As expected, single-leaf rates of net photosynthesis were greater for plants grown at elevated CO_2 than they were for those measured under ambient conditions. This photosynthetic response to elevated CO_2 was sufficient to elicit a 30% increase in whole-plant productivity, with the roots showing the greatest effect. Greater root growth at elevated CO_2 was associated with greater labile C in the rhizosphere, presumably from increased exudation or greater rates of root turnover, but there was no difference between ambient and twice-ambient CO_2 treatments in the labile C pool of the bulk soil. Microbial biomass C was also greater in soils collected from around CO_2-enriched plants compared to soils collected from the ambient treatments, with the greatest effect being observed for rhizospheric soils collected from the elevated CO_2 treatment. Microbial C in the rhizosphere was about 50% greater in the twice-ambient treatment than it was from the ambient-grown controls. And finally, rates of net nitrogen mineralization, as assayed by short-term laboratory incubations of the bulk soil, were greatest in soils collected from the elevated CO_2 treatment, suggesting greater nitrogen availability for plant uptake.

Although the results of Zak et al. (1993) will require verification in other studies, many of these observations argue for the existence of a positive feedback between atmospheric CO_2 enrichment, increased root growth, enhanced microbial activity, and greater nitrogen availability in the rhizosphere. While these results have important implications for how plants and soil biota respond to rising CO_2 concentrations, this study also illustrates how a model can be used to identify critical hypotheses and how a rather simple conceptualization exercise can help researchers focus their efforts on key questions. Application of models to areas where modeling has not traditionally played a strong role (i.e., symbiotic nitrogen fixation and mycorrhizal research) deserves further attention.

Mycorrhizal symbiosis and whole-plant carbon dynamics

Symbiotic associations between mycorrhizal fungi and roots are widely recognized for their important contributions to plant and ecosystem function (O'Neill, 1990; Reid et al., 1991). Mycorrhizae benefit plant growth in several ways, including increased root-absorbing area and enhanced nutrient uptake, particularly for the less mobile nutrient phosphorus. Through

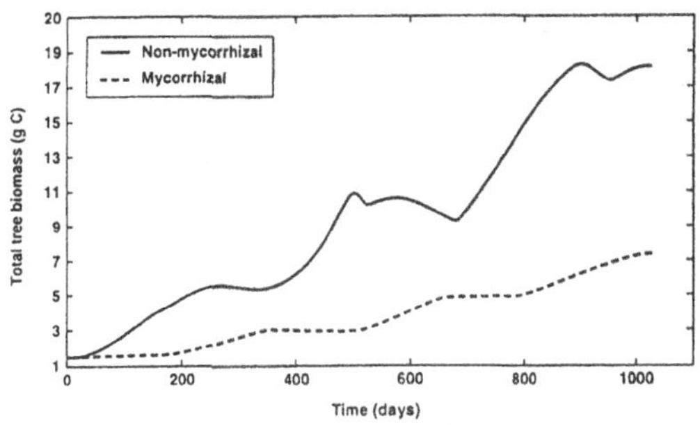

Fig. 4. Simulated growth of *Pinus ponderosa* seedlings growing either with or without the mycorrhizal fungus *Hebeloma crustulinforme* for three years. An approximate 65% reduction in whole-plant biomass for the mycorrhizal seedlings was due to the increased respiratory costs associated with fungal hyphae. Adapted from Andersen et al. (1992) with permission.

their beneficial effects on plant growth, mycorrhizae also influence ecosystem production. There is, however, a cost associated with the symbiosis, one in which C is diverted away from the plant and allocated instead to mycorrhizae. Quantifying this cost has proven difficult, especially in terms of reconciling the costs versus the benefits derived from the symbiotic association.

Frustrated with an inability to define the contribution of mycorrhizae to whole-plant growth, Andersen et al. (1992) noted that mycorrhizal fungi are rarely considered in process-level models of plant growth, but are instead included as some poorly characterized component of fine roots (e.g., Thornley, 1991). If mycorrhizae are indeed so important to our understanding of plant and ecosystem function, Andersen et al. (1992) argued that it may be necessary to represent these organisms in models as discrete plant "organs" due to their demonstrated effects on carbon allocation, nutrient and water uptake, root growth, root respiration, root turnover, and their distribution in the soil profile. This idea of a separate mycorrhizal component was pursued by Andersen et al. (1992) when they asked if a plant-growth model could be used to simulate the growth of mycorrhizal seedlings by simply changing various input parameters to reflect, among other things, hyphae instead of fine roots. Using the TREGRO model (Weinstein et al., 1991), ponderosa pine (*Pinus ponderosa* Douglas ex P. Lawson and Lawson) seedlings growing in association with the ectomycorrhizal fungus *Hebeloma crustulinforme* (Bull. ex St. Amans) were simulated by altering shoot photosynthesis and respiration to match rates observed in previous studies (Andersen and Rygiewicz, 1991; Miller et al., 1989), and by changing fine-root radius and respiration, and specific root length to approximate hyphae. These changes resulted in a substantial increase in initial root surface area and length, but did not appreciably alter initial belowground biomass or specific root volume.

Simulations with the TREGRO model indicated that when changes were made in the input parameters to reflect those of the symbiotic association, seedling growth after three years was reduced by almost 65% compared to that of non-mycorrhizal seedlings (Fig. 4). While this decrease in growth with mycorrhizal "infection" was surprising and occurred despite increased nutrient uptake by mycorrhizal seedlings, it was clear to Andersen et al. (1992) that the TREGRO model was highly sensitive to certain physiological variables, particularly the choice of fine root, or in this case, hyphal respiration. They concluded from this exercise that separate modules for mycorrhizal hyphae and fine roots would be needed in models due to their different rates of respiration and the relative strength of these two components as C sinks. What was equally apparent, was that given an unrestricted flow of C to meet the demands of fungal respiration, the mycorrhizal symbiosis could place a tremendous, bordering on unrealistic, C cost on the host plant. The question arises as to whether this characteristic reflects the energetic costs of growing and maintaining the mycorrhizal fungus or whether this represents a problem with how C is allocated in the TREGRO model to meet respiratory demands. Many models, for example, preferentially allocate C to satisfy rates of organ-specific respiration, with the remaining C allocated to growth. It is important to evaluate whether allocation strategies in existing plant-growth models will be sufficient for meeting the respiratory demands of mycorrhizal fungi or if new approaches to allocating C to mycorrhizae will be necessary.

Through these exploratory studies, Andersen et al. (1992) demonstrated several ways in which healthy feedback can occur between experimental research and model development. First, these authors recognized that growth models seldom include mycorrhizae, nor the costs and benefits explicitly associated with the symbiosis, and thus they logically raised the question as to how such an inclusion might alter model output. Having asked the question, Andersen et al. (1992) chose a model with sufficient physiological detail, made some simplifying assumptions, and then proceeded to examine whether simulated plant growth was sensitive to a separate mycorrhizal component. It was shown that the TREGRO model was indeed quite sensitive to the choice of fine-root or hyphal respiration, suggesting that the growth and maintenance costs of coarse-root, fine-root, and hyphal respiration, and the mechanisms regulating these rates, need to be better known in order to adequately simulate plant growth. Finally, it was concluded that maybe the manner in which C is allocated to mycorrhizae will be different from that currently used in models to meet the C demands of fine-root respiration. If such is the case, plant-growth models which incorporate a mycorrhizal component will have to make this distinction. Here we begin to see how a beneficial exchange of ideas can occur between modelers and experimentalists, one in which results from a simulation model contribute to the formation of new questions for future experiments, and experiments which provide new information for the development and verification of plant-growth models.

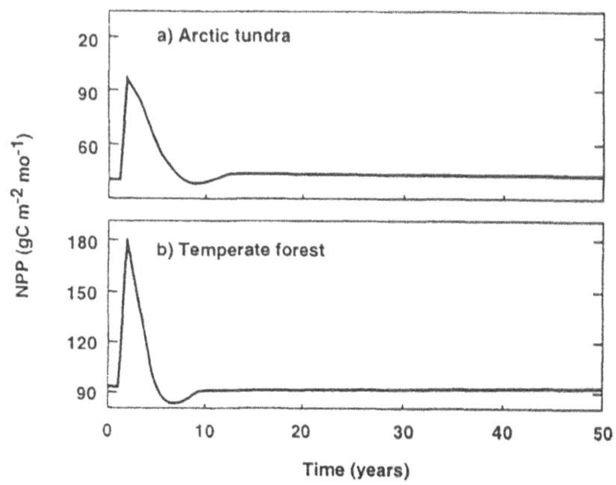

Fig. 5. Simulated net primary productivity (NPP) in (**a**) arctic tundra and (**b**) temperate hardwood forests exposed to a doubling of atmospheric CO_2 concentration. Net primary productivity increased initially with elevated CO_2 but declined in later years because of a nitrogen limitation. Adapted from Rastetter et al. (1991) with permission.

Ecosystem processes

At the ecosystem level, elevated CO_2 may cause changes in nutrient cycling through effects on litter quality, decomposition, and nitrogen mineralization. Although these effects are potentially quite important to the biogeochemical cycles of C and nitrogen, these processes are not equally represented in all models, nor do all models agree on how ecosystems will respond to atmospheric CO_2 enrichment. Some models suggest that there will be little or no long-term CO_2 fertilization effect in nutrient-limited ecosystems, while others indicate that net productivity can be expected to increase with rising CO_2 concentrations. Tempting as it might be to ask which generalization is correct, it is probably more useful to consider how these models differ from one another and to then ask whether these differences can help guide researchers to processes about which more research is needed. We shall see that in exploring how the biogeochemical cycles of C and nitrogen are linked within terrestrial ecosystems this can be a useful exercise.

Carbon and nitrogen cycles in terrestrial ecosystems

Quantifying the response of plants and ecosystems to rising CO_2 concentrations and associated climate change will require that the fate and transport of both C and nitrogen within the plant-soil system be known (Ågren et al., 1991; McGuire et al., 1992; Ojima et al., 1991). This understanding is critical because the linkage between C and nitrogen cycles will provide constraints and feedbacks to plant and ecosystem function. Consider, for example, Rastetter et al. (1991) who developed a general model for simulating C and nitrogen cycles in terrestrial ecosystems and who used this model to examine the response of arctic tussock tundra and temperate hardwoods to changes in atmospheric CO_2 concentrations. Simulations indicated that there was a large initial increase in net primary productivity for both ecosystems exposed to twice-ambient CO_2 (Fig. 5). With time, however, there was a substantial "redistribution" of nitrogen from vegetation to soils which caused a gradual nitrogen limitation to occur and a return of net primary productivity to pre-exposure levels. Although not stated as such, a critical assumption driving these simulated responses was that the growth of plants at elevated CO_2 produced litter with a C-to-N ratio higher than that of ambient-grown plants, which subsequently reduced nitrogen mineralization in the soil and decreased the amount of nitrogen available for plant uptake. Rastetter et al. (1991) concluded that interactions between the C and nitrogen cycles provided a negative feedback in both arctic tundra and temperate forest ecosystems exposed to atmospheric CO_2 enrichment. A similar conclusion was reached by Comins and McMurtrie (1993) who suggested that *if* reduced nitrogen concentrations in leaves growing at elevated CO_2 implies a correspondingly lower concentration of nitrogen in leaf litter, then this may lead to lower rates of nitrogen mineralization in the soil, imposing greater nutrient limitations on plants, and thereby canceling gains in productivity that would otherwise result from CO_2 enrichment.

Similarly recognizing that constraints associated with nutrient availability were important determinants of ecosystem productivity, Post et al. (1992) modified an existing individual-based gap model to include feedbacks mediated through the C and nitrogen cycles, and then used this model, LINKAGES, to examine the effects of elevated CO_2 on forest productivity. Unlike the model of Rastetter et al. (1991), however, which allowed litter C-to-N ratios to increase with CO_2 enrichment, the model LINKAGES places no such requirement on litter quality. As a result, limits imposed on productivity by a change in litter C-to-N ratios can be ignored and model simulations broadly compared with those of Rastetter et al. (1991) where productivity was limited by nitrogen availability. Such simulations indicate that when constraints on the C-to-N ratios of litter are reduced, net ecosystem productivity is higher at elevated CO_2 than it is at ambient CO_2 concentrations and that this difference can be sustained throughout a 500 year model run (Post et al., 1992). Contrary to the report of Rastetter et al. (1991), this example illustrates that a considerable potential exists for increasing C storage in forest ecosystems through the effects of elevated CO_2 on plant growth.

By presenting the contrasting results of Rastetter et al. (1991) and Post et al. (1992) our intention has not been to consider which scenario is correct, nor to suggest that the productivity of any one ecosystem will or will not change with rising CO_2 concentrations. Rather, it is hoped that by contrasting these two models, we could highlight the major assumptions explicit in each and illustrate how models can often be used to identify areas of uncertainty. Obviously, one of the primary differences between these two models was in whether elevated CO_2 alters litter C-to-N ratios. Here litter refers not only to foliage, but also to woody debris, coarse roots, fine roots, exudates, and conceivably microbial biomass. As shown through the two examples, whether or not elevated CO_2 alters litter quality is not a trivial distinction, but rather is one that makes the difference between whether a plant or ecosystem can sequester C in a higher-CO_2 world or not. Surprisingly, there are only meager data on the C-to-N ratios of leaf *litter* produced at elevated CO_2 and the effects of this on decomposition (Couteaux et al., 1991; Curtis et al., 1989; Norby et al., 1986b) and there exists no such information on the decomposition of roots. Given that ecosystem models could immediately benefit from data on decomposition, the gathering of such information should be given a high priority in future CO_2 enrichment studies.

Discussion

We have discussed several conceptual and quantitative models that illustrate some of the ways in which modeling can be used to improve our understanding of how plants, soil biota, and ecosystems might respond to edaphic and climatic change. As a complement to existing CO_2 enrichment studies, these models were shown to be useful for exploring numerically complex processes over time (e.g., C allocation to fine root biomass, mycorrhizae, and whole-plant C dynamics), spatially complex processes (e.g., root architecture and nutrient acquisition), and complex systems (e.g., biogeochemical cycles of C and nitrogen). Modeling was, in every case, a valuable tool for use both in guiding existing studies and in suggesting new hypotheses for future research. In this role, models serve a *heuristic* function; they are useful primarily for the insights they provide rather than for their ability to predict or explain natural phenomena.

As with any attempt to simplify a complex system, researchers interested in model development will find it necessary to make explicit assumptions concerning the nature of processes, pools, and interconnections among these processes and pools for their system of interest. Once identified, any of these assumptions can then be rigorously examined. During this examination, the following questions may be asked,

1. Do the parameters included within the model realistically delimit the system? In their analysis of C allocation to ectomycorrhizal fungi, Andersen et al. (1992) provided an excellent example of how this question can serve as a guide for model development. To reiterate, Andersen et al. (1992) observed that process-level models of plant growth seldom include the costs, nor the benefits associated with mycorrhizal fungi and concluded that such models are unrealistically delimited. They subsequently modified the TREGRO model to more realistically reflect the C costs and functioning of mycorrhizal infection. This exercise suggested several avenues for future research.
2. What are the effects of disturbances in individual pools and processes? Exploring this question gives insights into which aspects of the system might exert the most influence and are thus the most important foci for research. An example of this was

seen in the tradeoffs associated with C allocation to fine root biomass, wood production, and nutrient availability in conifer forests (King, 1993). The ability of models to direct research to key processes or hypotheses is especially relevant in ecosystem research, in which experimental investigation is often costly and laborious. Models therefore serve a useful function in identifying fruitful areas of investigation prior to extensive and possibly costly experimentation.

Once experimentation is underway models continue to be useful as vehicles for integration of experimental results. For example, respiratory costs associated with the growth and maintenance of mycorrhizal hyphae may have uncertain implications for whole-plant C allocation unless such data are incorporated into a model that integrates these processes over time with other C flows and resource acquisition functions within the plant. The same would be true for the cycling of C and nitrogen in terrestrial ecosystems, where it is doubtful that field research alone could capture the integrated effects of changing C-to-N ratios on the long-term processes of litter decomposition and nutrient availability. Experimental studies, nonetheless, will continue to provide new ideas about how plants respond to edaphic and climatic change, some of which will be used to develop new models, while others will be used to refine existing models. As a result, models will increasingly be better able to suggest promising avenues of further investigation, in an iterative interaction between modeling and experimentation. It is worth noting that the use of explicit simulation models in this context is merely an elaboration of the process of scientific inquiry that occurs in any discipline; that is, an iterative interplay between experimentation and our conceptual "model" of how nature works.

As the process of model development, validation, and targeted experimentation matures, models begin to assume more relevance as reliable representations of the actual system and can be used, often along with cautions and caveats, for prediction and mechanistic explanation of natural processes. Models are particularly useful in this regard when the natural system of interest is largely inaccessible to direct measurements. This is certainly true for processes occurring belowground, since many of these processes occur over long periods of time, across areas of large extent, and involve highly complex interactions between poorly understood components. As a predictor of natural systems, the utility of a model will in some measure be proportional to the degree that the model is process-oriented rather than empirical, since it is likely that fundamental processes such as photosynthesis or nitrification are similar across modeling scenarios, whereas the actual behavior of systems that we observe may vary greatly with species composition, temperature regime, soil type, and other model parameters.

Finally, we note that an obstacle to the optimal use of modeling in understanding the response of plants to edaphic and climatic change is that research efforts have so far concentrated on only a few model systems, while neglecting other systems such as the humid tropics. Although process-oriented models may help extrapolate research results across different environments, it is important that experimentation with natural systems include a reasonable breadth of parameters likely to be encountered in dominant terrestrial ecosystems, if such experimental studies are to be useful in developing reliable predictive models. A model may function quite well in describing and predicting system behavior in one environment characterized by a given set of constraints or processes, but may not be helpful at all in analyzing an environment characterized by a different set of constraints. Raich et al. (1991) recognized this when they concluded that a terrestrial ecosystem model that includes the biogeochemical cycles of C and nitrogen, but fails to include phosphorus may be of limited use in calculating net primary productivity for regions like the Amazon Basin of South America where soils are often phosphorus deficient. A thoughtful distribution of research and modeling efforts across a range of key environmental gradients would be more useful in the long run than excessive emphasis on a few ecosystems that may not be truly representative of other terrestrial environments.

Conclusions

Models of plant and ecosystem responses to global change provide conceptual frameworks which can yield insight into the effects of edaphic and climatic change not accessible from experimentation alone. We suggest, however, that both modelers and researchers suffer from working in isolation and note that a closer interaction between these groups should be encouraged. By so doing critical feedbacks associated with the response of terrestrial vegetation rising CO_2 concentrations and climate change can be identified and their significance to plant, ecosystem, and global processes more readily assessed.

Acknowledgements

We acknowledge with appreciation manuscript reviews by G S Edwards, E G O'Neill, and R J Norby. Special thanks to P R Kemp for discussions which led to the writing of this manuscript, and to C P Andersen, D A King, W M Post, E B Rastetter, and D R Zak for approving the use of their research in the present context. Partial financial support (JPL) was provided by the Root Biology Training Program, a unit of the DOE/NSF/USDA Collaborative Research in Plant Biology Program, funded by NSF grant BIR-9220330. Financial support (GMB) was provided from the Graduate Fellowships for Global Change Program administered by Oak Ridge Institute for Science and Education for the US Department of Energy, Office of Health and Environmental Research, Atmospheric and Climate Research Division. Research sponsored by the Carbon Dioxide Research Program, Office of Health and Environmental Research, U S Department of Energy, under contract No. DE-AC05-840R21400 with Martin Marietta Energy Systems, Inc. Publication No. 4299 Environmental Sciences Division, Oak Ridge National Laboratory.

References

Ågren G I, McMurtrie, Parton W J, Pastor J and Shugart H H 1991 State-of-the-art of models of production-decomposition linkages in conifer and grassland ecosystems. Ecol. Appl. 1, 118–138.

Andersen C P and Rygiewicz P T 1991 Stress interactions and mycorrhizal plant response: Understanding carbon allocation priorities. Environ. Poll. 73, 217–244.

Andersen C P, Laurence J A and Hogsett W E 1992 To what extent do mycorrhizae need to be considered in carbon budgets and process modeling efforts ? *In* Proceedings 12th North American Forest Biology Workshop, The Role of Physiology and Genetics in Forest Ecosystem Research and Monitoring. p 145. August 17–20, Sault ste. Marie, Ontario, Canada.

Bazzaz F A and McConnaughay K D M 1992 Plant-plant interactions in elevated CO_2 environments. Aust. J. Bot. 40, 547–563.

Berntson G M and Woodward F I 1992 The root system architecture and development of *Senecio vulgaris* in elevated CO_2 and drought. Func. Ecol. 6, 324333.

Comins H N and McMurtrie R E 1993 Long-term biotic response of nutrient-limited forest ecosystems to CO_2 enrichment; equilibrium behaviour of integrated plant-soil models. Ecol. Appl. (*In press*).

Couteaux M–M, Mousseau M, Celerier M–L and Bottner P 1991 Increased atmospheric CO_2 and litter quality: decomposition of sweet chestnut leaf litter with animal food webs of different complexities. Oikos 61, 54–64.

Curtis P S, Drake B G and Whigham D F (1989) Nitrogen and carbon dynamics in C3 and C4 estuarine marsh plants grown under elevated CO_2 in situ. Oecologia 78, 297–301.

Fitter A H, Strickland T R, Harvey M L and Wilson G W 1991 Architectural analysis of plant root systems. 1. Architectural correlates of exploitation efficiency. New Phytol. 118, 375–382.

Johnson I R (1990) Plant respiration in relation to growth, maintenance, ion uptake and nitrogen assimilation. Plant Cell Environ. 13, 319–328.

King D A 1993 A model analysis of the influence of root and foliage allocation on forest production and competition between trees. Tree Physiol. 12, 119–135.

Körner C and Arnone J A III 1992 Responses to elevated carbon dioxide in artificial tropical ecosystems. Science 257, 1672–1675.

McGuire A D, Melillo J M, Joyce L A, Kicklighter D W, Grace A L, Moore III, B and Vorosmarty C J 1992 Interactions between carbon and nitrogen dynamics in estimating net primary productivity for potential vegetation in North America. Global Biogeochem. Cycles 6, 101–124.

McMurtrie R E 1985 Forest productivity in relation to carbon partitioning and nutrient cycling: a mathematical model. *In* Attributes of Trees as Crop Plants, Eds. M G R Cannell and J E Jackson. pp 194–207. Institute of Terrestrial Ecology, Abbots Ripton, Huntingdon, England.

McMurtrie R E 1991 Relationship of forest productivity to nutrient and carbon supply — a modeling analysis. Tree Physiol. 9, 87–99.

Miller S L Durall D M and Rygiewicz 1989 Temporal allocation of 14C to extramatrical hyphae of ectomycorrhizal ponderosa pine seedlings. Tree Physiol. 5, 239–249.

Nielsen K L, Lynch J P, Jablokow A G and Curtis P S 1994 Carbon cost of root systems: an architectural approach. Plant and Soil (This volume).

Norby R J, O'Neill E G and Luxmoore R J 1986a Effects of atmospheric CO_2 enrichment on the growth and mineral nutrition of *Quercus alba* seedlings in nutrient-poor soil. Plant Physiol. 82, 83–89.

Norby R J, Pastor J and Melillo J M 1986b Carbon-nitrogen interactions in CO_2-enriched white oak: physiological and long-term perspectives. Tree Physiol. 2, 233–241.

Norby R J, O'Neill E G, Hood W G and Luxmoore R J 1987 Carbon allocation, root exudation and mycorrhizal colonization of *Pinus echinata* seedlings grown under CO_2 enrichment. Tree Physiol. 3, 203–210.

Ojima D S, Kittle T G F, Rosswall T and Walker B H 1991 Critical issues for understanding global change effects on terrestrial ecosystems. Ecol. Appl. 1, 316325.

O'Neill E G, Luxmoore R J and Norby R J 1987a Elevated atmospheric CO_2 effects on seedling growth, nutrient uptake, and rhizosphere bacterial populations of *Liriodendron tulipifera* L. Plant and Soil 104, 3–11.

O'Neill E G, Luxmoore R J and Norby R J 1987b Increases in mycorrhizal colonization and seedling growth in *Pinus echinata* and *Quercus alba* in an enriched CO_2 atmosphere. Can. J. For. Res. 17, 878–883.

O'Neill E G, O'Neill R V and Norby R J 1991 Hierarchy theory as a guide to mycorrhizal research on large-scale problems. Environ. Poll. 73, 271–284.

Post W M, Pastor J, King A W and Emanuel W R 1992 Aspects of the interaction between vegetation and soil under global change. Water Air Soil Poll. 64, 345363.

Raich J W, Rastetter E B, Melillo J M, Kicklighter D W, Steudler P A, Perterson B J, Grace A L, Moore III B and Vorosmarty C J 1991 Potential net primary productivity in South America: Application of a global mode. Ecol. Appl. 1, 399429.

Rastetter E B, Ryan M G, Shaver G R, Melillo J M, Nadelhoffer K J, Hobbie J E and Aber J D 1991 A general biogeochemical model describing the responses of the C and N cycles in terrestrial ecosystems to changes in CO_2, climate, and N deposition. Tree Physiol. 9, 101–126.

Reekie E G and Bazzaz F A 1989 Competition and patterns of resource use among seedlings of five tropical trees grown at ambient and elevated CO_2 . Oecologia 79, 212–222.

Reid C P P 1990 Mycorrhiza. In The Rhizosphere. Ed. J Lynch. pp 281–315. John Wiley & Sons.

Rogers H H, Prior S A and O'Neill E G 1992 Cotton root and rhizosphere responses to free-air CO_2 enrichment. Crit. Rev. Plant Sci. 11, 251–263.

Stulen I and den Hertog J 1993 Root growth and functioning under atmospheric CO_2 enrichment. Vegetatio 104/105, 99–115.

Thornley J H M 1991 A transport-resistance model of forest growth and partitioning. Ann. Bot. 68, 211–226.

Tolley L C and Strain B R 1984 Effects of CO_2 enrichment and water stress on growth of *Liquidambar styraciflua* and *Pinus taeda* seedlings. Can. J. Bot. 62, 21352139.

Van der Werf A, Visser A J, Schieving F and Lambers H 1993 Evidence for optimal partitioning of biomass and nitrogen at a range of nitrogen availabilities for a fast- and slow-growing species. Funct. Ecol. 7, 63–74.

Weinstein D A, Beloin R M and Yanai R D 1991 Modeling changes in red spruce carbon balance and allocation in response to interacting ozone and nutrient stresses. Tree Physiol. 9, 127–146.

Zak D R, Pregitzer K S, Curtis P S, Teeri J A, Fogel R and Randlett D L 1993 Elevated atmospheric CO_2 and feedbacks between carbon and nitrogen cycles. Plant and Soil 151, 105–117.

Plant and Soil **165**: 161-169, 1994.

Carbon cost of root systems: an architectural approach

Kai L. Nielsen[1], Jonathan P. Lynch[1,4], Andrei G. Jablokow[2] and Peter S. Curtis[3]
[1]*Department of Horticulture,* [2]*Department of Mechanical Engineering, The Pennsylvania State University, University Park, PA 16802, USA and* [3]*Department of Plant Biology, The Ohio State University, Columbus, OH 43210, USA.* [4]*Author for correspondence*

Key words: carbon cost, phosphorus, *Phaseolus vulgaris*, root architecture, root growth, root simulation model

Abstract

Root architecture is an important component of nutrient uptake and may be sensitive to carbon allocational changes brought about by rising CO_2. We describe a deformable geometric model of root growth, *SimRoot*, for the dynamic morphological and physiological simulation of root architectures. Using SimRoot, and measurements of root biomass deposition, respiration and exudation, carbon/phosphorus budgets were developed for three contrasting root architectures. Carbon allocation patterns and phosphorus acquisition efficiencies were estimated for *Phaseolus vulgaris* seedlings with either a dichotomous, herringbone, or empirically determined bean root architecture. Carbon allocation to biomass, respiration, and exudation varied significantly among architectures. Root systems also varied in the relationship between C expenditure and P acquisition, providing evidence for the importance of architecture in nutrient acquisition efficiency.

Introduction

The acquisition, allocation and utilization of resources by plants has many parallels with economic systems. Resources can be used in the maintenance of existing biomass or in the construction of new biomass, such as leaves and roots, which are then involved in the acquisition of additional resources (Bloom et al., 1985). One of the most important costs for roots is carbon loss by respiration both for growth and maintenance and carbon structurally incorporated into roots (Rundel and Nobel, 1991). Most cost-benefit analyses of vegetative growth have used carbon (C) to describe the cost of acquiring various resources (Mooney, 1972; Orians and Solbrig, 1977).

As atmospheric CO_2 rises, the C economy of many plants will be fundamentally altered. At elevated CO_2 levels, photorespiration is reduced in C_3 plants leading to increased N use efficiency (Wong et al., 1992) and quantum yield (Long and Drake, 1991), among other effects. Enhanced C assimilation at low nutrient or light availability may in turn allow enhanced allocation of C towards the acquisition of those limiting resources. Recent elevated CO_2 research has focused on changes in C allocation to roots and the effect such changes may have in long-term plant and ecosystem responses to global climate change (Rogers et al., 1993; Zak et al., 1993). Root architecture is an important element of whole plant nutrient uptake and several studies suggest that it may be responsive to changing CO_2 levels (Berntson and Woodward, 1992; Del Castillo et al., 1989; Rogers et al., 1992). Understanding the degree to which plant roots may respond architecturally to CO_2, however, requires not only a consideration of possible benefits to the plant, through altered patterns of soil exploration and nutrient uptake, but also an examination of the costs associated with different root growth patterns.

When carbon costs are considered along with nutrient acquisition, carbon/nutrient budgets can be used as an integrated view of the adaptive value of specific root traits involved in nutrient acquisition. We operationally define the relationship between belowground C expenditure and nutrient acquisition as the physiological efficiency of nutrient acquisition or 'NAE' (Nutrient Acquisition Efficiency) and predict that NAE will be influenced by the 3-dimensional structure of a root system over time. The complexity of numerical calculations required to estimate the integrated cost and benefit of various root system traits, and the large number of samples that must be taken to adequately describe the morphological and physiological development of actual root systems, make computer modeling a useful tool in evaluating different architectures. Most existing root system models, however, have either been intended for visualization only (Aono and Kunii, 1984), have had no time component, or have been limited to a narrow range of possible architectures (Diggle, 1988; Fitter, 1987; Langley, 1973; Pages et al., 1989). Fitter et al. (1991) worked with a model that was capable of considering a wide range of architectures. Carbon cost in this model was based on biomass of the root system.

We used *SimRoot*, a geometric model of root growth, (Davis, 1993), to simulate contrasting root systems. Based on the empirically measured spatial distribution of biomass C, C exudation, and respiration along root axes, a carbon cost function was incorporated in the model. Primary outputs from this model were visual projections of root growth in time and space, total root system length, volume, surface area, total respiration, total biomass C, total C exudation, number of meristems and volume of soil depleted of a specific nutrient. We evaluated phosphorus (P) acquisition because its uptake is very dependent on soil exploration by roots or their symbionts (Anuradha and Narayanan, 1991; Haynes et al., 1991) and because suboptimal P availability is a primary limitation to plant growth in many terrestrial ecosystems, including much of the humid tropics, Arctic regions, and areas with volcanic soils (Sanchez, 1976; Sanchez and Logan, 1992).

Materials and methods

Plant material

Seeds of common bean (*Phaseolus vulgaris* L. cv. G4000, CIAT) were surface sterilized, germinated for 24 hours in an incubator at 28°C between layers of paper moistened with 0.5 m*M* $CaSO_4$, and were then exposed to fluorescent light (12 hr day 25°C, 12 hr night 25°C, 50 μE m^{-2} s^{-1}). Seedlings used for exudation measurements were moved to Plexiglas root cylinders (described below) after the germination period. Root growth parameters of field grown *P. vulgaris* were taken from a study at the CIAT Palmira Research Station in Colombia, South America, Fall 1989, and used for simulation of the bean-like root architecture. Palmira (3° 30′ N lat., 76° 21′ W long.) is located at 965 m above sea level (Lynch and van Beem, 1993; Sadeghian, 1991)

Respiration

The primary root tips of 5 bean seedlings were placed in a 20 cm^3, humidified Plexiglas chamber and isolated from the rest of the seedling by foam gaskets. Air was circulated in the closed system through tubing connected with a closed system IRGA (Li-Cor 6200, LI-COR, Lincoln, Nebraska). The bottom of the chamber was covered with moist filter paper and the air flow bypassed the desiccant tube on the IRGA, ensuring a high relative humidity in the system and preventing the root tips from drying out. Roots were allowed to equilibrate in the system for 5 minutes prior to the measurement. Change in CO_2 concentration in the chambers over a 5 minute period was measured with the IRGA. To calculate respiration in 2 mm root length increments, new seedlings were mounted in the chamber, the length of the root tips in the chamber increased by 2 mm, and new measurements taken. Respiration from any given 2 mm section was calculated by subtracting respiration rates previously measured on sections closer to the apex.

Carbon deposition

Primary roots from 20 seedlings, 72 hours old, were cut into 2 mm pieces, dried at 58°C, and weighed. The spatial distribution of C deposition was estimated assuming a 41.3% C concentration in legume roots (Hansen et al., 1992).

Total organic carbon exudation

A Plexiglas chamber was used for seedling growth and collection of exudate. Initially, two 6 cm long Plexiglas cylinders were placed on top of each other and covered with aluminum foil to exclude light. The top of the upper cylinder was closed with a Plexiglas plate with two holes, one hole for the root to penetrate the chamber, the other for input of aerated 0.5 m*M* $CaSO_4$ solution. When the primary root in the lower chamber was approximately 7–8 cm long, the lower cylinder was removed and the outermost part of the attached primary root was placed on top of a series of small wells drilled in a Plexiglas block. Each well had a diameter of 10 mm and contained 1 mL 0.5 m*M* $CaSO_4$ and was isolated from neighboring wells by Silicon grease. Samples were taken from each well after 5 hours of exposure. Total organic C (TOC) was measured using a CO_2 Coulometer model 5010 (UIC Coulometrics Inc., Joliet, Illinois). The pH in the samples was acidified to below 5, then CO_2 was flushed by bubbling pure O_2 through the samples for 30 sec. prior to the measurement. A 0.2 mL sample was injected into the 1000°C combustion chamber, and after purification of the gas, the total amount of CO_2 was measured.

Statistics

The results of the measurements of spatial distribution of carbon costs were analyzed according to a one-way ANOVA.

Phosphorus acquisition

P acquisition was modeled as a function of diffusion to the root within zones of depletion. The depletion zone radius (*r*dz) was calculated from the expression: $r_{dz} = l + 2(Dt)^{\frac{1}{2}}$ (Nye and Tinker,1977) where l was root segment radius, D was the P diffusion coefficient (1.38×10^{-7} cm^2 s^{-1}, Nye and Tinker, 1977; Barber, 1984) for a moist loamy sand, and t was time in which diffusion occured, corresponding to the age of the link arising from the node.

Modeling

SimRoot, a 4-dimensional (i.e. space and time) data structure and visualization system for functional process modeling of botanical root systems (Davis, 1993) was employed to provide a simulation of root growth based on the following morphological and physiological parameters; types of root branches (e.g. primary tap root, lateral, basal roots etc.), branching angles, growth velocities of specific types of branches, spatial distribution of respiration, C exudation, and biomass C along root axes. Respiration, biomass deposition and exudation were measured on *Phaseolus vulgaris* seedlings under laboratory conditions. Three contrasting root architectures were modeled: herringbone, bean and dichotomous (Fig. 1). Root growth parameters for the bean root architecture were taken from field grown *Phaseolus vulgaris*, the herringbone and dichotomous root systems were theoretical extreme architectures, not derived from actual root systems (Fitter, 1991). The simulated growth period was 336 hours. Output from SimRoot simulating a *Phaseolus vulgaris* root at four time steps, illustrates the imitating capacities of this model (Fig. 2). Output was in the form of total respiration, C exudation and total C biomass of whole root systems. The spatial and temporal distribution of each root segment in the root system and the radius of the depletion zone of P around each segment was also calculated. The overlap of depletion cylinders between root axes (inter-root competition), was calculated by converting data from SimRoot into a volume element representation, where each element represented a cubiform volume (voxel) within the depletion cylinder. Duplicate voxels, representing inter-root competition were then eliminated and the depletion volume without cylinder overlap was estimated. This procedure was similar to that described by Fitter et al. (1991).

Results and discussion

Profiles of C cost along root axes

We observed high respiration rates in the first 4 mm of the primary root tip (Fig. 3). The apical part of the primary bean root contains mitotically active cells with small and sparse vacuoles, while the cells in more basal regions are more differentiated, contain larger vacuoles, have thicker cell walls and thereby a lower relative content of protoplasm and lower respiratory activity (Wanner, 1950). The increase in respiration 6–16 mm from the root tip was spatially correlated with the formation of root hairs. CO_2 influx was observed in the region 16–18 mm from the root tip. Amiro and Ewing (1992) found that uptake of ^{14}C by plants via the roots appears to be mostly a passive process that proceeds at a rate similar to transpiration. Earlier work by Arteca and Poovaiah (1982a) has shown that $^{14}CO_2$ uptake by roots can be substantial, and that both potato and bean plants respond to root zone application of CO_2 by increasing the amount of phosphoenolpyruvate carboxylase (Arteca and Poovaiah, 1982b), indicating a greater amount of fixation in the roots. This could lead to a net uptake of CO_2 in specific regions of the root. The large decrease in respiration observed between the region 14–16 mm from the root tip and the region 18–20 mm from the root tip is very drastic, and could be explained by variation between roots used for respiration measurements, although the roots were selected for uniformity. Regions further away from the root tip, 18–66 mm, had very low respiration rates. The tendency of tissue respiration to decrease with increased distance from the root tip agrees with early work using excised onion root tips (Berry and Brock, 1946; Wanner, 1950), although root excision may also

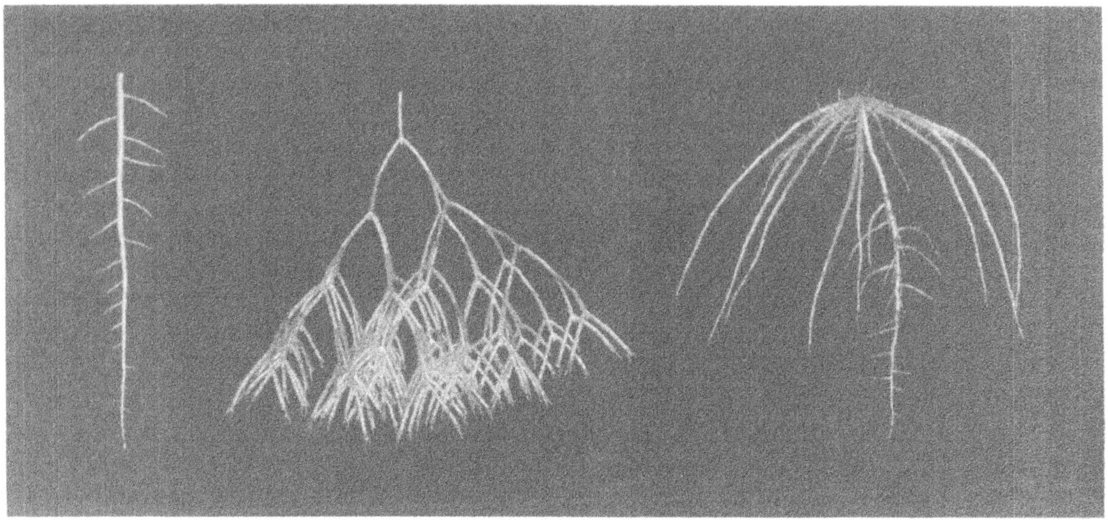

Fig. 1. Visual outputs on the Silicon Graphics computer showing the simulated growth of three contrasting root architectures. (**a**) Herringbone architecture. (**b**) Dichotomous architecture. (**c**) *Phaseolus vulgaris* cv. 'Carioca' architecture.

rapidly reduce root respiration (Bloom and Caldwell, 1988; Saglio and Pradet, 1980; Nielsen, unpublished data). In Figure 3 the range of standard errors (SE) is summarized. The SE indicates relatively stabile measurements of respiration, exudation, and deposition of carbon. The relatively large variations in root respiration of the different segments along the root axis can not be explained by large standard errors.

Carbon biomass density (μmoles C mm^{-1}) was higher in the root tip region (0–4 mm), but declined quickly in the elongation zone (4–15 mm), possibly as a consequence of cell expansion (Fig. 3). Carbon biomass density increased in the mature zone (> 16 mm) possibly because of root hair and cell wall formation and lignification.

Carbon exudation was also found to be highest in the 0–10 mm zone closest to the root tip (6.65 nmole C mm^{-1} hour^{-1}) falling to 5.245, 5.9 and 4.73 nmole C mm^{-1} hour^{-1} in the regions 13–23, 26–36 and 39–49 mm from the root tip respectively. In the region 53–63 mm from the primary root tip, exudation increased to 8.804 nmole C mm^{-1} hour^{-1} (Fig. 3). This increase coincided with the formation of lateral roots in this region.

Consideration of the spatial distribution of C cost along root axes makes it possible to draw conclusions regarding the effect of architecture on partitioning of below ground carbon resources between total respiration, exudation and deposition, and gives a more precise estimate of total carbon costs of contrasting root architectures and thereby nutrient acquisition efficiency. In this model we assume that the spatial distribution of carbon cost is independent of branching order. Future modifications of the model may have to include consideration of possible physiological differences between root types (e.g. basal vs. taproot) and root branching orders (Waisel and Eshel, 1991). Because respiration is highest in the meristematic zone, C loss in the form of CO_2 will be highest from root systems with rapidly increasing number of meristems over time. The dichotomous root system has the most rapidly increasing number of meristems of the three architectures simulated (Fig. 4). By definition, the rate of increase in number of meristems by apical bifurcation is geometric. Respiratory costs will accumulate faster and the amount of C spent on respiration will therefore be greater in the dichotomous as opposed to the bean root architecture. That the herringbone architecture used a relatively larger fraction of C for respiration was a consequence of the relatively lower carbon deposition over time. The herringbone architecture had lower carbon deposition because it did not branch extensively, and had few higher order roots.

Inter-root competition

Fitter (1991) stated that because lateral roots of high developmental order develop within the depletion zone of their parents, the exploitation efficiency (NAE) will be reduced in root systems with many lateral roots

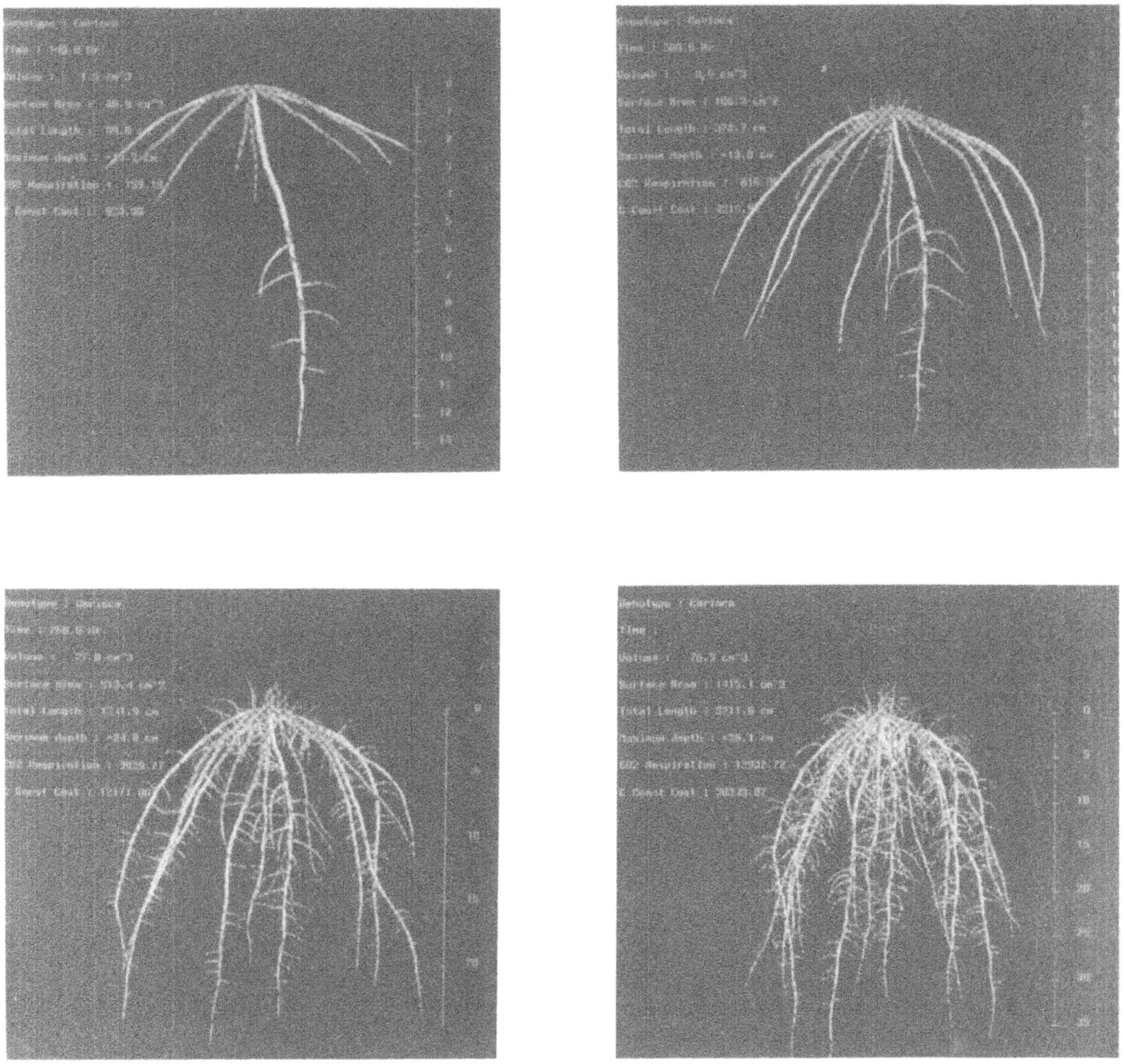

Fig. 2. Visual outputs on the Silicon Graphics computer showing the simulated growth of a *Phaseolus vulgaris* L. cv. Carioca root 140, 200, 260, and 336 hours after germination.

of high developmental order such as the dichotomous root system. In his model (Fitter, 1991), the soil volume depleted by the entire root system was estimated considering overlaps between depletion zones of adjacent links.

In our analysis, recharge of the solution phosphate availability in the depleted volume around root axes was not considered. It has been indicated (Paul and Clark, 1989) that such recharge of P in soil solution takes place rapidly in the rhizosphere, and as an effect of this remobilization, inter-root competion might be reduced to some extent. We also note that our analysis of inter-root competition may change if nutrients with other mobility than P were considered, since mobility of the nutrient would influence both the size and intensity of depletion zones. Baldwin et al. (1972) found that competition between two roots does not start until $(D \times t)^{\frac{1}{2}}$ is greater than the average distance between roots. If the inter-root distance is greater than $(D \times t)^{\frac{1}{2}}$ a root system can be regarded as composed of isolated single roots. In the simulations of root growth shown here, especially in the dichotomous root system, the inter-root distance became very small and there was intense competition between single roots.

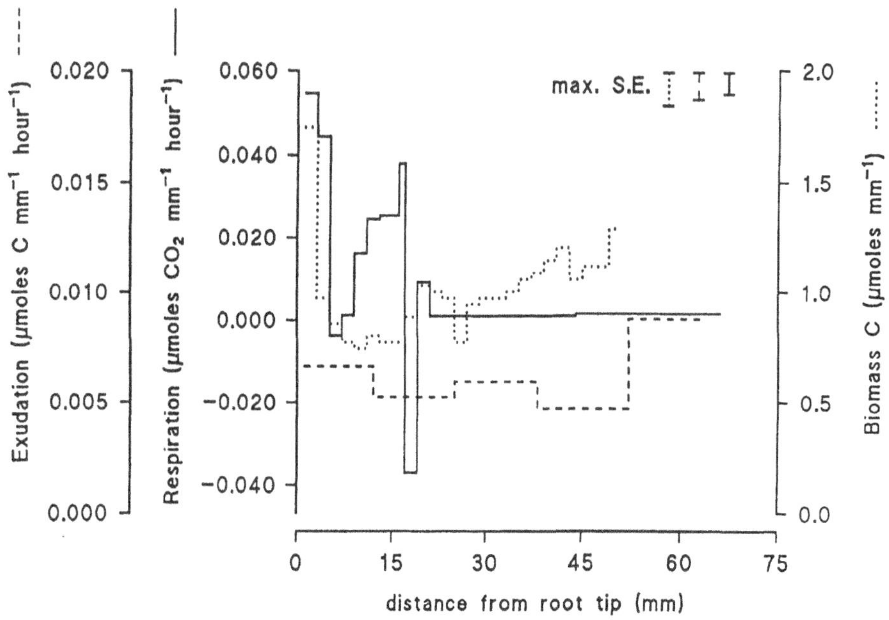

Fig. 3. Spatial distribution of respiration (μmoles CO_2 mm^{-1} $hour^{-1}$), exudation (μmoles total organic C mm^{-1} $hour^{-1}$), and biomass (μmole C mm^{-1}) as found in primary root tips of *Phaseolus vulgaris* L cv. G4000. SE for respiration varied between 0.000256 and 0.00478 μmoles CO_2 mm^{-1} $hour^{-1}$, for exudation between 0.000546 and 0.00141 μmoles organic C mm^{-1} $hour^{-1}$, and for carbon deposition between 0.0165 and 0.1516 μmole C mm^{-1}.

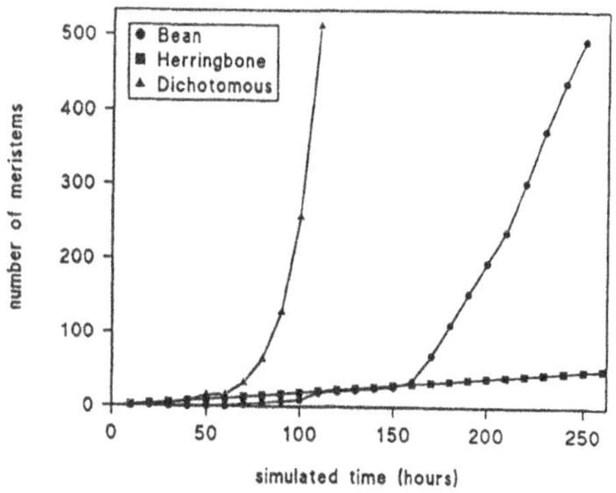

Fig. 4. Estimates of number of meristems formed in the three contrasting root architectures over time. Data based on output from SimRoot.

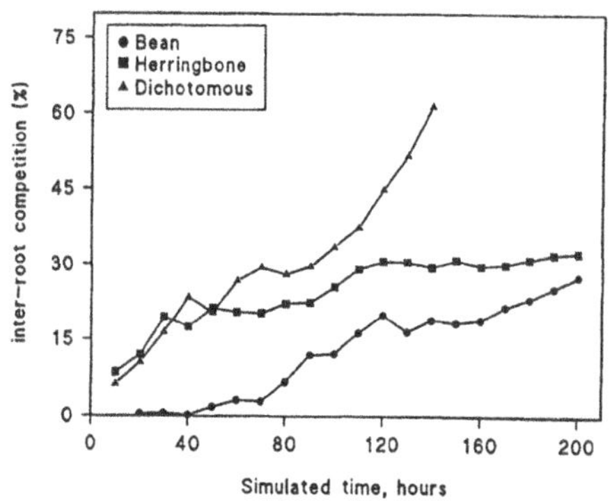

Fig. 5. Percent inter-root competition (overlap of depletion cylinders around root axes) over time for the three contrasting root architectures. Data based on output generated by SimRoot and DepZone.

The inter-root competition of the two extreme root systems, herringbone and dichotomous, was quite similar in the first 50 hours of simulation. Competition then stabilized for the herringbone root architecture, but increased drastically for the dichotomous root systems as a result of increased root densities (Fig. 5). Inter-root competition was low in the bean root system in the first 70 hours of simulation. Lateral roots then began forming in soil zones already depleted by parent roots leading to a rapid increase in inter-root competition from 80–120 hours. After 100 hours, basal roots formed and competition declined as the basal roots entered soil unexplored by their parent root after 120 hours. Continued lateral branching of basal roots then led to increased inter-root competition. As root growth continued, competition between basal roots would be expected to intensify.

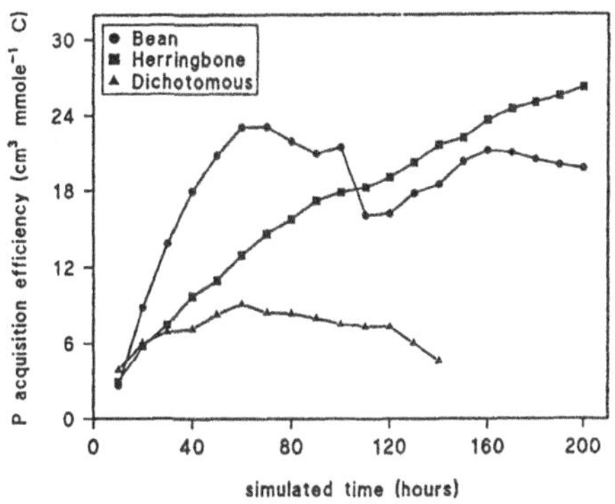

Fig. 6. Phosphorus acquisition efficiency (cm^3 soil depleted per mmole C spent by the root) over time for the contrasting root architectures. Data based on output generated by SimRoot and DepZone.

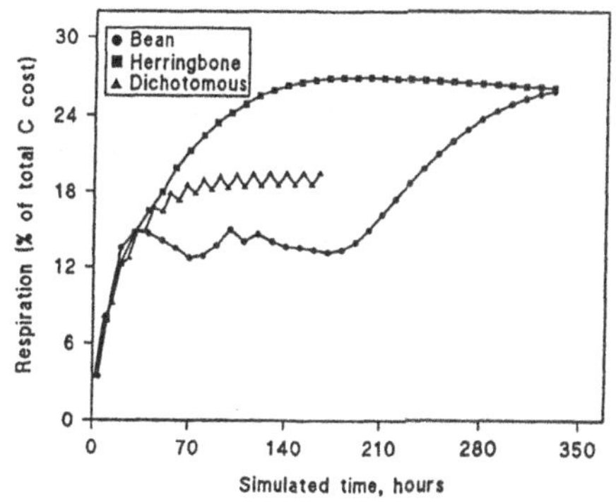

Fig. 8. Percent respiration of total below ground C cost over time in the three contrasting root architectures. Data based on output generated by SimRoot.

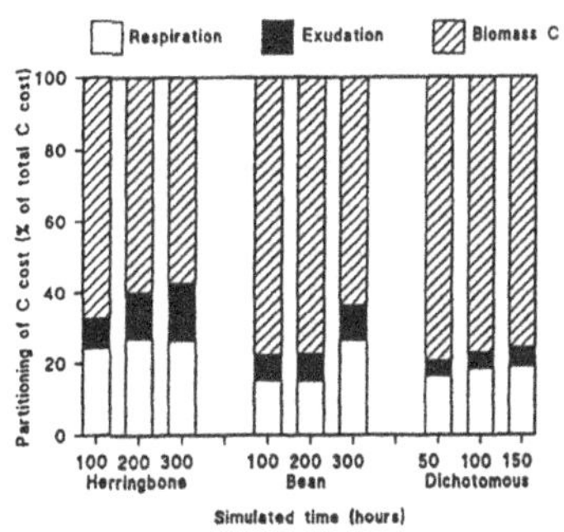

Fig. 7. The effect of root architecture and simulation time on partitioning of belowground carbon resources between *biomass C, exudation* and *respiration*. Data based on output generated by SimRoot and DepZone.

Phosphorus acquisition efficiency

High inter-root competition within the dichotomous root system led to declining PAE over time. A steady increase of PAE over time was observed in the herringbone architecture. The high efficiency of this architecture was due to its topology. No lateral branches of high developmental order were formed, leading to low root densities and relatively low inter-root competition. The PAE of the bean root system was first higher than dichotomous or herringbone, but when lateral and basal roots began forming after 60–100 hours, the efficiency declined to a level parallel to the PAE of herringbone. Increased inter-root competition as a result of lateral branching of the basal roots led to further decline after 160 hours of growth (Fig. 6). Although there were a larger number of roots of high developmental order in the bean root system, P acquisition in relation to C cost of the bean root system was very close to the relationship found in the herringbone system. The bean root system can essentially be considered a collection of herringbone root systems. The slightly lower PAE in bean root architecture is due to competition between the branches in the architecture. It has been found that respiratory losses increase with decreasing inherent growth rate of a species and with decreasing nitrate supply (Van der Werf et al., 1992). It would be expected that a newly formed root segment entering a zone with high phosphorus availability will have a lower rate of respiration than newly formed segments penetrating already depleted zones. In addition to that it might be expected that the formation of phosphorus releasing compounds would increase the amount of exudates in low phosphorus zones. The effects of the availability of nutrients on respiration were not considered in this model.

These results show the importance of explicitly considering the different components of C cost, such as respiration, biomass C, and exudation when evaluating different root architectures (Fig. 7). Very different proportions of C were allocated to biomass deposition, respiration, and exudation among the three root systems examined. The amount of C used for biomass deposition varied from 55 – 80 % of the total amount of carbon allocated to the root system, depending on root architecture and time. Percentage of total C exuded and respired increased over time in all three architectures. Respiration and exudation increased steadily with time in both the herringbone and dichotomous root systems, whereas the bean root system used a smaller fraction of C for respiration from 50–200 hours (Fig. 8). A large increase in respiration was observed after 200 hours, due to a massive increase in the number of meristems. (Figs. 4, 7).

We conclude that in simple root systems, volume and length density measurements are insufficient for the calculation of carbon costs associated with phosphorus acquisition, because variation in root architecture can affect the quantity of C allocated to structural biomass, exudation, or respiration. These different C fractions have very different residence times in soil and could differentially affect microbial populations and mineralization of soil organic matter. Efficiency of nutrient acquisition may also be an important component of plant fitness and species competition The large differences in PAE observed among the contrasting architectures indicate the importance of considering root architecture in relation to nutrient acquisition efficiency and environmental factors, such as rising CO_2 that affect belowground allocation of carbon resources.

Acknowledgements

This work was partially supported by the Root Biology Training Program, a unit of the DOE/NSF/USDA Collaborative Research in Plant Biology Program, funded by NSF grant BIR–9220330 to Dr Lynch. We thank Mr Robert D Davis III for his assistance in modifying SimRoot and developing DepZone and Mr Joe Bodkin, Material Characterization Laboratory, Penn State, for assistance during TOC analysis.

References

Amiro B D and Ewing L L 1992 Physiological conditions and uptake of inorganic carbon–14 by plant roots. Environ. Exp. Bot. 32, 203–211.

Anuradha M and Narayanan A 1991 Promotion of root elongation by phosphorus deficiency. Plant and Soil 136, 273–275.

Aono M and Kunii T L 1984 Botanical Tree Image Generation. IEEE Comput. Graphics Appl. 4, 10–34.

Arteca R N and Poovaiah B W 1982a Absorption of $^{14}CO_2$ by potato roots and its subsequent translocation. J. Am. Soc. Hort. Sci. 107, 398–401.

Arteca R N and Poovaiah B W 1982b Changes in phosphoenolpyruvate carboxylase and ribulose–1,5-bisphosphate carboxylase in *Solanum tuberosum* L. as affected by root zone applications of CO_2. HortScience 17, 396–398.

Barber S A 1984 Soil nutrient bioavailability. A mechanistic approach. John Wiley and Sons, New York. 398 p.

Baldwin J P, Tinker P B and Nye P H 1972 Uptake of solutes by multiple root systems from soil II. The theoretical effects of rooting density and pattern on uptake of nutrients from soil. Plant and Soil 36, 393–708.

Berry L J and Brock M J 1946 Polar distribution of respiratory rate in the onion root tip. Plant Physiol. 21, 242–249.

Berntson G M and Woodward F I 1992 The root system architecture and development of *Senecio vulgaris* in elevated CO_2 and drought. Func. Ecol. 6, 324–333.

Bloom A J and Caldwell R M 1988 Root excision decreases nutrient absorption and gas fluxes. Plant Physiol. 87, 794–796.

Bloom A J, Chapin III F S and Mooney H A 1985 Resource limitation in plants – an economic analogy. Ann. Rev. Ecol. Syst. 16, 363–392.

Davis III R D 1993 Geometric modeling and computer visualization of botanical root systems. Master Thesis The Pennsylvania State University.

Del Castillo D, Acock B, Reddy V R and Acock M C 1989 Elongation and branching of roots on soybean plants in a carbon dioxide-enriched aerial environment. Agron. J. 81, 692–695.

Diggle A J 1988 ROOTMAP: A Root Growth Model. Math. Comput. Simulation 30, 175–180.

Fitter A H 1991 Characteristics and functions of root systems. *In* Plant Roots: The Hidden Half. Eds. Y Waisel and U Kafkafi. pp 3–25. Marcel Dekker, Inc., New York.

Fitter A H 1987 An architectural approach to the comparative ecology of plant root systems. New Phytol. 106, 61–77.

Fitter A H, Stickland T R, Harvey M L and Wilson G W 1991 Architectural analysis of plant root systems I: Architectural correlates of exploitation efficiency. New Phytol. 118, 375–382.

Hansen A P, Yoneyama T and Kouchi H 1992 Short term nitrate effects on hydroponically-grown soybean cv. Bragg and its supernodulating mutant. J. Exp. Bot. 43, 1–7.

Haynes B, Koide R T and Elliott G 1991 Phosphorus uptake and utilization in wild and cultivated oats (*Avena* spp.). J. Plant Nutr. 14, 1105–1118.

Langley D R 1973 The growth of root systems – A numerical computer simulation model. Plant and Soil 38, 145–159.

Long S P and Drake B G 1991 Effect of the long-term elevation of CO_2 concentration in the field on the quantum yield of photosynthesis of the C_3 sedge, *Scirpus olneyi*. Plant Physiol. 96, 221–226.

Lynch J and van Beem J 1993 Growth and architecture of seedling roots of common bean genotypes. Crop Science 33(6), 1253–1257.

Mooney H A 1972 The carbon balance of plants. Ann. Rev. Ecol. Syst. 3, 315–346.

Orians G and Solbrig O 1977 A cost-income model of leaves and roots with special reference to arid and semiarid areas. Am. Nat. 111, 677–690.

Nye P H and Tinker P B 1977 Solute Movement in the Soil-Root System. Blackwell Scientific Publications, Oxford. 342 p.

Pages L, Jordan M O and Picard D 1989 A simulation model of the three-dimensional architecture of the maize root system. Plant and Soil 119, 147–154.

Paul E A and Clark F E 1989 Soil Microbiology and Biochemistry. pp 222–231. Academic Press, Inc. San Diego, California, USA.

Rogers H H, Runion G B and Krupa S V 1993 Plant responses to atmospheric CO_2 enrichment with emphasis on roots and the rhizosphere. Env. Poll. 83, 155–159.

Rogers H H, Peterson C M, McCrimmon J N and Cure J D 1992 Response of plant roots to elevated atmospheric carbon dioxide. Plant Cell Environ, 15, 749–752.

Rundel P W and Nobel P S 1991 Structure and function in desert root systems. *In* Plant Root Growth. An Ecological Perspective. Ed. D Atkinson. pp 349–378. Blackwell Scientific Publications, London.

Sadeghian S 1991 Influencia de algunas caracteristicas de las semillas y plantulas de frijol *Phaseolus vulgaris* L. sobre la tolerancia a la baja disponibilidad de fosforo en el suelo. Univ. Nacional de Colombia. Fac. Ciencias Agropecuarias, Palmira, Colombia.

Saglio P H and Pradet A 1980 Soluble sugars, respiration, and energy charge during aging of excised maize root tips. Plant Physiol. 66, 516–519.

Sanchez P A 1976 Properties and Management of Soils in the Tropics. pp 254–294. John Wiley and Sons, USA.

Sanchez P A and Logan T J 1992 Myths and science about the chemistry and fertility of soils in the tropics. *In* Myths and Science of Soils in the Tropics. SSSA Special Publication No. 29. pp 35–46. Soil Science Society of America and American Society of Agronomy, Madison, WI, USA.

Van der Werf A, Welschen R and Lambers H 1992 Respiratory losses increase with decreasing inherent growth rate of a species and with decreasing nitrate supply: a search for explanations for these observations. *In* Molecular, Biochemical and Physiological Aspects of Plant Respiration. Eds. H Lambers and L H W van der Plas. pp 421–432. SPB Academic Publishing bv, The Hague, The Netherlands.

Waisel Y and Eshel A 1991 Multiform behavior of various constituents of one root system. *In* Plant Roots: The Hidden Half. Eds. Y Waisel and U Kafkafi. pp 39–52. Marcel Dekker, Inc., New York.

Wanner H 1950 Histilogische und Physiologische Gradienten in der Würzelspitze. Ber. Schweiz. Bot. Ges. 60, 404–412.

Wong S C, Kriedemann P E and Farquhar D 1992 CO × nitrogen interaction on seedling growth of four species of eucalypt. Aust. J. Bot. 40, 457–472.

Zak D R, Pregitzer K S, Curtis P S, Teeri J A, Fogel R and Randlett D L 1993 Elevated atmospheric CO_2 and feedbacks between carbon and nitrogen cycles. Plant and Soil 151, 105–117.

Developments in Plant and Soil Sciences

1. J. Monteith and C. Webb (eds.): *Soil Water and Nitrogen in Mediterranean-type Environments.* 1981 ISBN 90-247-2406-6
2. J. C. Brogan (ed.): *Nitrogen Losses and Surface Run-off from Landspreading of Manures.* 1981 ISBN 90-247-2471-6
3. J. D. Bewley (ed.): *Nitrogen and Carbon Metabolism.* 1981 ISBN 90-247-2472-4
4. R. Brouwer, I. Gašparíková, J. Kolek and B. C. Loughman (eds.): *Structure and Function of Plant Roots.* 1981 ISBN 90-247-2510-0
5. Y. R. Dommergues and H. G. Diem (eds.): *Microbiology of Tropical Soils and Plant Productivity.* 1982 ISBN 90-247-2624-7
6. G. P. Robertson, R. Herrara and T. Rosswall (eds.): *Nitrogen Cycling in Ecosystems of Latin America and the Caribbean.* 1982 ISBN 90-247-2719-7
7. D. Atkinson, K. K. S. Bhat, M. P. Coutts, P. A. Mason and D. J. Read (eds.): *Tree Root Systems and Their Mycorrhizas.* 1983 ISBN 90-247-2821-5
8. M. R. Sarić and B. C. Loughman (eds.): *Genetic Aspects of Plant Nutrition.* 1983 ISBN 90-247-2822-3
9. J. R. Freney and J. R. Simpson (eds.): *Gaseous Loss of Nitrogen from Plant-Soil Systems.* 1983 ISBN 90-247-2820-7
10. United Nations Economic Commission for Europe (ed.): *Efficient Use of Fertilizers in Agriculture.* 1983 ISBN 90-247-2866-5
11. J. Tinsley and J. F. Darbyshire (eds.): *Biological Processes and Soil Fertility.* 1984 ISBN 90-247-2902-5
12. A. D. L. Akkermans, D. Baker, K. Huss-Danell and J. D. Tjepkema (eds.): Frankia *Symbioses.* 1984 ISBN 90-247-2967-X
13. W. S. Silver and E. C. Schröder (eds.): *Practical Application of* Azolla *for Rice Production.* 1984 ISBN 90-247-3068-6
14. P. G. L. Vlek (ed.): *Micronutrients in Tropical Food Crop Production.* 1985 ISBN 90-247-3085-6
15. T. P. Hignett (ed.): *Fertilizer Manual.* 1985 ISBN 90-247-3122-4
16. D. Vaughan and R. E. Malcolm (eds.): *Soil Organic Matter and Biological Activity.* 1985 ISBN 90-247-3154-2
17. D. Pasternak and A. San Pietro (eds.): *Biosalinity in Action.* Bioproduction with Saline Water. 1985 ISBN 90-247-3159-3
18. M. Lalonde, C. Camiré and J. O. Dawson (eds.): Frankia *and Actinorhizal Plants.* 1985 ISBN 90-247-3214-X
19. H. Lambers, J. J. Neeteson and I. Stulen (eds.): *Fundamental, Ecological and Agricultural Aspects of Nitrogen Metabolism in Higher Plants.* 1986 ISBN 90-247-3258-1
20. M. B. Jackson (ed.): *New Root Formation in Plants and Cuttings.* 1986 ISBN 90-247-3260-3
21. F. A. Skinner and P. Uomala (eds.): *Nitrogen Fixation with Non-Legumes* (Proceedings of the 3rd Symposium, Helsinki, 1984). 1986 ISBN 90-247-3283-2
22. A. Alexander (ed.): *Foliar Fertilization.* 1986 ISBN 90-247-3288-3
23. H. G. v.d. Meer, J. C. Ryden and G. C. Ennik (eds.): *Nitrogen Fluxes in Intensive Grassland Systems.* 1986 ISBN 90-247-3309-X
24. A. U. Mokwunye and P. L. G. Vlek (eds.): *Management of Nitrogen and Phosphorus Fertilizers in Sub-Saharan Africa.* 1986 ISBN 90-247-3312-X
25. Y. Chen and Y. Avnimelech (eds.): *The Role of Organic Matter in Modern Agriculture.* 1986 ISBN 90-247-3360-X
26. S. K. De Datta and W. H. Patrick Jr. (eds.): *Nitrogen Economy of Flooded Rice Soils.* 1986 ISBN 90-247-3361-8
27. W. H. Gabelman and B. C. Loughman (eds.): *Genetic Aspects of Plant Mineral Nutrition.* 1987 ISBN 90-247-3494-0
28. A. van Diest (ed.): *Plant and Soil: Interfaces and Interactions.* 1987 ISBN 90-247-3535-1

Developments in Plant and Soil Sciences

29. United Nations Economic Commission for Europe and FAO (eds.): *The Utilization of Secondary and Trace Elements in Agriculture.* 1987 ISBN 90-247-3546-7
30. H. G. v.d. Meer, R. J. Unwin, T. A. van Dijk and G. C. Ennik (eds.): *Animal Manure on Grassland and Fodder Crops.* Fertilizer or Waste? 1987 ISBN 90-247-3568-8
31. N. J. Barrow: *Reactions with Variable-Charge Soils.* 1987 ISBN 90-247-3589-0
32. D. P. Beck and L. A. Materon (eds.): *Nitrogen Fixation by Legumes in Mediterranean Agriculture.* 1988 ISBN 90-247-3624-2
33. R. D. Graham, R. J. Hannam and N. C. Uren (eds.): *Manganese in Soils and Plants.* 1988 ISBN 90-247-3758-3
34. J. G. Torrey and J. L. Winship (eds.): *Applications of Continuous and Steady-State Methods to Root Biology.* 1989 ISBN 0-7923-0024-6
35. F. A. Skinner, R. M. Boddey and I. Fendrik (eds.): *Nitrogen Fixation with Non-Legumes* (Proceedings of the 4th Symposium, Rio de Janeiro, 1987). 1989 ISBN 0-7923-0059-9
36. B. C. Loughman, O. Gašparíková and J. Kolek (eds.): *Structural and Functional Aspects of Transport in Roots.* 1989 ISBN 0-7923-0060-2; Pb 0-7923-0061-0
37. P. Plancquaert and R. Haggar (eds.): *Legumes in Farming Systems.* 1990 ISBN 0-7923-0134-X
38. A. E. Osman, M. M. Ibrahim and M. A. Jones (eds.): *The Role of Legumes in the Farming Systems of the Mediterranean Areas.* 1990 ISBN 0-7923-0419-5
39. M. Clarholm and L. Bergström (eds.): *Ecology of Arable Land – Perspectives and Challenges.* 1989 ISBN 0-7923-0424-1
40. J. Vos, C. D. van Loon and G. J. Bollen (eds.): *Effects of Crop Rotation on Potato Production in the Temperate Zones.* 1989 ISBN 0-7923-0495-0
41. M. L. van Beusichem (ed.): *Plant Nutrition – Physiology and Applications.* 1990 ISBN 0-7923-0740-2
42. N. El Bassam, M. Dambroth and B.C. Loughman (eds.): *Genetic Aspects of Plant Mineral Nutrition.* 1990 ISBN 0-7923-0785-2
43. Y. Chen and Y. Hadar (eds.): *Iron Nutrition and Interactions in Plants.* 1991 ISBN 0-7923-1095-0
44. J. J. R. Groot, P. de Willigen and E. L. J. Verberne (eds.): *Nitrogen Turnover in the Soil-Crop System.* 1991 ISBN 0-7923-1107-8
45. R. J. Wright, V.C. Baligar and R. P. Murrmann (eds.): *Plant-Soil Interactions at Low pH.* 1991 ISBN 0-7923-1105-1
46. J. Kolek and V. Kozinka (eds.): *Physiology of the Plant Root System.* 1992 ISBN 0-7923-1205-8
47. A. U. Mokwunye (ed.): *Alleviating Soil Fertility Constraints to Increased Crop Production in West Africa.* 1991 ISBN 0-7923-1221-X; Pb 0-7923-1222-8
48. M. Polsinelli, R. Materassi and M. Vincenzini (eds.): *Nitrogen Fixation* (Proceedings of the 5th Symposium, Florence, 1990). 1991 ISBN 0-7923-1410-7
49. J.K. Ladha, T. George and B.B. Bohlool (eds.): *Biological Nitrogen Fixation for Sustainable Agriculture.* 1992 ISBN 0-7923-1774-2
50. P.J. Randall, E. Delhaze, R.A. Richards and R. Munns (eds.): *Genetic Aspects of Plant Mineral Nutrition.* 1993 ISBN 0-7923-2118-9
51. K.S. Kumarasinghe and D.L. Eskew (eds.): *Isotopic Studies of Azolla and Nitrogen Fertilization of Rice.* 1993 ISBN 0-7923-2274-6
52. F.A. Bliss and G. Hardarson (eds.): *Enhancement of Biological Nitrogen Fixation of Common Baen in Latin America.* 1993 ISBN 0-7923-2451-X
53. M.A.C. Fragoso and M.L. van Beusichem (eds.): *Optimization of Plant Nutrition.* 1993 ISBN 0-7923-2519-2
54. N.J. Barrow (ed.) *Plant Nutrition - From Genetic Engineering to Field Practice.* 1993 ISBN 0-7923-2540-0
55. A.D. Robson (ed.): *Zinc in Soils and Plants.* 1993 ISBN 0-7923-2631-8
56. A.D. Robson, L.K. Abbott and N. Malajczuk (eds.): *Management of Mycorrhizas in Agriculture, Horticulture and Forestry.* 1994 ISBN 0-7923-2700-4

Developments in Plant and Soil Sciences

57. P.H. Graham, M.J. Sadowsky and C.P. Vance (eds.): *Symbiotic Nitrogen Fixation*. 1994 ISBN 0-7923-2781-0
58. F. Baluška, M. Čiamporová, O. Gašparíkova and P.W. Barlow (eds.): *Structure and Function of Roots*. 1994 ISBN 0-7923-2832-9
59. J. Abadía (ed.): *Iron Nutrition in Soils and Plants*. 1994 ISBN 0-7923-2900-7
60. P.S. Curtis, E.G. O'Neill, J.A. Teeri, D.R. Zak and K.S. Pregitzer (eds.): *Belowground Responses to Rising Atmospheric CO_2*. Implications for Plants, Soil Biota, and Ecosystem Processes. 1994 ISBN 0-7923-2901-5
61. P.C. Struik, W.J. Vredenburg, J.A. Renkema and J.E. Parlevliet (eds.): *Plant Production on the Threshold of a New Century*. 1994 ISBN 0-7923-2903-1

Kluwer Academic Publishers – Dordrecht / Boston / London

SPRINGER NATURE

GPSR Compliance

The European Union's (EU) General Product Safety Regulation (GPSR) is a set of rules that requires consumer products to be safe and our obligations to ensure this.

If you have any concerns about our products, you can contact us on ProductSafety@springernature.com

In case Publisher is established outside the EU, the EU authorized representative is:

Springer Nature Customer Service Center GmbH
Europaplatz 3
69115 Heidelberg, Germany

The manufacturer's authorised representative in the EU is Springer Nature Customer Service Centre GmbH, Europaplatz 3, 69115 Heidelberg, Germany. If you have any concerns regarding our products, please contact ProductSafety@springernature.com

Printed and bound by CPI Group (UK) Ltd, Croydon, CR0 4YY
23/09/2026
02221726-0002